Industrial Composites, Optimization Methods, and Tribological Considerations

This book recapitulates the significance of metal composites in modern sustainable industry, along with the critical review of process optimization techniques and tribological aspects for industrial and automotive applications.

The book aims to cover all the topics relevant to the mechanical characteristics of metal matrix composites (MMCs) such as reinforcement types (particles, short fiber, and long or continuous fibers), their effect on the microstructure of composites, and significance in enhancing the materials' mechanical and physical properties. The first two chapters aim to provide detailed learning about recent industrial advancements of diverse metal composites: aluminum, copper, and magnesium, and their significance in the modern workable industry. The next two chapters highlight a detailed analysis of particle, short-fiber, and continuous-fiber materials used for composite fabrication. Further chapters discuss the behavior of composites under various testing conditions, process optimization techniques, tribological technologies, lubrication knowledge including additives and lubricants, and improvements to the tribological properties of MMCs for industrial and automotive applications.

Industrial Composites, Optimization Methods, and Tribological Considerations is an essential book for experts, scientists, and engineers who are involved in the domain of composite materials, tribology, and process optimization.

Manufacturing Design and Technology

Series Editor
J. Paulo Davim

This series will publish high-quality references and advanced textbooks in the broad area of manufacturing design and technology, with a special focus on sustainability in manufacturing. Books in the series should find a balance between academic research and industrial application. This series targets academics and practicing engineers working on topics in materials science, mechanical engineering, industrial engineering, systems engineering, and environmental engineering as related to manufacturing systems, as well as professions in manufacturing design.

Non-Conventional Hybrid Machining Processes: Theory and Practice
Edited by Rupinder Singh and J. Paulo Davim

Functional Materials and Advanced Manufacturing: 3-Volume Set
Edited by Chander Prakash, Sunpreet Singh, and J. Paulo Davim

Industry 4.0: Challenges, Trends, and Solutions in Management and Engineering
Edited by Carolina Machado and J. Paulo Davim

Sustainable Manufacturing for Industry 4.0: An Augmented Approach
Edited by K. Jayakrishna, Vimal K.E.K., S. Aravind Raj, Asela K. Kulatunga, M.T.H. Sultan, and J. Paulo Davim

Industrial Tribology: Sustainable Machinery and Industry 4.0
Edited by Jitendra Kumar Katiyar, Alessandro Ruggiero, TVVLN Rao and J. Paulo Davim

Innovative Development in Micromanufacturing Processes
Edited by Pawan Kumar Rakesh and J. Paulo Davim

Tribology in Sustainable Manufacturing
Edited by Jitendra Kumar Katiyar, TVVLN Rao, Ahmad Majdi Abdul Rani, Mohd Hafis Sulaiman and J.P. Davim

Sustainable Manufacturing: An Emergence in Industry 4.0
Edited by Kamalpreet Sandhu, Sunpreet Singh, Ranvijay Kumar, JP Davim and Seeram Ramakrishna

Industrial Composites, Optimization Methods, and Tribological Considerations
Edited by Gurpreet Singh Saini, Jai Inder Preet Singh, Pankaj Kumar Singh, Raghu Echempati, and J. Paulo Davim

For more information about this series, please visit: https://www.routledge.com/Manufacturing-Design-and-Technology/book-series/CRCMANDESTEC

Industrial Composites, Optimization Methods, and Tribological Considerations

Edited by
Gurpreet Singh Saini
Jai Inder Preet Singh
Pankaj Kumar Singh
Raghu Echempati
J. Paulo Davim

CRC Press
Taylor & Francis Group
Boca Raton London New York

CRC Press is an imprint of the
Taylor & Francis Group, an **informa** business

First edition published 2026
by CRC Press
2385 NW Executive Center Drive, Suite 320, Boca Raton FL 33431

and by CRC Press
4 Park Square, Milton Park, Abingdon, Oxon, OX14 4RN

CRC Press is an imprint of Taylor & Francis Group, LLC

ISBN: 978-1-041-02268-8 (hbk)
ISBN: 978-1-041-02684-6 (pbk)
ISBN: 978-1-003-62057-0 (ebk)

DOI: 10.1201/9781003620570

Typeset in Times
by SPi Technologies India Pvt Ltd (Straive)

Contents

Preface

The rapid advancement of engineering technologies and the growing demand for high-performance, durable, and lightweight materials in industrial and automotive sectors have intensified research into advanced composite materials. Among these, MMCs have emerged as promising candidates due to their superior mechanical, thermal, and tribological properties. MMCs, reinforced with ceramics or other materials, offer a compelling combination of strength, wear resistance, and thermal stability—making them ideal for high-stress environments such as engines, brake systems, and aerospace components.

This book, *Industrial Composites, Optimization Methods, and Tribological Considerations*, provides a comprehensive exploration of the tribological behavior of MMCs, with a special focus on their applications in demanding industrial and automotive environments. Tribology—the study of friction, wear, and lubrication—is a critical discipline in materials science, as it governs the reliability, efficiency, and longevity of components in motion. Understanding the tribological response of MMCs under various operational conditions is vital for designing next-generation engineering systems.

This was driven by the fact that there is a dire need to solve the material degradation challenges that affect the properties of mechanical components and their life cycle. Conventional materials do not usually work in the harshest environments, such as those of high load, temperatures, and speeds. Researchers have managed to substantially improve the wear resistance, minimize the friction, and optimize the mechanical performance by embedding reinforcements, that is, silicon carbide (SiC), boron carbide (B_4C), graphite, or carbon nanotubes, into metal matrices, that is, aluminum, magnesium, or titanium.

In this book, the order of information has been designed in such a way that it provides both introductory and progressive knowledge. It starts by giving a dose of what the MMCs are, their members, and their methods of fabrication. This is followed by several chapters that explore various elements of tribology, described as dry wear and lubricated wear behavior, thermal effect, load dependence, speed dependence, surface engineering techniques, and the contribution of hybrid multi- and nano-reinforcement. Experimental techniques and application in real-life scenarios in industry-critical sectors like automotive/engines, aerospace/structure, and heavy machinery attract special attention.

Years have passed, and many works by various researchers have been conducted and contributed before this book on tribology in MMCs has been finally published. It would fill the gap between the science theory and engineering practice by providing a more detailed understanding of the effect on the wear, friction, and surface performance of MMCs of various reinforcement materials, composition of the matrix, processing methods, and service conditions.

The main aim of this book would be to offer a one-stop reference book to researchers, material scientists, mechanical engineers, and the industry fraternity that are involved in the development and fabrication of high-performance materials used in applications that are challenging in nature. It provides extensive research on the dynamics of MMCs and their behavior under different conditions, especially in dry sliding, lubricated conditions, under different heat loading, under different loads, and under different sliding velocities. Such knowledge will be required to maximize the life of the components, minimize the expenses of maintenance, and improve the performance of the system.

We have made a conscious effort to balance theoretical explanations with experimental data, visual diagrams, and literature reviews, making this book both a foundational text for newcomers and a technical reference for seasoned professionals. In preparing this book, we owe a great deal to the global community of researchers and engineers whose pioneering work forms the backbone of this field. We extend our deepest appreciation to academic mentors, laboratory teams, peer reviewers, and institutional collaborators who supported the development of this manuscript through constructive feedback, critical discussions, and data contributions. We also acknowledge the students and research scholars whose enthusiasm and curiosity have continually inspired us to dig deeper and think more broadly. Their inquisitiveness has played a key role in shaping the structure and depth of this book.

It is our sincere hope that this volume will not only serve as a resource for learning and application, but also ignite further interest and innovation in the realm of tribology and composite materials. As industries move toward lightweighting, sustainability, and high-performance engineering, we believe MMCs and their tribological optimization will play a pivotal role in shaping the next generation of technologies.

Dr. Gurpreet Singh Saini
Dr. Jai Inder Preet Singh
Dr. Pankaj Kumar Singh
Prof. Raghu Echempati
Prof. J. Paulo Davim

About the Editors

Dr. Gurpreet Singh Saini is an assistant professor-III in the Department of Automobile and Aeronautical Engineering, Amity Institute of Technology, Amity University Uttar Pradesh (AUUP), Noida, India. He received his Ph.D. in Mechanical Engineering (Composite Materials) from J. C. Bose University of Science and Technology, YMCA in 2017. He earned his M.E. from PEC Chandigarh (2011), India and B.Tech from Kurukshetra University, India (2006) respectively. He has more than 15 years of teaching and research experience with core areas of research includes composite materials, material characterization, process optimization, wear and Tribology. He is an active life member of Science and Engineering Institute (SCIEI). He has also contributed in more than 25 research papers in various international journals of repute.

Dr. Jai Inder Preet Singh is currently working as professor and department head in School of Mechanical Engineering, Lovely Professional University, Jalandhar. His area of research is green composites, polymer composites, self-healing composites and mechanical characterization. He has published more than 20 research articles in peer reviewed international journals. He guided 1 Ph.D., 12 M.Tech Thesis and 1 Ph.D. thesis. He has published five book chapters and is editorial board of some of the reputed Scopus indexed journals. He is also a guest editor of one of the Scopus indexed international peer reviewed journals. He has granted 4 international patents, 7 national patents and published 50+ patents.

Dr. Pankaj Kumar Singh is currently working as campus director, Delhi Skill and Entrepreneurship University (Government of NCT of Delhi) Wazirpur-I Campus, Delhi, an eminent faculty, researcher, administrator, and a dynamic motivator for students. He did his B.Tech (M.E.) from VTU & M.Tech and Ph.D. (Mechanical Engineering) from Birla Institute of Technology, Mesra, Ranchi, India. He was awarded 1st Prize Dronacharya Award (Best Faculty Advisor) in BAJA SAE INDIA 2016 and 2nd Prize for Technology (Go-Green- Emissions) in BAJA SAE INDIA 2013. He has a teaching experience of more than 18 years and published many international research papers and has actively participated in workshops and international and national conferences.

Dr. Raghu Echempati holds a Ph.D. in Mechanical Engineering from IIT Kharagpur, India, and has over 40 years of academic experience, having taught at prestigious institutions including IIT Delhi, the University of Florida, Ohio State University, and Kettering University, where he retired as a full professor in 2022. He has also served as a part-time professor at Wayne State University and as a visiting professor in countries such as Germany, India, New Zealand, and Thailand. Dr. Echempati's teaching expertise spans a wide range of subjects, including solid mechanics, vibrations, machine design, and automotive lightweighting. In addition to his academic roles, he has held various industry internships with Bosch, General Motors, and GEMA, providing expertise in areas like airbag injury studies and die design. He is a registered professional engineer in Mississippi, a fellow of ASME, and a member of ASEE and SAE. His contributions to engineering education and research have earned him multiple awards, including a Fulbright Specialist Award and the ASME Dedicated Service Award. Dr. Echempati has published over 160 papers and has delivered keynote lectures at numerous international conferences.

Prof. (Dr.) J. Paulo Davim is a full professor at the University of Aveiro, Portugal. He is also distinguished as honorary professor in several universities/colleges/institutes in China, India and Spain. He received his Ph.D. in Mechanical Engineering, M.Sc. in Mechanical Engineering (materials and manufacturing processes) and Mechanical Engineering degree (5 years), from the University of Porto (FEUP), the Aggregate title (Full Habilitation) from the University of Coimbra and the DSc (Higher Doctorate) from London Metropolitan University. He is Fellow (FIET) of IET-London, Fellow (FIMF) of IMF-Birmingham and Eur Ing by Engineers Europe FEANI-Brussels. He is also senior specialist engineer by the Portuguese Order of Engineers with an MBA and Specialist titles in Industrial Management as well as in Metrology. He has more than 35 years of teaching and research experience in manufacturing, materials, mechanical and industrial engineering, with special emphasis in machining, tribology & surface engineering. He has also interest in design, management, sustainability, Industry 5.0, engineering education and higher education for sustainability. He has guided large numbers of postdoc, Ph.D. and master's students as well as has coordinated and participated in several financed research projects. He has received several scientific awards and honors. He has worked as evaluator of projects for ERC-European Research Council and other international research agencies as well as examiner of Ph.D. thesis for many universities in different countries. He is the editor-in-chief of several international journals, book series editor and scientific advisory for many conferences.

Contributors

Gaurav Arora
Tool and Production Engineering Department
Delhi Skill and Entrepreneurship University
New Delhi, India

Kunal Arora
Mechanical Engineering Department
Shandong University of Technology
Zibo, Shandong, P. R. China

Rishu Arora
Department of Mathematics, Amity Institute of Applied Science
Amity University
Noida, India

Bharat Bajaj
Cleveland State University
Cleveland, Ohio, USA

Indira Bhati
Maharishi Markandeshwar Institute of Management
Maharishi Markandeshwar (Deemed to be University)
Mullana, India

Vaibhav Chandra
Tool and Production Engineering Department
Delhi Skill and Entrepreneurship University
New Delhi, India

J. Paulo Davim
University of Aveiro
Campus Universitario de Santiago
Aveiro, Portugal

Raghu Echempati
Division of Engineering Technology
Wayne State University
Detroit, Michigan, USA

Neeraj Kumar Gahlot
Amity Institute of Aerospace Engineering
Amity University Uttar Pradesh
Noida, India

Piyush Gulati
Mechanical Engineering Department
Lovely Professional University
Phagwara, India

Shaman Gupta
Department of Mechanical Engineering
Maharishi Markandeshwar Engineering College, Maharishi Markandeshwar (Deemed to be University)
Mullana, India

Ismail Hossain
Department of Nuclear and Renewable Energy
Ural Federal University
Yekaterinburg, Russia

Amarjeet Kaur
PGT (Chemistry)
St. Soldier Divine Public School
Jalandhar, India

Umesh Khandey
Department of Manufacturing Engineering
National Institute of Advanced Manufacturing Technology (NIAMT)
Ranchi, India

Praveen Kumar
Tool and Production Engineering Department
Delhi Skill and Entrepreneurship University
New Delhi, India

Gurpreet Singh Saini
Amity Institute of Technology
Amity University Uttar Pradesh
Noida, India

Rajesh Kumar Saluja
Amity Institute of Aerospace Engineering
Amity University Uttar Pradesh
Noida, India

Harpinder Singh Sandhu
School of Mechanical Engineering
Lovely Professional University
Phagwara, India

Ashok Kumar Sharma
Manipal University Jaipur
Jaipur, India

Neeraj Sharma
Production engineering Department
National Institute of Technology Agartala
Tripura, India

Sachin Sharma
Department of Mechanical Engineering
School of Advanced Engineering, UPES
Dehradun, India

Vikas Sharma
University Center for Research and Development (UCRD)
Chandigarh University
Mohali, India

Harjit Singh
School of Mechanical Engineering
Lovely Professional University
Phagwara, India

Jai Inder Preet Singh
Mechanical Engineering Department
Lovely Professional University
Phagwara, India

Kanwarjeet Singh
Tool and Production Engineering Department
Delhi Skill and Entrepreneurship University
New Delhi, India

Maninder Singh
School of Mechanical Engineering
Lovely Professional University
Phagwara, India

Narender Singh
Amity Institute of Aerospace Engineering
Amity University Uttar Pradesh
Noida, India

Pankaj Kumar Singh
Campus Director, Mechanical Engineering Department
Delhi Skill and Entrepreneurship University
New Delhi, India

Lagouge Tartibu
Department of Mechanical and Industrial Engineering Technology
University of Johannesburg
Johannesburg, South Africa

1 An Overview of Metal Matrix Composites and Significance in Modern Sustainable Industry

Harpinder Singh Sandhu, Piyush Gulati, and Sachin Sharma

1.1 INTRODUCTION

The unending quest for industrial progress indeed faces a dilemma: how to balance performance requirements against the sustainable utilization of resources. The quest for materials with better attributes (stronger, lighter, and more durable, with better functionality) has been the driver for innovation from the Stone Age through to the Silicon Age. Composite materials, engineered for specific properties by combining two or more constituents (a matrix and a reinforcement) to create properties not achievable with either of the constituents alone, represent perhaps the pinnacle of this quest. The range of composite materials is broad, but Metal Matrix Composites (MMCs) are a critical species of composites. MMCs combine the durability, heat transfer characteristics, and manufacturability of metals (e.g., aluminum, magnesium, titanium, or copper) with the better stiffness, strength, wear properties, or function of a ceramic (e.g., silicon carbide, alumina, or boron carbide) or carbon-based (e.g., graphite and carbon nanotubes (CNTs)) reinforcement.

The importance of MMCs is tremendous and growing rapidly, given the urgent need for sustainable technology solutions. As Khanna et al. (2024) note, MMCs are being recognized as key enablers for performance and sustainability. To be clear, the idea of MMCs has been around for millennia, but the modern era of MMCs began in earnest in the mid-20th century, dictated by aerospace and defense drivers. The early developments were focused on continuous fiber reinforcements for niche applications, for example, boron fiber/aluminum struts in the Space Shuttle (Mortensen and Llorca, 2010; Miracle, 2005). There was much progress on fundamental processing science and interfacial interactions in the 1970s and 1980s. Real advancement of industrial relevance occurred with the advancement and maturation of discontinuously reinforced MMCs (DRMMCs) – primarily particles, whiskers, or short fibers included into the metal matrix (Makhesana et al., 2025; Khanna et al., 2024). DRMMCs offered an attractive combination that has managed to retain a lot of the isotropic characteristics and traditional processability of metals (e.g., casting,

DOI: 10.1201/9781003620570-1

forging, and extrusion) and gained substantial improvements in specific stiffness, specific strength, elevated temperature performance, and, importantly, tribological properties (e.g., wear). Expansion into markets outside of aerospace opened up possibilities in automotive, thermal management, and industrial wear.

Currently, the focus on sustainability has completely reshaped the MMC story. Gupta et al. (2024) and Madhu et al. (2025) demonstrate compellingly how MMCs are important for sustainability in manufacturing. Their durability and wear resistance extend product life, therefore limiting material consumption. The impacts that lightweighting can have on energy efficiency in transportation (lower fuel consumption and emissions) and machinery with the ability to achieve similar or better performance for less mass are time-tested, relevant today, and paramount going forward. In addition, the effort is being made to find ways to utilize recycled matrix materials and to develop recycling pathways of MMCs within circular economy focuses (Gupta et al., 2024; Pooja et al., 2025; Miracle, 2005).

The material science community has yet to comprehensively explore the advantages of the engineering flexibility that MMCs offer in relation to tailoring the MMC properties based on matrix/reinforcement selection, architecture, and processing optimization to solve engineering problems in which performance, weight, longevity, and environmental impact must be considered simultaneously. As Pooja et al. (2025) explain, MMCs are "revolutionary materials for shaping the future," which are now progressing to engineered solutions involving sustainable industry. This chapter discusses this evolving area and includes materials, production, the significance of optimization methods, and the tribological aspects that will determine the success of implementing them into a sustainability-focused industrial structure.

1.2 CLASSIFICATION OF METAL MATRIX COMPOSITES

Building upon the foundational understanding of MMCs as pivotal enablers of sustainable industrial progress, we now explore their systematic classification. This is a direct consequence of how MMCs can be used, which will guide the selection of materials and the selection of processing route and work toward the successful implementation of MMCs into performance-enhancing yet sustainability-driven applications. The classification scheme is multilayered, based on four significant dimensions that will likely have a huge impact on the properties, cost, manufacturability, sustainability, and tribological performance of the composite (Miracle, 2005).

1.2.1 Classification of Matrix Materials

Like all engineering materials, the choice of metal matrix governs the baseline properties such as density, melting point, ductility, thermal/electrical conductivity, and corrosion resistance, while also influencing recyclability potential and contribution to lightweighting. The systems most commonly used, and successful as a commercial development, simply demonstrate some flavor of collaborative performance, processing, and cost.

- **Aluminum Matrix Composites (AMCs)**: AMCs are the most common and commercially successful of MMCs (Makhesana et al., 2025; Khanna et al., 2024). Aluminum alloys, as a metal matrix system, offer low density, high strength, good corrosion resistance, and good casting and machining ability. If explicitly reinforced with SiC, Al_2O_3, B_4C, or graphite particles, AMCs exhibit considerable improvements in regard to specific stiffness, specific strength, potential wear resistance, and elevated temperature capabilities. Their main objective is to enable weight loss (i.e., lightweighting, which is essential for reducing energy consumption and emissions) in automotive and aerospace components and to improve thermal management in electronic products (Madhu et al., 2025; Yan & Xu, 2025; Gupta et al., 2024).

- **Magnesium Matrix Composites (MMgCs)**: MMgCs represent the ultimate lightweighting material (Pooja et al., 2025; Madhu et al., 2025), since magnesium alloys are ~35% lighter than aluminum. Composites with Mg matrices reinforced with SiC, Al_2O_3, or CNTs exhibit remarkable specific strength and stiffness, making them highly attractive for use in aerospace, defense, and portable electronics. However, limited ductility, corrosion vulnerability, and higher manufacturing costs have limited broader use. Recent research has focused on improving corrosion resistance and developing inexpensive ways to promote lighter-weight composites in a sustainable manner.
- **Copper Matrix Composites (CMCs)**: CMCs are intended to provide strength and wear resistance while possessing ideally superior thermal and electrical conductivity. Reinforcements are typically in the form of graphite (to provide self-lubrication), SiC, Al_2O_3, diamond, or carbon fibers. CMCs find common use in high-performance electrical contacts, welding electrodes, heat sinks, and sliding contacts (Makhesana et al., 2025; Khanna et al., 2024). Development of improvements in energy efficiency associated with power transmission and thermal management is significant in terms of sustainability.
- **Titanium Matrix Composites (TMCs)**: TMCs are intended for the most demanding applications and high specific strength, with ideally exceptional creep and corrosion resistance at higher temperatures (jet engines/hypersonic vehicles). Continuous SiC fibers are readily available in Ti alloys, but there are also (discontinuous) reinforced TMCs (Makhesana et al., 2025; Pooja et al., 2025). Currently, their high cost limits applications primarily to aerospace. The net sustainability benefits of TMCs arise from component longevity in extreme conditions, as well as decreased fuel consumption due to weight savings.

1.2.2 Classification Based on Reinforcement Type and Architecture

Composite classification by reinforcement design can be based on the structure, morphology, and distribution of the reinforcement phase. This will govern mechanical, thermal, and even tribological properties (e.g., wear resistance).

- **Particulate-Reinforced Composites**: This is the most commercially relevant type, featuring ceramic particles (SiC, Al_2O_3, B_4C, and TiB_2) typically in the 1 to 100 micron size. They contain isotropic qualities, adequate ductility, and formability based on stir casting and powder metallurgy. Volume fractions range from 5% to 30%. The main benefits are higher strength, higher stiffness, and higher wear resistance at a reasonable price, making these composites better suited for applications in automotive components, brake systems, or fretting wear parts (Khanna et al., 2024; Makhesana et al., 2025; Madhu et al., 2025).

- **Continuous Fiber-Reinforced Composites**: Continuous fiber-reinforced composites have long, aligned fibers (SiC, Al_2O_3, carbon, and boron) embedded in the matrix. Continuous fibers give the best specific strength and stiffness (i.e., anisotropic) characteristics, along with higher creep resistance and fatigue resistance. Also, continuous fibers can be difficult and expensive to produce, have low transverse strength, and exhibit variable workability. So these composites are largely limited to ultra-high-performance aerospace applications (Mortensen & Llorca, 2010; Makhesana et al., 2025) where weight matters the most.
- **Whisker-Reinforced and Chopped Fiber-Reinforced Composites**: Whisker-reinforced composites utilize single crystal whiskers (SiC and Si_3N_4) or chopped fibers. These composites are in the middle. They can provide similar or stronger/stiffer properties than particulates (especially if aligned), while maintaining isotropic properties and manufacturability advantages compared to continuous fibers. There are also health (whiskers) or stable orientation disadvantages. The composites can be used for structures where more stiffness than particulates is needed, typically done through powder metallurgy (Makhesana et al., 2025).

1.2.3 Hybrid Composites: Synergistic Multi-functionality

Hybrid MMCs are another sophisticated development in materials engineering, distinct from advancing single-reinforcement systems. Hybrid MMCs contain two or more distinct types of reinforcement in a single metallic matrix (Al, Mg, Cu, and Ti) (Gupta et al., 2024; Pooja et al., 2025). By establishing hybridization, users can directly target very particular property combinations to achieve very particular objectives that would otherwise be unattainable or contradictory. The ultimate goal is to achieve synergistic multi-functionality and optimize the composite for simultaneously accommodating competing requirements of paramount importance in sustainable design (Khanna et al., 2024; Pooja et al., 2025). Hybridization can be designed to exploit complementary functionalities; for instance, hybridization can be designed to mitigate disadvantages (e.g., adding solid lubricants such as graphite to combat the high friction of hard ceramic reinforcements such as SiC) or aim to blend capabilities (e.g., combining strength with neutron absorption or thermal management with electrical insulation) (Gupta et al., 2024; Madhu et al., 2025; Adi & Malik, 2025). This hybridization strategy can indeed improve cost-effectiveness by the use of small amounts of expensive, high-performance reinforcement (e.g., CNTs) blended with larger amounts of inexpensive reinforcement (e.g., SiC particulates). It specifically targets sustainability goals across multiple performance parameters, such as lowering friction/energy loss, prolonging component lifetime, improving efficiency with lightweight functionality, or enabling sustainable designs requiring resources. The design of effective hybrid MMCs requires careful consideration across many interacting factors. Reinforcement compatibility is critical and includes chemical compatibility – both in the matrix environment and relative to each other, to prevent detrimental reactions leading to brittle phases; thermal compatibility – to control coefficient of thermal expansion (CTE) mismatches leading to residual stresses and thus inducing microcracking or distortion; and physical compatibility – to avoid complex separation based on density leading to segregation, and particle size/morphology impacting dispersion. Interfacial interactions are critically challenging because multiple faces exist (Matrix–Reinforcement A, Matrix–Reinforcement B, and potentially Reinforcement A-Reinforcement B). Each interface must be designed for optimal bonding to facilitate effective load transfer and synergy of properties; weak interfaces would be sites of failure, and interfacial reactions represent a continued significant concern, especially with liquid-state processing. Processing challenges are compounded: achieving effective uniformity of homogeneous dispersion of different types of reinforcement (especially if they impair any type, shape, or density) is extremely difficult; agglomeration negates all benefits; and low-viscosity reinforcements (whiskers, short fibers, and CNTs) are fragile and therefore can easily be damaged in mechanical processing steps. Process selection is also limited; stir casting cannot accommodate long fibers, which must instead be introduced alongside particles during p/m or infiltration (Parikh et al., 2023; Mussatto et al., 2021).

Several case studies explain the advantages of hybridization in Al/SiCp + Gr flake composites, which is due to the synergistics of the extra reinforcement imparted via the combined effects of SiC and graphite, and particularly so in the solid state. SiC supplies strength and resistance to abrasive wear, and graphite comes with solid lubrication properties, which reduce the friction coefficient significantly. For use in bearing applications, bushings, or piston skirts, this hybrid has improved energy and component life (Gupta et al., 2024; Madhu et al., 2025). Al/SiC whiskers + SiC particles combine the ultra-high stiffness and strength of whiskers with the lower cost, improved isotropy, and better fracture toughness from particles. This inexpensive combination is aimed at high-performance automotive (suspension and brake calipers) and aerospace structural parts (Makhesana et al., 2025). Al/B_4Cp + CNTs is a wonderful blend of the great neutron absorption of B_4C with the awesome strength and stiffness of CNTs, improving the mechanical integrity and the radiation stability of the components for nuclear shielding, control rods, and fusion reactor structures (Pooja et al., 2025). The Cu/Al_2O_3 Fibers + Gr Particles offer copper-like components, with the reinforcement of Al_2O_3 fibers for wear resistance and thermal stability, while providing graphite particles for some essential self-lubrication. This hybrid is invaluable for durable and reliable electrical sliding contacts, like pantographs and high-current brushes, minimizing downtime and energy losses (Khanna et al., 2024; Gupta et al., 2024).

New trends suggest an exciting direction forward. One such trend is multi-scale hybridization, which involves combining reinforcements of different/distinct length scales (e.g., microscale particles + nanoscale particles/tubes) to take advantage of synergistic effects that may improve strength, toughness, and fatigue resistance. Functionally Graded Hybrids (FGHs) are enabled by Additive Manufacturing (AM), which allows the type, size, or concentration of reinforcements to vary spatially inside a single component. This enables optimization of properties such as hardness or thermal conductivity, whether in a deeper part of the part or a thinner or more complex part of the same component (Chen et al., 2024). The exploration of novel reinforcement combinations could include metallic glasses (for strength/corrosion resistance), Shape Memory Alloys (SMAs) (damping/crack closure), and 2D materials (graphene and h-BN for lubrication/barrier properties). There is also the idea of "Smart" hybrids introducing sensory (e.g., piezoelectric particles) and/or self-healing systems. Lastly, hybridization is already expanding to matrix systems themselves, with the use of Multi-Principal Element Alloys (MPEAs)/High-Entropy Alloys (HEAs) and/or intermetallic matrices (e.g., TiAl and NiAl) together with combinations of tailored hybrid reinforcements to formulate combinations of properties that have not been previously accomplished (Pooja et al., 2025). Hybrid MMCs exemplify smart conceptual design for materials, optimizing synergies to address the complexity of aiming for sustainable engineering, often complicated by conflicting priority objectives. Partially fulfilled packages offer great promise for producing lower-weight, more durable, highly energy-efficient, and better-performing components, which, together with the expanded modes available, strongly position MMCs as a foundational engineering material of future industrial ecosystems.

1.2.4 Classification by Processing

The processing method used to manufacture MMCs is of paramount significance in establishing the microstructure and its interfaces, defect density, final properties, cost implications, manufacturing scalability, and the lifecycle environmental impact (Mortensen & Llorca, 2010; Miracle, 2005). Moreover, each processing method will provide unique advantages, disadvantages, and consequential performance and sustainability implications of interest to manufacturers, regulators, and consumers of products constructed with composites.

- **Stir Casting (Liquid-State Processing)**: It consists of stirring molten matrix metal until a vortex is created, and the reinforcement particles (usually) may be added/dispersed and cast off into the fused phase. Variations of stir casting methods, such as compocasting (stirring in the semi-solid state), allow for improved reinforcement distribution, reducing settling and reactions. The foremost advantages of stir casting are its low cost (uses existing foundry capability), response to scalability requirements at higher levels of mass and spatial scale component processing, demonstration of excellent speed of production to facilitate near-net-shape complex geometries, and workability with castable alloys (Al, Mg, Zn, and Cu). The sustainability advantages of stir casting include lower energy input than powder capabilities and excellent compatibility with recycling matrix melts to maximize circularity (Gupta et al., 2024; Makhesana et al., 2025; Parikh et al., 2023; Mussatto et al., 2021). However, some major limitations exist: it restricts the use of long fibers, imposes challenges when processing materials with a high suspended load of particles/whiskers or short fibers (as long fibers are damaged), and makes it difficult to obtain uniform dispersion – agglomeration/sedimentation is often prevented only with new, novel techniques (e.g., ultrasonic cavitation). Furthermore, low wettability means fluxes/coatings need to be employed; highly offensive interfacial reactions (e.g., Al4C3 in Al/SiC) require protection/control; porosity and weak interfaces limit achievable properties; interfacial volume fraction is generally low (for Vf 50%). It provides excellent microstructural control of reinforcement morphology, allows the production of near-net-shape parts (sometimes complex shapes), and minimizes interfacial reactions due to lower temperature. It also permits the use of alloys that are otherwise difficult to cast, and it literally yields a recyclable scrap – including recyclable MMC swarf – all of which can be delivered to optimize the use of property per unit mass and limits machining scraps (Khanna et al., 2024; Makhesana et al., 2025; Madhu et al., 2025; Parikh et al., 2023; Mussatto et al., 2021). Disadvantages include the cost of entry (expensive powders and complex processing); the use of MMCs often produces porosity, which usually requires a refinement/HIP, size, and shape limitations; powders during handling require attention (to limit contamination, oxidation, and human health hazards); and the expense/energy input for powder manufacturing and processing (compaction and sintering/HIP). P/M yields interface quality, density, and reinforcement distribution (may be anisotropic), and higher strength/stiffness composite products are possible. P/M is the dominant route for TMCs, CMCs, whisker-reinforced composites, nanocomposites, and high-performance automotive/aerospace (e.g., connecting rods and valve seats).

- **Infiltration Techniques**: These techniques rely on a porous reinforcement preform (fibers, a particle mix, and foam) that the matrix metal is forced into when in molten form. Infiltration with pressure (e.g., Squeeze Casting) forces the molten metal into the die with preform under 50–150 MPa, thereby overcoming poor wettability, minimizing porosity, and enabling quick solidification period to produce a fine microstructure and good dimensional accuracy. Spontaneous infiltration (e.g., DIMOX) applies no external pressure; instead, it makes use of chemical reaction mechanisms to promote wetting and capillary force infiltration. Infiltration works quite well with continuous fiber preforms complex enough to allow for high Vf and good alignment/preservation of the reinforcement. Infiltration also allows near-net shape production with good potential along various matrices: Al, Mg, Cu, Ni, and intermetallics. It also has limitations: equipment (pressure) costs, controlled atmosphere/chemistry (spontaneous infiltration), fabrication costs and complexity of making the preform, opportunities for preform instability, interfacial reactions (spontaneous infiltration), limited size and shape constraints, and energy demands (particularly high pressure). Infiltration is required for continuous fiber-reinforced aerospace structures (i.e., tubes and members), high-performance thermal management substrates (i.e., SiC fiber/Al), and other complex architecture such as brake discs (Parikh et al., 2023; Mussatto et al., 2021).
- **Spray Deposition**: A stream of inert gas injects the reinforcements and molten matrix metal to atomize the matrix metal; then, the resulting solidified mixture forms into a preform bound to a substrate that rapidly solidifies into

a bulk or near-net shape component. The advantages of spray deposition include rapid solidification that creates very fine and homogeneous microstructures, reduced segregation, the ability to form near-net shapes, compatibility with many matrix and particulate choices, and limited interfacial reactions due to the short contact time with the vertices. The disadvantages of spray deposition include unavoidable porosity (2–5%) that requires a subsequent consolidation; developing precise, complicated process control; difficulty in achieving a completely uniform and random reinforcement dispersion; high equipment cost; and limitation to simpler geometries. Spray deposition has property effects associated with the fine grain size, limited segregation, and limited interfacial reactions (including porosity limiting-assemblies). The principal utilization for spray deposition is as billets for forging/extrusion high-performance parts and overlays for wear-resistant phases (Parikh et al., 2023; Mussatto et al., 2021; Xie et al., 2021).

- **Additive Manufacturing**: AM of MMCs builds parts from a digital model using MMC feedstock material and allows building parts layer by layer. This can occur predominately through Powder Bed Fusion (SLM and EBM), Directed Energy Deposition (LMD and WAAM), or Friction Stir Additive Manufacturing (FSAM). AM is revolutionizing the way we design and produce parts with unprecedented potential benefits: 1) Design freedom is unparalleled: AM allows for many complex features to be manufactured and designed for topology-optimized lightweight structures; 2) Unique capability to manufacture Functionally Graded Materials (FGMs) – spatially varying the reinforcement of a component; 3) Intricately architected materials (lattices and cellular structures) can be rapidly fabricated; 4) Waste of material is reduced (as indicated in terms of the appropriate maximum weight to initial weight ratio (purchase to fly)); 5) Rapid prototyping and customization; 6) Less tooling, as many parts can be manufactured directly; and 7) Sustainability as a result of large reductions in material waste and the possibility of lightweight products creating less operational energy (Pooja et al., 2025; Adi & Malik, 2025; Lakshmikanthan et al., 2025; Gupta et al., 2024).

AM is becoming increasingly popular and has garnered a great deal of interest, but the process has many significant obstacles, such as extreme process complexity in melting and solidification in conjunction with reinforcements, consistent distribution that assures no agglomerations or clustering of reinforcements and degradation of reinforcement (e.g., damaging CNT), many defects (e.g., porosity, lack-of-fusion, cracking, and residual stresses), limiting reinforcements primarily to particulates (fibers, especially continuous are very difficult), expensive, difficult-to-manufacture feedstock material (namely composite powders/wire), slow build rates, expensive equipment/materials, frequent post-processing (HIP, machining, etc.), poor surface finish (i.e., dimensional accuracy), and high energy intensity required to complete the build. Current applications focus on prototyping, complex MMC parts, lightweight structurally optimized component parts (e.g., aerospace/defense applications), customized biomedical implants, functionally graded devices for thermal management, specialized tooling (AM), and repair applications. There are research initiatives and pilot projects working to expand the technology further (Lakshmikanthan et al., 2025; Chen et al., 2024).

Choosing a processing route for an MMC is ultimately a sustainability decision considering the entire system (i.e., performance, cost, scalability, design freedom, and lifecycle sustainability). Performance is inclusive of the microstructure and interface properties, defects, and whether materials and reinforcement behave as desired properties. Cost is cumulative between capital costs and materials costs (direct cost), energy, scrap yield (indirect cost), and costs associated with post-processing, such as machining and surface finishing, and e-waste, which are not insignificant. Scalability to deliver a design with a production scaling from prototype to a volume may dismantle some factories as they scale their production size, pace, and part-building system for greater speed. Is design freedom critical when the performance is optimum and sustainability is determined by the lifecycle energy cost per part, material efficiency (including kilograms in the part and kilograms sold), compatibility with used recycled materials, the recyclability of any scrap, or the potential for lightweighting/longevity? In summary, there is no single best process for producing every part; this decision is very much influenced by the specific matrix or matrices, type of reinforcement, desired properties or performance, processing method, the part geometry, production volume, and eligible cost targets, with sustainability priorities at the forefront.

1.3 METAL MATRIX COMPOSITES FOR INDUSTRIAL APPLICATIONS: ENGINEERING PERFORMANCE

The systematic classification of MMCs provides the necessary framework for their utilization across different industrial segments. It is clear that since there have been some global movements toward sustainability, MMCs are moving from special-purpose materials to industrial staples, offering solutions to demanding engineering issues (Gupta et al. 2024). This section examines how certain classes of MMCs influence major industries by utilizing their engineered features.

- **Automotive Industry**: The automotive industry is the largest consumer of DRMMCs, predominantly AMCs, because of the constant pressure for weight reduction and more efficient powertrains. AMCs – made with Al alloys (such as A356 and 6061) and containing either SiC or Al_2O_3 particles (normally 10–25 vol%) and produced by stir casting or powder metallurgy (P/M) – have improved unique properties. Engine components such as piston crowns and cylinder liners take advantage of wear resistance, together with elevated temperature strength and lower thermal expansion of AMCs, with SiC/Al pistons indicating a 5–7% reduction in weight and a 15% reduction in wear compared to standard cast iron (Sharma et al. 2020). AMCs used in brake systems offer great advantages due to their high thermal conductivity relative to specific heat capacity, and improved wear resistance (often with a higher SiC fraction), resulting in enhanced performance. Braking stability is also enhanced with hybrid composites using graphite (Gupta et al. 2024; Pooja et al. 2025). The driveline and chassis parts (e.g., 6061/SiCp driveshaft – up to 50% weight saving) reduce the overall weight compared to steel and lower rotational inertia. The safety and part performance improvement by using AMCs indicates greater strength and time savings. Suspension components, such as springs, require high specific stiffness and strength, and AMCs can also be used in hybrid composites (Khanna et al., 2024; Makhesana et al., 2025; Rudrakotti & Chandrasekaran, 2025; Yan & Xu, 2025).

- **Aerospace Industry**: Aerospace configurations are the most advanced use of MMCs. Aerospace heat-resistant and extreme-lightweighting MMC applications exemplify both challenges. For example, SiC-reinforced AMCs are used in airframe and structural components like ventral fins. Discontinuously reinforced AMCs are commonly employed (e.g., 2009/SiC/15p) in longerons and transmission housing. These applications provide weight savings of 20–30% compared to monolithic alloys while having similar stiffness and damage tolerance properties. Because of the requirements of aerospace applications, particularly with mechanical properties, structural and airframe applications are typically produced using powder metallurgy (P/M) methods to improve consumption rates of raw materials and maximum performance (Madhu et al., 2025; Pooja et al., 2025). In propulsion applications, we see the effect of high-temperature MMC surfaces. TMCs can employ continuous SiC fibers (e.g., SCS-6/Ti-6Al-4V) and operational temperatures above 500°C and are used for component systems such as compressor rings, blades, and shafts. Discontinuously reinforced TMCs and monolithic nickel-superalloy MMCs are being developed for turbine applications (Makhesana et al., 2025). For hypersonics and space applications, where materials are required to perform under extreme thermo-mechanical loads, scientists and engineers are testing ultra-high-temperature ceramic (UHTC) reinforced refractory metal matrices for leading edges and components of engine systems. Examples of UHTC materials include C/C-SiC, and they are fiber-reinforced with ZrB_2/SiC fibers (Pooja et al., 2025).
- **Defense and Military**: The defense industry/application is a key marketplace for tailored MMCs – specifically as it relates to performance, survivability, and longevity in austere, rugged environments. Armor systems include lightweight aluminum and magnesium MMCs that are reinforced with high-hardness ceramics such as Al_2O_3 and B_4C. These systems have been developed for improved ballistic performance per unit weight compared with steel, and for demonstrating hybrid capabilities (Madhu et al., 2025). Track components and other wear parts are other examples of specialty use of MMCs subjected to extreme wear and impact. High wear-resistant AMCs, or Steel Matrix Composites (SMCs), with TiC/WC particles are specified for use in tank tracks, suspension, and recoil (Makhesana et al., 2025).
- **Electronics and Thermal Management**: AMC-based lightweight, stiff, and hybrid CMC heat management systems are also used for thermal

management and aerospace applications on lightweight electronic systems like optical systems and drones (Khanna et al., 2024). Electrical and thermal applications: MMCs for energy management. This sector is largely influenced by CMCs while also claiming a place for specialized AMCs, with engineered properties being the core of these sectors. CMCs for high-power electronics and thermal management applications consist of CMCs with high thermal conductivity reinforcement materials (diamond, graphite flakes, etc.) that meet or exceed the thermal conductivity of copper with the same CTE as semiconductors, greatly reducing thermal fatigue of contacts, coolers, heat sinks, and heat management packages like electronic systems. Similarly, AMCs with SiC and graphite serve as heat spreaders that reduce weight (Madhu et al., 2025; Makhesana et al., 2025). CMCs and functionalized CMCs have been used for electrical contacts and sliding contact applications, where the requirements for conductivity, strength, and wear resistance are conferred in some CMCs. Conventionally, Cu/Gr (graphite-reinforced copper) has been used for self-lubricating brushes in motors and pantographs. Hybrid CMCs (Al_2O_3/SiC particles) are used where wear resistance is essential, and AMCs with CNTs are being tested as lightweight conductors (Khanna et al., 2024; Gupta et al., 2024). Welding and resistance heating applications also involve the composite space with materials like Cu/W or Cu/Mo composites, combining together materials with high conductivity (Cu) alongside the high temperature strength and erosion resistance of W/Mo electrodes, making the electrode much more efficient and long-lasting (Makhesana et al., 2025). These varied applications illustrate the diverse strategies involved in selecting MMC classes (matrix, reinforcement, and processing) to solve specific industrial needs, which naturally leads toward keeping sustainability issues in mind for their consolidated impact.

1.4 RELEVANCE TO TODAY'S MODERN SUSTAINABLE INDUSTRY

The performances of MMCs across many industrial fields reveal significant potential as enablers of industrial sustainability (Gupta et al., 2024). Its inherent characteristics, designed from matrix, reinforcement, and processing, address the core columns of Sustainable Development: resource efficiency, energy efficiency, emission reduction, and lifespan extensibility. Transitioning to using MMCs is a purposeful step toward a resilient and sustainable industry.

- **Lightweight**: The actuating element of energy efficiency is weight. Without question, the most immediate contribution of MMCs is mass reduction. By replacing heavier metals (steel and cast iron) with lighter MMCs (Al, Mg), it will reduce moving component and structure mass substantially, resulting in lower energy demand and energy reductions. The need for lighter materials is especially potent in the world of transportation. Industry studies overwhelmingly confirm that a 10% mass reduction on a vehicle equates to a 6–8% fuel efficiency improvement (Khanna et al., 2024).

MMC components (pistons, brakes, driveshafts, and airframe) will result in mass reductions, and thus reduced fuel requirements and lifetime greenhouse gas (GHG) emissions. The weight savings from MMCs can similarly reduce fuel consumption per flight hour in aerospace applications (Madhu et al., 2025; Pooja et al., 2025). In industrial machinery, there are many opportunities for reduced mass on robotic arms, parts of equipment, and handling systems, which will all reduce acceleration/deceleration energy requirements, ultimately improving overall system efficiency.

- **Enhanced Durability**: MMCs will often provide superior wear resistance, fatigue strength, creep resistance, and corrosion resistance as compared to monolithic alloys, which fundamentally alters the lifecycle of a product. Longer service life is a direct benefit of using MMC in the case of components like cylinder liners, brake rotors, wear plates, cutting tools, and electrical contacts. An example of MMC's components is that they can deliver significantly longer service life. For example, SiC/Al brake rotors have a wear rate of 50% lower than cast iron (Makhesana et al., 2025). This results in less replacement during use, reducing material and energy used during extraction and manufacturing, as well as minimizing the waste during manufacturing and waste disposal of the replacement parts. Less maintenance and downtime are achieved through long-lasting components, which have implications for operational efficiency through more resource-efficient use while reducing the environmental impact of maintenance activities (transport, lubricants, etc.) (Gupta et al., 2024). Moreover, the high-temperature stability of the MMC provides the potential for better life expectancy under harsh conditions.
- **Operational Efficiency**: The operational efficiency associated with lightweighting and reduced friction (one major tribological advantage of many MMCs) provides immediate operational benefits. The lower fuel/power demand is a net result of a decrease in mass and frictional losses (e.g., MMCs in engines or bearings) and by using less energy (fuel/electrical) for the same work, resulting in a permanent reduction in operational carbon footprint. Emissions foreground simply, and as a result of reduced fuel combustion, there are fewer CO_2, NO_x, and particulate emissions. Moreover, MMCs allow for the technology of efficiency upgrades (e.g., advanced turbocharging or heat recovery) using their high service temperature capacity (Madhu et al., 2025), and potential strategies to help achieve aggressive emissions regulations worldwide (Pooja et al., 2025).
- **Recycling and End-of-Life**: In order to achieve true sustainability, it is necessary to address end-of-life (EoL) management, which creates challenges and even opportunities, both determined by matrix choice and process. However, there are challenges, as the value of old recyclables is poorly understood by traditional metal recycling streams, especially for materials with ceramic or carbon reinforcements. Rapid processing can risk melt contamination for recycled metal; hybrid composite EoL is even worse in traditional metal markets since they have two or more materials, highlighting the need for efficient, sensible separation processes (Gupta et al., 2024).

Opportunities and progress in the space include material recycling processes such as salt-flux separation, as well as specific remelting to recover metal from the scrap for lower-grade use or even recovery with controlled contamination for the formation of new MMCs (Makhesana et al., 2025; Khanna et al., 2024). Direct reuse or remanufacturing should also be real opportunities for components lasting as long as MMCs. The design for recycling (DfR) concepts also focus on reducing reinventing Mc components upon separation (e.g., introduction of certain types of reinforcements), or systems with inert matrices, as P/M will make scrap recyclable, etc. Recovery of reinforcements for high-value materials (SiC or carbon fibers) is being given some degree of consideration compared to the recovery of the matrix; however, they are still less mature.

The role of MMCs toward a sustainable industry is clear. They generate lighter structures, they provide lower operational energy and emissions, and they can achieve significant lifespans, thus promoting circularity in the industry. Their effect is compounded throughout the lifecycle – they impact material selection, facilitate optimized manufacture, extend service life, and can promote more responsible EoL management for sustainability. As costs reduce and technologies develop, MMCs can progress from merely high-performance methods of achieving the desired outcome to vital enablers for a sustainable industry future, while the optimization of their properties and their excellent tribological performance (discussed next) are essential for maximizing their potential for sustainable industry.

1.5 STRUCTURE AND PROPERTIES OF METAL MATRIX COMPOSITES

The sustainability advantages made possible by MMCs – lightweighting, durability, emission reduction, and lifecycle management – are based fundamentally on the unique structure-property relationships for MMCs. The complex interactions between the metallic matrix, the ceramic or carbon reinforcement, interfacial features, and the controlled processing and classification provide the performance characteristics for various sustainable industrial applications. This section discusses the microstructural origins and properties that make MMCs essential to a resource-efficient future, and provides the necessary science-based rationale for the optimization and tribological behavior discussed later.

1.5.1 Microstructural Features: Governing Performance

The microstructure of an MMC is the dominant factor governing the overall behavior of an MMC, and is conditioned by the following three components, which are interconnected by the characteristics of the matrix, the type of reinforcement, and the processing route (Rudrakotti & Chandrasekaran, 2025; Mortensen & Llorca, 2010; Miracle, 2005). The matrix microstructure (the continuous metallic phase: Al, Mg, Cu, Ti) dominates the baseline properties of the composite; this may include ductility, toughness, thermal and electrical conductivity, and corrosion resistance. The features of the matrix microstructure (e.g., grain size, phase distribution [e.g., precipitates in age-hardenable alloys], and dislocation density) are also influenced by the presence of the reinforcement and the processing route. An example of this includes the rapid solidification that occurs during spray deposition, or through AM, which may produce extremely fine-grained structures, which may lead to a much stronger structure (Rudrakotti & Chandrasekaran, 2025; Lakshmikanthan et al., 2025; Makhesana et al., 2025; Xie et al., 2021). The characteristics of the reinforcement (including type, size, shape, volume fraction, and spatial distribution (SiC, Al_2O_3, B_4C, graphite, CNTs, etc.)) remain the most controlled levers of performance. Uniform dispersion is particularly important in particular composites to avoid stress concentration to achieve the desired isotropic properties, while fiber alignment ultimately dictates the amount of anisotropic behavior in continuous fiber MMCs. In particular, nanoscale reinforcement (nanoparticles and CNTs) is the most effective in terms of strength development and also the most difficult to obtain a stable, agglomerate-free dispersion (Pooja et al., 2025). The matrix–reinforcement interface is arguably the most critical microstructure because this interface governs the critical transfer of load from the relatively ductile matrix to the stiffened, strengthened reinforcement. A chemical bond at the interface is essential for reinforcing, yet this area can be adversely affected by several negative phenomena: this interface can react chemically during processing using liquid-state processing, generating brittle intermetallic compounds (e.g., Al_4C_3 in the Al/SiC composite), which will severely reduce mechanical properties and corrosion resistance. Some methods to inhibit those reactions are processing time and temperature, controlled inert atmosphere, and protective coatings (e.g., SiO_2 on SiC particles) (Khanna et al., 2024; Madhu et al., 2025). Thermal mismatch stresses at the interface arise due to differences in CTE between the matrix and reinforcement. Residual stresses occur at cooling room temperature from processing or service temperatures. If residual stresses occur, this will dramatically change yield strength behavior, fatigue crack initiation and propagation, and dimensional stability, which aids with material selection and process control (Gupta et al., 2024). Most importantly, this interface also controls wear resistance and friction coefficient – important tribological properties, which are the focus of this book.

1.5.2 Mechanical Properties: Strength, Stiffness, and the Ductility Challenge

In general, MMCs develop improved mechanical properties over the original alloy matrix using microstructural interaction from varying processing conditions. Tensile strength and stiffness are dramatically improved through several mechanisms: load transfer is accomplished by having the stiff, high-modulus reinforcement take a disproportionately larger fraction of the applied load. Orowan strengthening takes place when small particles serve as barriers to dislocation slip in the matrix, increasing the stress required to deform plastically. Grain refinement is frequently caused by grain boundaries being pinned by the reinforcements during solidification, leading to a finer matrix microstructure (Hall-Petch strengthening). A greater dislocation density is developed through geometry around the reinforcements due to CTE mismatch in cooling, and this also inhibits dislocation slip. The increase in elastic modulus (stiffness) can be considerable, especially when the reinforcement is a high-modulus ceramic or continuous fibers aligned to the load direction. Predictive models range from the Rule of Mixtures (in theory, the upper bound) to modified rules such as Halpin-Tsai for short-fiber composites. The influence of strength and stiffness increases with the percentage volume fraction of the reinforcements and decreases with the particle size of the reinforcements for particulate MMCs (Makhesana et al., 2025; Madhu et al., 2025; Rudrakotti & Chandrasekaran, 2025). Hardness is defined in this context as an indicator of resistance to surface deformation, indentation, abrasion, and wear, which is also greatly improved through the incorporation of hard ceramic reinforcements such as SiC, B_4C, and TiB_2. The additional hardness supports the greatly improved durability benefits required for sustainability. A negative relationship exists between toughness and ductility. While strength and hardness increase, ductility (percentage elongation to failure) and fracture toughness (KIC) decrease relative to the matrix. Brittle reinforcements can act as crack initiators, and the high level of constraint imposed on thermal and mechanical expansion or contraction inhibits plastic deformation and promotes crack tip blunting. Ways to reduce this trade-off include using a bigger ductility matrix alloy, using lower volume fractions of reinforcement, using finer and more uniform reinforcement distribution, preferably using particles that are spherical in morphology rather than angular to limit stress concentrations, and developing a strong but non-embrittled interface. Hybrid composites may form a pathway to achieve a balance of competing requirements, such as using hybrid particle strength with a ductile phase or solid lubricant (Rudrakotti & Chandrasekaran, 2025; Gupta et al., 2024; Khanna et al., 2024). These enhancements to mechanical properties are the basis for lightweighting and structural efficiency.

1.5.3 Thermal and Electrical Properties: Enabling Efficient Systems

MMCs offer unique pathways to customize thermal and electrical properties, which are integral to sustainable energy systems and modern electronic applications. Thermal conductivity (k) is extremely sensitive to matrix and reinforcement properties, volume fractions, distribution, and most importantly, the quality of the interface. CMCs with high-k reinforcements such as diamond particles or graphite flakes can have thermal conductivities greater than pure copper. This means they have potential for high-heat-flux applications such as heat sinks. AMCs with SiC are favorable, as they provide a good trade-off of thermal conductivity and low density. In contrast, poor interfacial bonding, porosity, or thermal barrier layers at interfaces can all significantly reduce the effective thermal conductivity of the composite and act as bottlenecks for the effective heat flow (Madhu et al., 2025; Rudrakotti & Chandrasekaran, 2025; Makhesana et al., 2025; Miracle, 2005). Good thermal management offers additional benefits by reducing waste energy expenditure in electronics and power for engineering systems. CTE is another very important property because ceramic reinforcements possess a lower CTE than metals. Utilizing combinations can diminish the composite CTE to suit other constituent materials within an assembly (e.g., semiconductors or ceramics within electronic packages). This results in reduced deleterious thermal stresses developed during temperature cycling, and provides much better reliability to your devices, whether it's in longevity or lower rates of failure – things that go toward sustainable electronics and high-precision engine parts (Rudrakotti & Chandrasekaran, 2025; Gupta et al., 2024). Electrical conduction is mainly controlled by the matrix phase in ceramic-reinforced MMCs. Copper/graphite composites have reasonable conductivity (since copper is only partially replaced with graphite), and can benefit from graphite's solid lubricity. Introducing insulating ceramic reinforcements (SiC and Al_2O_3) cuts down on bulk electrical conductivity (the reduction in conductivity may be advantageous for certain applications that require controlled resistivity, e.g., grounding straps, or electromagnetic interference (EMI) shielding may be useful to control ESD or EMI entrances or exits) (Khanna et al., 2024). Modifying these properties is crucial to defining the role of MMCs in energy-efficient systems – sustainable electronics or high-precision machinery.

1.6 EVOLUTION OF MMCS: FROM AEROSPACE CURIOSITY TO INDUSTRIAL NECESSITY

It is not a coincidence that MMCs are nearly universally superior in specific strength, stiffness, thermal stability, wear resistance, and a host of functional properties compared to conventional materials. MMCs are not just laboratory enhancement stories, but part of a decades-long actual evolution, demonstrating how conventional materials are being replaced and the role MMCs play as mission-critical components within demanding engine and industrial applications (Khanna et al., 2024). The success of these applications, based on demands for enhanced performance and efficiency, but even more so, precise sustainability criteria, relies on the transition in how design paradigms developed toward material choices. MMCs are being viewed as expensive items from the aerospace industry and design processes. This view is misleading: they should not be seen merely as expensive or difficult to use but rather as effective industrial material solutions worthy of high value.

1.6.1 Wear and Corrosion Resistance: Pillars of Component Longevity

Excellent tribological and corrosion characteristics are at the core of MMCs being able to tell a sustainability story through prolonged product life, resulting in less consumption of resources and turnover. Wear resistance is a dominant characteristic for many MMCs, particularly those that exhibit hard ceramic particles of some type (SiC, Al_2O_3, or B_4C) or solid lubricants (graphite, MoS_2) and in general exhibit a better performance in abrasive, adhesive, or erosive wear when compared to monolithic alloy materials. They have better resistance to penetration and abrasion from counterfaces or particles that are entrained. In comparison to a monolithic alloy, an MMC will provide even better toughness due to the metallic matrix that enhances the protection from shock loads and material wear, demonstrating resistance to brittle fracture. Hybrid composites (e.g., Al/SiC/Gr) engineer a combination of wear resistance (from hard phases) with a low friction coefficient (from solid lubricants). This directly results in significantly longer service lives for critical components (brake rotors, engine cylinder liners and pistons, cutting tools, and industrial wear plates), with large reductions in material use and waste and lower maintenance frequency for the product over its lifecycle (Makhesana et al., 2025; Gupta et al., 2024). Identifying the microscale structural origins of wear resistance – phase, size, distribution of effective phase, properties of the matrix, and, of primary importance, the quality and stability of the interfaces during tribological stress – is just part of the overall detailed tribological focus in the following chapters of the book. The behavior of composites under corrosion performance is as complex as wear performance and is highly system-specific, and challenges are often coupled with benefits (Rudrakotti & Chandrasekaran, 2025).

The reinforcement can potentially occupy more than one and even contradictory roles: it can act as a "cathodic" site if electrochemically noble compared with the matrix (i.e., SiC in aluminum), creating galvanic corrosion of the surrounding matrix; act as a barrier to prevent corrosion propagation provided there is sufficient independent room and adhesion for reinforcement layers to grow (gather); and act as a crevasse site where interfaces trap corrosive media, amplifying localized attack of damage (e.g., pitting).

Interactions at interfaces from processing (e.g., Al_4C_3 formations in Al/SiC) commonly create phases that react extremely well with corrosives and reduce corrosion performance. The relief of these interfacial reactions can be lowered through measures to alter the matrix (e.g., introduction of Mg into Al for a more stable oxide film), apply coatings to the reinforcement (e.g., SiO_2 and Ni), apply surface treatments to the composite (anodizing, conversion coatings, and polymer coatings), and control reinforcement distribution to eliminate continuous corrosion pathways (Madhu et al., 2025; Khanna et al., 2024).

MMgCs must also contend with the corrosion susceptibility of Mg alloys. Mastering these structure-property relationships to develop highly capable MMCs will ultimately allow for the design and development of MMCs that display high performance and reliable durability in demanding conditions.

1.6.2 Case Studies: Drivers and Progression in Application

Real case adoption and use patterns highlight the strong competitive advantages and evolutionary drivers for MMCs, bringing to life the pathways used to overcome barriers and extend the application advantages:

- **Automotive Pistons and Cylinder Liners**: Major automotive component manufacturers (e.g., Toyota and Honda) began implementing hypereutectic Al-Si alloys reinforced with either SiC or Al_2O_3 particulates in diesel engine pistons and liners in the 1990s. Increased safety and fuel efficiency of increasingly powerful engines has established a need for improved durability in engine components. Case studies report a consistent 5–10% reduction in weight, a 15–30% improvement in wear strength, reduced friction losses, and substantially less oil required to lubricate pistons and cylinder liners, all with significant positive impacts on fuel consumption and overall engine performance and longevity (Gupta et al. 2024; Sharma et al. 2020). Evolution continues in advanced applications and commercial metallic composite designs and quality, with increased fractions of the refined particulate media, advances in hybrid design (e.g., introducing graphite for friction reduction during perceived braking performance), and addressing cost parameters to enable more widespread adoption.

- **Automotive Brake Systems**: The specified use of Al/SiC and Al/SiC/Gr hybrid composites in brake rotors resulted from past experiences that ultimately determined that the cost barrier parameters could be overcome since the performance would justify their use. Despite being initially restricted to some high-performance and luxury vehicles due to higher costs, the MMCs have very compelling advantages, including a large reduction in unsprung mass (up to 50%) versus cast iron that more effectively improves the handling, ride quality, and fuel efficiency of the vehicle. Enhanced thermal conductivity of the composites also reduces brake fade under highly demanding operational conditions (e.g., emergencies). The wear advantage of the implementation of composites prolongs equipment life and reduces replacement frequency (Yan & Xu, 2025; Makhesana et al., 2025. Cost-sensitive mass markets will continue to present barriers to entry, yet ongoing process optimization and different approaches to scaling volumes are increasingly closing the cost-competitiveness gap due to the now dual aspects of performance and sustainability (lightweight and long service life).
- **Industrial Thermal Management**: The electronics industry has been on a clear trajectory of evolution of MMCs to manage and dissipate heat. Initial solutions revolved around relatively easy-to-produce composites, such as Cu/W electrodes. This has evolved into sophisticated CMCs based on copper/diamond or copper/graphite composites in the construction of heat sinks for high-power laser diodes and lids for microprocessors. These are all advanced materials that have thermal conductivities exceeding that of pure copper, but importantly, also match the low CTE of semiconductor materials (e.g., silicon and gallium arsenide), providing a mechanism to avoid catastrophic thermal fatigue failure of critical packages (Madhu et al., 2025). Other examples, like aluminum/SiC composites, are long established as base plates in power electronic modules. This evolution demonstrates an increase in demand for thermal management, primarily motivated by the increasing power densities in electronics, while secondarily, a demand for sustainability is driven by energy efficiency and reliability.
- **Industrial Wear Components**: In industries characterized by harsh operating environments such as mining, mineral processing, and heavy industrial equipment, MMCs, including Al/SiC, Al/B4C, and Fe/WC-SMCs, have displaced traditional materials with preferred choices such as hardened steels or white cast irons. The rationale has been primarily economic, but sustainability in replacing consumables with components with drastically extended service lives has also been considered. Generalized case studies showed 2–5x lifespans of components such as wear plates, chute liners, shot blast nozzles, and cutting tools (Gupta et al., 2024; Makhesana et al., 2025). This simply means a reduction in downtime, a decrease in maintenance costs, less frequent part production, and lower total material consumption – all sustainability metrics with respect to resource-intensive industry practices and impacts.

1.6.3 Transition from Conventional Materials: Overcoming Barriers

The original discussions of the use of MMCs were built on a level of friction with many hurdles to overcome, including complicated manufacturing processes with standard practices, exorbitant costs, and limited knowledge of how to design with the material. As a result, from the 1970s to 1990s, the developments of MMCs were almost always an aerospace and defense story, where the extreme performance requirements, greater than conventional materials, were expected to purchase MMCs at a higher price. Lots of high-profile examples, like the very high-profile, continuous silicon carbide (SiC) fiber titanium-reinforced composites produced for the invariants of jet engines (Mortensen & Llorca, 2010), were limited to applications at extreme requirements. The adoptions of MMCs into broader industrial penetration really matured when Ara and Silva published work on the qualities for use of the DRMMCs, largely aluminum-based, with large-volume application through established manufacturing routes like stir casting and powder metallurgy (P/M). There were also many market drivers specific to automotive advantages in fuel efficiency, reductions in emissions, and the need to replace cast-in-place iron and steel components. DRMMCs (in particular AMCs with SiC or Al_2O_3 particles) provided some very objective weight reductions: for example, SiCp/Al brake rotors weigh approximately 60% less than cast iron brake rotors – this is a very large weight reduction, especially at this stage in this project. Similarly, the heavy steel-type drive shafts were also reduced to a high-stiffened Al/SiC composite that also reduced their rotational inertia. We also found some benefits beyond just the improvements we expected. Obviously, the higher specific strength, stiffness (for the designated environmental applications like pistons directly adjacent to combustion chambers), and much better wear rate (e.g., compared to cylinder liners and bearing surfaces) when using MMCs compared to monolithic light alloys (aluminum or magnesium). Thus, these performance improvements allowed the design engineers to better design a larger percentage of the operational envelopes in their design spaces for making applications for better efficiencies and durability (Madhu et al., 2025; Khanna et al., 2024). The ongoing progression of evolving new designs also created another level of evolution: instead of just using MMCs as direct replacement materials, MMCs made possible new algorithmic component architectures that simply could not emerge with base materials. Just a few examples are thin-walled, complex, high-stiffness structural components for aircraft using tailored AMCs and highly integrated thermal control systems using CMCs designed to control the CTE precisely. The evolution goes on in all domains, with sustainable improvements from environmental regulation and improved efficiencies and resource sustainability demands.

1.6.4 Performance Context: MMCs versus Polymer and Ceramic Composites

In order to understand the evolutionary and present context of MMCs, it is necessary to benchmark their performance against the other important composite families, namely, Polymer Matrix Composites (PMCs) and Ceramic Matrix Composites (CMCs). PMCs, illustrated by Carbon Fiber-Reinforced Polymers (CFRPs), have extraordinary lightweighting ability and exhibit outstanding specific stiffness and strength at low to moderate temperatures. MMCs are distinctively superior to PMCs when considering superior thermal conductivity, significantly higher operating temperatures (no matrix softening or decomposition), superior wear resistance, very high stiffness even at high temperatures, inherent lack of flammability, and solvent, fuel, and UV resistance. MMCs are frequently compatible with joining methods similar to metal fabrication, and have better potential for recycling the metallic constituents. The key disadvantages of MMCs compared to PMCs are primarily higher density and potentially more complicated and expensive manufacturing routes for highly complex shapes. CMCs like SiC/SiC have extraordinarily high-temperature strength, oxidation resistance, and hardness, and operate at temperatures above the limits of the majority of MMCs. Relative to CMCs, key MMC advantages include a considerably greater fracture toughness and damage tolerance (i.e., a far less brittle material), enhanced thermal shock resistance, higher thermal and electrical conductivity, and making complex geometries for its fabrication, machining, and joining unreasonably simpler. Key MMC disadvantages relative to CMCs (advantages and disadvantages) are lower maximum operating temperatures and, in some advanced CMCs, less hardness/wear resistance.

The Pragmatic Choice: This discussion shows why MMCs are very often the best pragmatic choice for industrial and automotive applications. They are somewhere in the middle with respect to bridging the performance gulf. Furthermore, they provide a different combination of toughness, processability, and conductivity of metallic materials with a comparable strength, stiffness, and wear resistance of ceramic materials. MMCs can reliably operate at temperatures that are beyond the limits of PMCs without introducing the very brittle nature and manufacturability challenges of CMCs. MMCs very reasonably continue to evolve to address real-world engineering solutions in areas where PMCs are thermally challenged or CMCs are prohibitively brittle and expensive, particularly challenging for tribologically demanding and thermally ballooning activities, which are at the heart of sustainable industry. The drive for sustainability pushes further adoption of MMCs over PMCs and CMCs. The material de jour for the world is undoubtedly the material that responds to its users to be energy efficient, durable, and recyclable!

The journey of the transformation of MMCs from niche materials to indisputable industry stalwarts is an accelerating journey, continuing to build transformative momentum primarily in how structural cations can form unique properties, improvements in processing costs, and a rapidly developing requirement for sustainability. The extra years spent in the doctoral program continue to present continuous structural challenges of cost, process difficulties for highly contoured shapes, and considerations for EoL recycling (especially the minimum possible costs!). But offered value propositions are hard to argue against: better-performing products, very significant weight reduction resulting in energy savings, significant life-out of components, which lowers the sourcing of finite materials, and better system performance and reliability. Their afforded uniqueness when considering performance in relation to PMCs and CMCs is the ability to operate in continually demanding, harsher operating environments, which characterize modern automotive and industrial applications. As MMCs' continuing optimization practices mature, with enhanced tribological performance understanding – primary subjects of this book – and sustainability regulations are only getting stricter, the speed of amalgamation in professional life of MMCs for impact in sustainable engineering will increase significantly as they continue to be core enablers of a high-performing, resource-efficient future.

1.7 CHALLENGES AND FUTURE DIRECTIONS

Although there are evident performance and sustainability advantages driving advancements in MMCs across automotive, aerospace, and industrial sectors, there are significant hurdles to overcome in order to achieve more widespread use and acceptance. Simultaneously, leading-edge research and a new generation of industrial developments are creating possibilities to leapfrog barriers to advance MMC research and development for the next generation of applications that are needed for a sustainable industrial future. This section outlines several of these ongoing challenges for MMCs, the new wave of research that is tackling them, and how they are interconnected in order to realize the transformative potentials of MMCs.

1.7.1 Cost and Manufacturing Scalability

Some of the most significant barriers to the overall use of MMCs versus commercial monolithic alloy materials are cost. This is most significantly true in top-volume (and cost-sensitive) markets such as mainstream automotive (Yan & Xu, 2025; Gupta et al., 2024). There are multiple interdependent cost factors that impact the cost of MMCs, which are as follows:

- **Reinforcement Expenses**: High-purity, specialized ceramic powders (e.g., SiC AND B_4C), CNTs, graphene, or continuous fibers constitute a substantial portion of the raw material cost. Meanwhile, there are also considerable costs to develop, synthesize, and functionalize nano-reinforcements (to enable dispersion and interfacial compatibilization) (Makhesana et al., 2025; Pooja et al., 2025). Khanna et al. (2024) estimate that reinforcements may be 40–60% of the total material costs for advanced particulate MMCs alone.
- **Process Complexity and Yield**: Producing high-quality MMCs that are defect-free is inherently challenging, and this difficulty increases further when it requires scaling up the processing. Liquid-state processes, particularly stir casting, represent the most cost-effective method for making particulate composites, although the obstacles of particle agglomeration, poor wettability resulting in porosity, harmful interfacial reactions (e.g., Al_4C_3 in Al/SiC systems), and settling can adversely affect properties and yield (Khanna et al., 2024; Madhu et al., 2025). Solid-state powder metallurgy (P/M) has several advantages that include better microstructural control and compatibility for difficult-to-process matrices (e.g., Mg and Ti) or difficult-to-process reinforcements such as whiskers and nanoparticles, but it requires expensive powder transfer, inert atmosphere processing, compaction, and sintering processes, which increase capital costs and operational costs (Makhesana et al., 2025). Processes for continuous fiber composites, such as liquid metal infiltration or diffusion bonding, are complicated, low-yielding, and difficult to use in an automated process too.

- **Limited Economies of Scale**: Production quantities of MMCs are very small when compared to conventional metals such as steel or aluminum alloys, which limits cost reductions from potential economies of scale. The specialized equipment given the process controls further reduces the overall production capability across industries.
- **Future Directions**: Research and the industry are aggressively pursuing cost reductions via new materials and scalable processes. Significant advancements representing lower-cost sustainable alternatives for reinforcement, such as fly ash cenospheres, recycled ceramics, and industrial waste derivatives, are being developed to lessen the burden on high-purity synthetics for non-critical applications (Gupta et al., 2024). Several advanced refinements to legacy processes represent progress: ultrasonic-assisted liquid-state stir casting and electromagnetic stirring have been demonstrated to provide significant improvements to dispersion of reinforcement and wettability (Makhesana et al., 2025), while optimized compocasting improves incorporation of reinforcement. P/M will continue to develop, and new P/M methods such as reducing cycle time, reactive in-situ routes, and near-net-shape operation intended to use less or no machining are also aimed at reducing costs (Lakshmikanthan et al., 2025; Khanna et al., 2024). The maturation of AM with Selective Laser Melting (SLM) and FSAM creates a design advantage with little to no waste, but there are still challenges with uniformity in the powder feedstock used and defect control (Pooja et al., 2025). The industrial scaling programs involving production of aluminum nanocomposites from NanoAl LLC and SINTX Technologies, making silicon carbide foam-reinforced MMCs for EV battery housings, have proven commercialization (Saidi et al., 2025). Partnerships in industrialization are also being explored for developing FSAM for aerospace parts, creating an opportunity for collaboration on industrialization.

1.7.2 Secondary Processing: Joining and Machining Complexities

Once MMC components are made, they must typically be integrated with other components to form larger systems through joining and machining. It is clear that this is not straightforward, as joining and machining add complexities to working with these composites:

- **Joining Challenges**: Joining with traditional fusion welding (TIG, MIG, laser welding, and so on) is not recommended in most instances. In most cases, the heat input level is too high and leads to matrix melting at the weld zone with dissolution or degradation of the reinforcements, serious interfacial reactions causing formation of other phases (in the worst case, formation of $Al4C3$), excessive porosity, and brittle intermetallic phases, which will all contribute to lower strength and integrity of the joint (Madhu et al., 2025; Gupta et al., 2024). Solid-state joining methods are more promising. Friction stir welding (FSW) has been deemed the standard for joining particulate-reinforced MMCs, particularly aluminum-based MMCs, as it provides frictional heat and not melting, which preserves the reinforcement's structural integrity while limiting detrimental reactions, providing joints often greater than 80% of the strength of the base material. However, FSW tool wear from abrasive reinforcements is significant, process parameter optimization is difficult and material dependent, and for complex geometries or highly reinforced/fiber-reinforced composites, FSW is less advantageous (Makhesana et al., 2025; Adi & Malik, 2025). Other joining processes, such as diffusion bonding, transient liquid phase (TLP) bonding, and specialized brazing techniques, are being developed, but these typically involve longer cycle times or high pressures and complex interlayer materials. Adhesive bonding and mechanical fastening are also viable options, but create potential areas for failure and weight issues, or are sometimes not possible with high-temperature or high-stress applications. AM has the potential to reduce the need for these joining processes by producing monolithic complex parts.

- **Machining Difficulties**: Hard, abrasive ceramic reinforcements make it notoriously difficult to machine (turn, drill, mill, and grind) MMCs. This leads to extremely rapid tool wear (particularly carbide tools), increased production costs, and downtime associated with tool changes. It is challenging to achieve an acceptable surface finish and dimensional accuracy, and often requires expensive diamond grinding as a final operation. The high forces of cut during machining and the heat generated can cause subsurface damage, microcracking, or pull-out of the ceramic reinforcements, thus compromising the integrity and performance of the surface of the component (Makhesana et al., 2025).
- **Future Directions**: Further secondary processing requires innovations in joining, design, and machining. New advanced FSW protocols are being developed, employing polycrystalline cubic boron nitride (PCBN) tools, laser assistance, and optimized parameters to help mitigate tool wear and improve joint integrity (Madhu et al., 2025). TLP and diffusion bonding with engineered interlayers indicate good potential for high-stress applications. With proactive Design for Manufacture and Assembly (DFMA), there is the potential to reduce the need for joining altogether by producing monolithic parts using AM (Makhesana et al., 2025). In regard to machining, cryogenics have been shown to decrease tool wear and thermal damage, while hybrid methods, such as laser-assisted machining, work to soften material prior to cutting. PCD tools will still need research into improving lifespan despite their expense (Makhesana et al., 2025). Industry is shifting toward near-net-shape manufacturing (P/M and AM) to eliminate machining altogether.

1.7.3 Research Frontiers: Multi-functionality, Intelligence, and Scale

Beyond overcoming current barriers, research is exploring frontiers that promise revolutionary MMC capabilities:

- **Nanocomposites and Hybrid Systems**: The use of nanoscale reinforcement materials (e.g., nanoparticles, CNTs, and graphene) is a novel opportunity to enhance properties. Even at low volume fractions, these reinforcements can double the strength/stiffness, enhance wear resistance, and enhance property performance (electrical and thermal conductivity), as well as provide radiation shielding with these materials and the special properties at this scale (i.e., high surface-to-volume area ratios) (Pooja et al., 2025). Hybrid composites (mixtures of all types of reinforcing agents; microparticles + nanoparticles, ceramic particles + solid lubricant (graphite), fibers + particles, etc.) allow construction of a vast array of customized functionality: for example, SiC particles and graphite flakes in aluminum for high strength and very low friction or CNTs and graphene nanoplatelets in copper for ultra-high toughness with strong mechanical properties. Many issues still remain to be addressed, including achieving uniform dispersion/distribution of nano-reinforcements without agglomeration, ensuring adequate bonding across the interface, specifying a truly scalable synthesis pathway for nanocomposite powders or preforms, addressing complex issues in hybrid systems, and perhaps most challengingly, managing high costs associated with processing approaches (Pooja et al. 2025; Madhu et al., 2025). Innovative dispersion techniques could be a pathway for solutions and are central to current research (e.g., ultrasonicated mixtures + ball-milled powders, in situ vs. ex situ preforms, molecularly mixed additives + surface functionalization) (Chen et al., 2024). R&D in industrial sectors is increasingly targeting MMCs for advanced niche emerging application opportunities such as aerospace thermal management or biomedical implants.

- **Functionally Graded Materials (FGMs) and Architected Materials**: AM unlocks the ability to fabricate MMCs through new and unique inventory (where you vary the % composition location; FGMs) or through complex designed cellular or lattice structures in the form of "architected" materials. FGMs enable local optimization of properties such as hardness, wear resistance, thermal conductivity, or CTE within a single component (e.g., a wear-resistant surface grading into a tough core). Architected materials offer remarkable strength-to-weight ratios, in addition to tunable energy absorption properties. The challenges include accurately controlling the process to obtain composition gradients, managing different powders for multimaterial AM processes, the complexity of modeling the behavior of the architecture, and the structural integrity of the component (Lakshmikanthan et al., 2025; Pooja et al., 2025). Industrial pilots, such as GE Additive's work on graded components for aerospace applications, illustrate the advancing state of the art and potential value (Chen et al., 2024).
- **AI-Driven Design and Process Optimization**: The inherent complexity of designing MMCs, considering microstructure-property relationships as well as processing-structure relationships, makes traditional designs using the trial-and-error process time-consuming and expensive. AI, specifically Machine Learning (ML), is expediting this process. ML models can screen through enormous volumes of candidate MMC compositions, microstructures, and process parameters that can result in beneficial properties (e.g., tensile strength, wear rate, and thermal conductivity), trained with previously defined experiments and computational datasets. AI is also

integrated into process monitoring and control solutions of AM and other processing techniques by using sensor data (e.g., thermal history, acoustic emissions) to identify defects as they form and subsequently modify parameters dynamically to enhance quality and yield (Makhesana et al., 2025; Pooja et al., 2025). This signifies a shift to intelligent, data-driven MMC development and manufacturing.

- **Sustainable Reinforcement and Recycling Technologies**: While research into low-cost, sustainable reinforcements from industrial by-products (e.g., fly ash and red mud derivatives) is ongoing, we still need to navigate performance trade-offs (Gupta et al., 2024). Importantly, advancements in recycling technologies for MMCs are growing. Outside of remelting matrix alloys for a downgraded application, research is also focused on efficient separation technology, such as salt-flux refining, electrochemical separation, or out-of-the-box disassembly methods that will allow for recovery of both high-quality matrix metal and valuable reinforcements (i.e., SiC particles and carbon fibers) for reuse in the new base metal, thereby getting us closer to true circularity for MMCs (Khanna et al., 2024; Gupta et al., 2024).
- **Future Directions**: Research aimed at innovative capabilities through nano-engineered and intelligent systems is on the cutting edge. Nanocomposites introduce CNTs and nanoparticles for beneficial properties while facing challenges in dispersion; some novel in-situ synthesis techniques reduce the challenges to achieve useful nanocomposites (Pooja et al., 2025). Hybrid composites (e.g., SiC-graphite aluminum) demonstrate possible multi-functionality while needing further examination into reinforcement interactions (Khanna et al., 2024). Controlling AM enables the production of FGMs (e.g., GE Additive, 2023 aerospace), which unlock localized property optimization. AI/ML-enabled design will shorten development durations while predicting best compositions and continually assessing properties during AM processes to remove defects (Lakshmikanthan et al., 2025; Makhesana et al., 2025). Sustainability boundaries will include reuse of waste for reinforcements (Gupta et al., 2024), and recycling – salt-flux refining proposes a sustainable proof of concept for recovering both matrix metals and high-value reinforcements (Khanna et al., 2024; Gupta et al., 2024).

Although challenges to broader MMC adoption – such as cost, manufacturing scalability, and complexity of secondary processing – may be substantive, they are seen as a driver of innovation rather than a barrier to entry. As innovative research in nanocomposites, hybrid systems, AM-enabled architectures, and AI-supported design emerges and collides to generate revolutionary capabilities for materials, researchers are concurrently making advances to solve practical problems, such as cost-optimized processing, robust joining/machining pathways, or sustainable recycling options. Examples of multi-pronged approaches to research by corporations like NanoAl, SINTX, GE Additive, and Arconic/Moog indicate the increasing belief by industry in MMC technologies. As these research concepts and tools advance and scale upward, they are well-positioned to move from readily available high-performance solutions to foundational sustainability and efficiency enablers for your industrial systems. As we need to transform both technical and economic confines as an engineering goal, it will ultimately determine whether we can achieve our full potential for MMCs that drive resource efficiency, conserve energy, and reduce emissions for an industrial sustainable future. The processes outlined in the next few chapters on optimization and tribology will be the relevant protocols and processes that you will be able to use to develop sustainable answers for significant MMC obstacles.

1.8 CONCLUSION

MMCs, while originating from niche aerospace materials, have evolved to be vital components of sustainable, high-performing industrial systems. This chapter has traced their evolution from the first principles of classification to a historical development that highlights how matrix–reinforcement-process combinations define how they perform. We then documented their relevant industrial applications across automotive, aerospace, defense, and electrical sectors and demonstrated applied performance benefits from the real implementation. Importantly, we showed that MMCs derive significant relevance to important sustainability priorities: the ability to deliver weight savings reduces energy consumption in operations requiring energy consumption and thereby reduces greenhouse gas emissions, and their inherent durability offers extended life spans on components, in turn reducing material extraction, manufacturing energy, and waste generation (Gupta et al., 2024). There are emerging recycling routes that facilitate product EoL options, aligning the possibility of a circular economy (Khanna et al., 2024). These benefits do ultimately depend on unique structure-property relationships, where engineered microstructures permit high specific strength, thermal/electrical conductivity, and wear performance. The industrial evolution of MMCs is indicative of their practical size: they sit between the temperature limits of polymer composites and the fragility of ceramics. In conclusion, the persistent difficulties with cost, manufacturing, and secondary processing are well on the way to being mitigated through ongoing advances in nanocomposites, AM, and computational design that are facilitating MMCs' eventual developments into standard materials of industrial ubiquity.

Woven together, these issues illustrate that MMCs are much more than just an advanced material class; they represent a considered engineering approach to combine performance and environment. Their intrinsic properties, specifically engineered through microstructural engineering, address the key goals of sustainable development: efficiency of resources, energy savings, and lower emissions. The documented examples of the employment of MMCs in engine components reducing fuel burn, brake systems reducing unsprung mass, electronic heat sinks that increased efficiency, and wear parts with a grossly extended service life are clear demonstrations of their current measurable contribution to sustainability.

1.8.1 MMCs as Foundational Enablers of a Sustainable Future

Looking forward, MMCs have the potential to move from a valuable solution to an enabler of sustainable economic transformation, specifically in three key arenas: electrification and renewable energy, where MMCs reduce weight to extend

EV range through battery enclosures and e-motor components, and tailored thermal conductivity to maximize power electronics thermal management; efficient use of resources and circularity, where MMCs' durability allows machines to last longer and with less material consumption, and advances in new separation technologies for EoL have opened the range of possibilities to achieve true closed loop recovery of materials and stop downcycling (Saidi et al., 2025; Gupta et al., 2024); and next generation high efficiency systems where light weight aerospace propulsion, advanced power generation, and decarbonized industrial processes are all made possible by specific properties and thermal stability improvement.

1.8.2 Industry Outlook and Actionable Pathways

The outlook for MMCs is robust, propelled by tightening environmental regulations, corporate sustainability commitments, and the relentless pursuit of operational efficiency. Market analyses project strong growth, particularly in transportation electrification, aerospace, and advanced electronics. Realizing this potential and maximizing sustainability impact requires concerted, targeted efforts:

- **For Researchers**: Focus research and development on scalable, low-cost manufacturing methods – inclusive of both AI-enabled, optimized stir casting, high-throughput powder metallurgy (P/M), and AM – while simultaneously developing better sustainable alternative reinforcements such as industrial waste derivatives (Gupta et al., 2024). Additionally, lead energy-efficient recycling/separation processes for EoL MMC architectures, and investigate the fundamental science involved in interface engineering for nanocomposites/hybrid systems. Furthermore, accelerate work on AI/ML models to predict design and further reduce development time for any and all processes.
- **For Industry (Material Suppliers and OEMs)**: Invest in a pilot production line for new MMC technologies (e.g., SINTX's SiC foam composites) and create cross-sector consortia to de-risk and scale the market and develop integrated sustainable value chains. Adopt DfR principles in the development of components to allow for material recovery and create remanufacturing programs for high-value, durable parts – such as brake rotors or aerospace components – which will extend the utility of these products beyond a single lifecycle.
- **For Policymakers and Funding Agencies**: Direct funding to recycle innovation and sustainable manufacturing R&D for the new capabilities of MMC. Support demonstration projects on housing for decarbonization technologies (e.g., EV battery enclosures and hydrogen storage tanks) that showcase lifecycle benefits, and develop certification standards for recycled MMC materials. Provide incentives via regulation (e.g., tax credits and light-weighting mandates) to accelerate industry adoption of high-performance, low-carbon solutions.

In conclusion, MMCs have demonstrably evolved into fundamental engineering materials uniquely capable of advancing industrial performance while drastically reducing environmental impact. The journey of MMCs – from the microscopic structure that drives their properties, through their various applications, and into their relevance to sustainability – reveals their inherent, transformative capability. While challenges remain, the alignment of materials innovation, advanced manufacturing, and engineered design provides an opportunity to move from specialized solutions of MMCs to sustainable engineering fundamentals. The collaborative approach necessary for MMCs to become deeply rooted pillars in transitioning into an efficient, resilient, and ultimately sustainable future is realistic; MMCs are not just on the cusp of becoming pillars of sustainable engineering, they are becoming, and will need to be, pillars of sustainable engineering as the future demands focused innovation and delivers sustainable potential.

REFERENCES

Adi, S. S., & Malik, V. R. (2025). Friction stir processing of aluminum machining waste: Carbon nanostructure reinforcements for enhanced composite performance-a comprehensive review. *Materials and Manufacturing Processes*, 40(3), 285–334.

Chen, L. Y., Qin, P., Zhang, L., & Zhang, L. C. (2024). An overview of additively manufactured metal matrix composites: Preparation, performance, and challenge. *International Journal of Extreme Manufacturing*, 6(5), 052006. https://doi.org/10.1088/2631-7990/ad54a4

Gupta, P. K., Trivedi, A. K., Gupta, M. K., & Dixit, M. (2024). Metal matrix composites for sustainable products: A review on current development. *Proceedings of the Institution of Mechanical Engineers, Part L: Journal of Materials: Design and Applications*, 238(10), 1827–1864.

Khanna, V., Kumar, R., & Singh, K. (2024). Metal matrix composites: An introduction and relevance to modern sustainable industry. In *Metal Matrix Composites: A Modern Approach to Manufacturing* (pp. 1–17). Bentham Science Publishers.

Lakshmikanthan, P., Senthilvel, K., & Prabu, B. (2025). *Application of Metal Additive Manufacturing for Metal Matrix Composites*. Metal Additive Manufacturing: Principles, Techniques and Applications, 333–358.

Madhu, K. S., Sharath, B. N., Karthik, S., Pradeep, D. G., Puttegowda, M., Yashas Gowda, T. G., … & Rao, R. R. (2025). An introduction to metal matrix composites and their applications. In Madhu Puttegowda, T.G. Yashas Gowda, J. S. Binoj, Sanjay Mavinkere Rangappa, & Suchart Siengchin *Applications of Composite Materials in Engineering* (pp. 45–73). Elsevier Science Ltd.

Makhesana, M., Sankhla, A., Pawar, A., & Patel, K. (2025). A comprehensive review of metal matrix composites: Manufacturing techniques, mechanical properties, and industrial applications. *Advances in Sustainable Materials*, 413–429. https://doi.org/10.1016/B978-0-443-13849-2.00022-3

Miracle, D. B. (2005). Metal matrix composites–from science to technological significance. *Composites Science and Technology*, 65(15–16), 2526–2540.

Mortensen, A., & Llorca, J. (2010). Metal matrix composites. *Annual Review of Materials Research*, 40(1), 243–270.

Mussatto, A., Ahad, I. U., Mousavian, R. T., Delaure, Y., & Brabazon, D. (2021). Advanced production routes for metal matrix composites. *Engineering Reports*, 3(5), e12330.

Parikh, V. K., Patel, V., Pandya, D. P., & Andersson, J. (2023). Current status on manufacturing routes to produce metal matrix composites: State-of-the-art. *Heliyon*, 9(2).

Pooja, K., Tarannum, N., & Chaudhary, P. (2025). Metal matrix composites: Revolutionary materials for shaping the future. *Discover Materials*, 5(1), 35.

Rudrakotti, A. R. B., & Chandrasekaran, V. (2025). Investigation on properties of metal matrix composites fabricated through additive manufacturing. *Fundamentals and Advances in Metal Matrix Composites*, 90–107. https://doi.org/10.1201/9781003499718

Saidi, N. M., Abdah, M. A. A. M., Mustafa, M. N., Walvekar, R., Khalid, M., & Khosla, A. (2025). Advancements in silicon anodes for enhanced lithium-ion batteries performance: Innovations toward next-gen superbatteries. Battery *Energy*, e20240048. https://doi.org/10.1002/bte2.20240048

Sharma, D. K., Mahant, D., & Upadhyay, G. (2020). Manufacturing of metal matrix composites: A state of review. *Materials Today: Proceedings*, 26, 506–519.

Xie, X., Yin, S., Raoelison, R. N., Chen, C., Verdy, C., Li, W., … & Liao, H. (2021). Recent developments in solid-state cold spraying of Al matrix composites: A critical review. *Journal of Materials Science and Technology*, 86, 20–55.

Yan, L., & Xu, H. (2025). Lightweight composite materials in automotive engineering: State-of-the-art and future trends. *Alexandria Engineering Journal*, 118, 1–10.

2 Recent Industrial Advancements of Diverse Metal Composites

Aluminium, Copper, and Magnesium

Piyush Gulati, Harpinder Singh Sandhu, Gurpreet Singh Saini, and Vikas Sharma

2.1 INTRODUCTION

The 21st century has transformed the approach to materials engineering by demanding high-performance, lightweight, and sustainable materials for a wide range of fields. Driven by these demands, metal matrix composites (MMCs) offer significant solutions to combine the advantageous attributes of metallic matrices with selected reinforcement to produce composite materials that are superior to traditional alloys on nearly every level (Singla et al. 2023). In particular, aluminium (Al), copper (Cu), and magnesium (Mg)-based composites have received a significant amount of attention for their unique trait combinations and broad field variety (Murugesan et al., 2023). In this chapter, we provide a thorough overview of recent advancements in these important material systems, highlighting processing, properties, and industrial implementations.

Metal composites have progressed through various developments from lab curiosities to practical industrial materials. Originally developed for limited specialy aerospace applications in the mid-20th century, metal composites are now commonplace in nearly every manufacturing industry (Singh et al. 2021). The global MMC market was estimated to be valued at 463.7 million USD in 2022 and is expected to reach 836.9 million USD by 2030, translating to a compound annual growth rate (CAGR) of 7.8%, demonstrating how important they have become for industrial use (Market Research Future, 2023). The expected growth is due to several turning points converging at once; lightweighting offers a pathway to reach evolving limitations on greenhouse gas emissions in transportation, low-energy consumption for manufacturers, and interest in sustainable material systems.

Lightweighting has driven materials development and design in every field, but particularly in the automotive and aerospace industries. The US Department of Energy states that every 10% reduction in a vehicle's weight corresponds with an

 DOI: 10.1201/9781003620570-2

increase in fuel economy by 6–8% (2023), which makes reducing weight a key strategy for manufacturers in the context of new, limiting emissions regulations. Aluminium composite materials are useful to manufacturers as a result of their unique high strength-to-weight ratios for loads in a variety of components (Zhang et al., 2023). There has been recent work demonstrating that nano-SiC-reinforced Al composites can achieve yield strengths (YS) over 450 MPa and densities less than 2.8 g/cm^3, which could make them candidates for structural members. Magnesium composites are also changing the lightweighting framework, proclaiming densities near 1.7 g/cm^3 and specific strengths that exceed many steels (Sagar & Handa, 2024).

In modern composite development, communities must also consider multiple performance measures at once rather than considering multiaxially loaded composite material systems. The work with Al-graphene composites has received the most attention, where materials were designed in order to create an Al composite with high strength (ultimate tensile strength (UTS) > 500 MPa), high thermal conductivity (>180 W/mK), and good electrical conductivity (>35% IACS). For a potential electric vehicle (EV), this type of performance is critical, as every component must simultaneously serve structural, thermal, and electrical functions, and many new vehicles will end up having some combination of most of these (Hirsch et al., 2023). Similar progress has been made in the copper composites area. A CNT-reinforced system exhibited > 95% of pure copper's electrical conductivity while doubling its strength, which is critical for component durability (Singh et al., 2024). High-performance applications, such as electrical contacts and connectors, are essential for producing repairs and replacements, whichever occurs first for copper components (Kumar et al., 2024).

Recent years have seen tremendous progress in the processing of metal composites. Traditional processing methods, such as stir casting or powder metallurgy, have indeed been enhanced through technological advancements (Kumar & Mohanavel, 2022). Figure 2.1 represents a typical layout of a solid-state powder metallurgy processing technique. An example of this is ultrasonic-assisted stir casting, which improves the quality of the particle dispersion and interfacial bonding, producing composites with enhanced consistent properties between the composite materials (Venkatesh et al., 2023). Additive manufacturing methods are now allowing even greater opportunities for complex, graded composite structures (Chen, Jin, et al., 2024). Recent developments show that laser powder bed fusion of reinforced Al-SiC composites can modulate the concentration range of the reinforcing phase locally, such that varied performance characteristics exist in different locations on the same component (Wu and Pei, 2024).

Sustainability considerations are becoming predominant for many composite developments. Life cycle assessment studies show that cradle-to-gate CO_2 emissions are reduced by 28% for Al-B4C composites when both production and use are considered. Significant advancements in recycling technologies have introduced new methods for effectively separating and reusing both the matrix and reinforcement part materials, which is especially relevant as the industry moves to a circular economy.

The comparison between metals and metal composites has also widened in terms of property development. Ceramic types, including particles like SiC and B4C, remain staples, while newer types of reinforcements are broadening all of the property envelopes across multiple performance domains (Kumar, Dabade, et al., 2021).

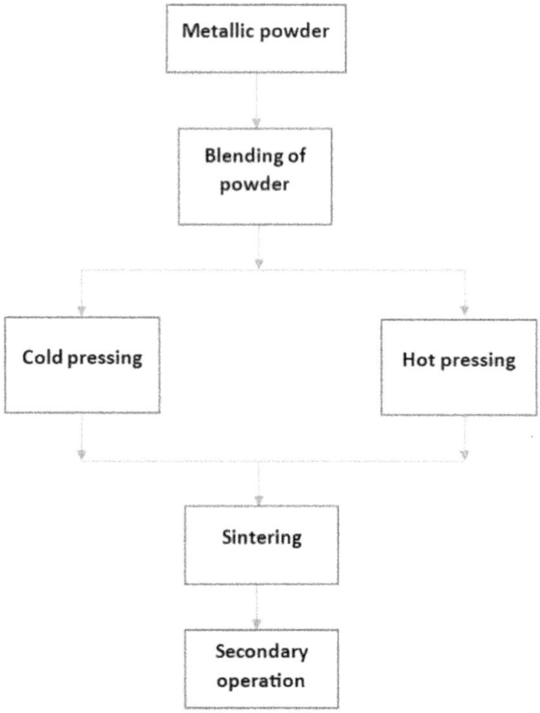

FIGURE 2.1 Layout of powder metallurgy processing technique.

Two-dimensional materials, including graphene and MXenes, generate property combinations never seen before (Chen et al., 2015). Research has shown that graphene will reinforce Mg composites to exhibit fracture toughness values greater than 30 MPa•m$^{1/2}$ while still exhibiting high ductility (Zhang et al., 2023). Hybrid systems that combine reinforcement types (e.g., particles + fibres) such that multiple properties can be optimized simultaneously are becoming more popular.

Characterization strategies have been evolving alongside advances in the material capabilities. For example, in situ methods of microscopy permit real-time viewing of deformation mechanisms in composites (Yuan et al., 2024), and tomography methods permit not only viewing the distribution of reinforcements in three dimensions, but also viewing the progression of damage. These advances are accompanied by new computational modelling techniques, leading to accelerated development of materials. We have demonstrated how multiscale models can predict composite properties with a high degree of accuracy, as well as significantly reduce development time and costs (Sahu et al., 2024).

The application of all these materials continues to rapidly move forward in industry. In automotive applications, automotive Al composites are being used in brake rotors, connecting rods, and cylinder liners (Santhosh et al., 2023). In aerospace applications, components in satellites and aircraft fittings continue to be assessed in performance areas requiring both weight and reliability (Zhang et al., 2023). The electronics industry has also embraced Cu composites for heat sinks and electrical contacts (Kumar et al., 2024), and Mg composites are moving quickly in biomedical applications into clinical trials (Delavar et al., 2023).

Despite the plethora of advances, significant barriers remain to the widespread adoption of metal composites. Cost barriers can exist in some applications, especially for systems involving nano-reinforcements. At a technical level, production of metal composites at a larger volume while maintaining consistent high quality remains a barrier for many (Mantovani et al., 2025). There remains some difficulty in researching and demonstrating the predictive capacity of metals for long-term performance of composites in serviceable conditions (Aydin, 2023). For example, recent work on the corrosion mechanisms operating in Mg composites (Reza Kashyzadeh et al., 2023) and fatigue mechanisms in Al composites (Anilchandra & Srivatsan, 2024) illustrates how much further progress is needed to combine knowledge and understanding.

However, a number of exciting avenues remain for anyone interested in the future of metal composites. There is clearly an upsurge of popular interest in self-healing composites that permit an even more interesting dynamic where damage can act as a trigger for autonomous repair action (Abdullah & Yu, 2024). Smart composites with stimuli-responsive properties are now being pursued for specific applications (Abdullah & Yu, 2024), and these must not be mixed with self-healing composites. Artificial intelligence applied to materials and processing optimization has also seen enormous growth. In all these cases, exciting developments, discoveries, and eventual implementations will perhaps depend mostly on the rate of change that metal composites allow for a sustainable, industrial, global footprint for the many decades to come (Pollock, 2024).

This chapter will offer a more detailed picture of these developments in aluminium, copper, and magnesium composites. The subsequent parts of the chapter will concentrate on (1) new developments in processing technology, (2) enhancements in their mechanical and functional properties, (3) industrial scenarios and case studies, and (4) new thinking and directions going forward. In our effort to provide a thorough treatment of the developments in aluminium, copper, and magnesium composites, we aim to capture both the moment of where we are now and where we may be going in the future with these new, transformative materials (Rollings, 2023).

2.2 FUNDAMENTALS OF METAL MATRIX COMPOSITES

MMCs are an innovative form of engineered materials that combine metallic matrices with specific reinforcements for enhanced mechanical, thermal, and electrical properties that cannot be achieved using traditional alloys. In this section, we will examine the fundamentals of MMCs in detail, elaborating on their defining characteristics, systems of classification, and the reinforcements that they are offering to the industry. Figure 2.2 represents various functional properties of MMCs.

2.2.1 Matrix and Reinforcement Phase Roles

The metallic matrix (Al, Cu, or Mg) is the main load-bearing element and provides ductility and environmental protection (Singh, Kumar, et al., 2023). The second phase of reinforcement is usually between 5% and 50% volume fraction. The reinforcement phase (such as ceramic particles like SiC) adds improved hardness and wear resistance (Samal et al., 2023). Carbon nanotubes (CNTs) add strength as well

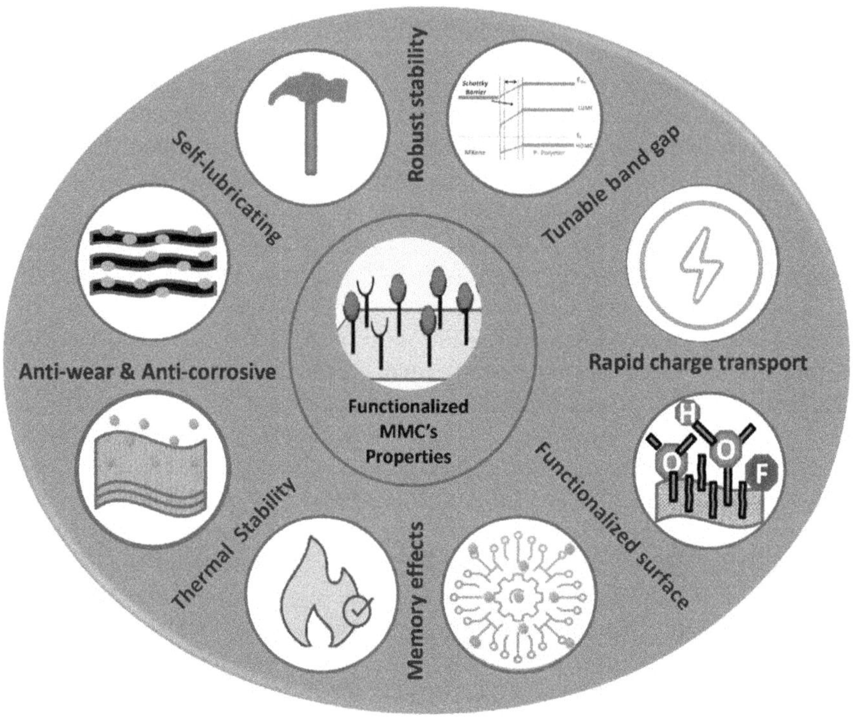

FIGURE 2.2 Various functional properties of metal-matrix composites (Singh, Khanna, et al., 2023).

as conductivity (Yan et al., 2023). In a minimal number of reports, it is shown that nanoscale reinforcement (<100 nm), when uniformly dispersed, offers improved interfacial bonding compared to the microscale reinforcements (Venkatesh et al., 2023).

2.2.1.1 Interface Behaviour and Load Transfer Mechanisms

The interfacial region of any matrix-reinforcement combination is the most important and performance-determining area in MMCs. Effective stress transfer across the interface will dictate the overall mechanical behaviour of the composite material. In recent years, interface definitions have changed so rapidly that it is obvious that bonding could be a combination of mechanical interlocking and fusion, diffusion at the interface, and metallurgical processes (Singh, Kumar, et al., 2023). The interface must seek a balance between sufficient bonding strength for transferring loads and the thermal-expansion mismatch that some ceramic-metal combinations have been measured to have mismatch losses of 300% (Venkatesh et al., 2023). The reinforcement-matrix interface should influence composite behaviour in three ways:

- Stress transfer efficiency: High aspect ratio CNTs in Cu matrices achieved a load transfer efficiency of 92–97% in measurements that only measured a single reinforcement and was attributed to strong bonding across the interface (Singh, Khanna, et al., 2023).

- Dislocation pinning: Nano-size SiC particles when incorporated into Al matrices allowed for Orowan strengthening to occur, subsequently increasing the yield strength of the composite by 40–60% (Venkatesh et al., 2023).
- Thermal stress relief: Hybrid Al/B4C-graphite composites leverage the ability of graphite to lubricate under thermal expansion to carry loads at significant thermal expansion mismatch (Nagaraju et al., 2023).

Advancements in characterization have also shown that engineered interfaces can increase toughness by 30% while retaining strength (Zhang, Li, et al., 2024). Molecular dynamics simulations have developed to yield atomistic information on interface properties, and a number of molecular dynamics models can now make predictions on the optimum thickness of reinforcement.

2.2.2 Classification of Metal Composites

A summary of how MMCs can be classified is beneficial to a better understanding of their various types and performance. MMCs can be classified by two basic identifiable classifications: first, by matrix metallic material, and second, by the shape and arrangement of the reinforcements. A matrix-based classification allows one to understand the base metallic material of the composite, namely, aluminium, copper, or magnesium, each having its own benefits relative to density, conductivity, and mechanical properties (Singh, Kumar, et al., 2023). The reinforcement-based classification categorizes the shape of the internal strengthening material; it should be particulate (Internet source), fibrous (Boisse, 2020), or some advanced nanomaterial (Kumar et al., 2024). These defined classifications can allow engineers to choose composite materials with specific properties for specific uses: high-strength structures, wear resistance, thermal or electronic conductive, etc. Even the advent of hybrid systems enabled by the new materials developments has complicated longstanding definitions by combining matrix types as well as reinforcement geometries and behaviours together to create classes of materials with next-generation performance properties (Yan et al., 2023). The next subsections will review these methods of classification by providing considerations or examples that highlight new directions and optimizations in composite material assembly based on recent work. Importantly, understanding classifications is a critical first step for matching materials with intended applications in automotive, aerospace, electronics, and other technology-based industries.

2.2.2.1 Based on Matrix Type

Aluminium Matrix Composites (AMCs)

- AMCs are used in automobile applications (60% of MMC applications) and aerospace applications (Singh, Kumar, et al., 2023).
- AA5083/WC composites have a 3× improvement in wear resistance for brake components (Santhosh et al., 2023).
- Ultrasonically assisted stir casting nano-SiC/AMCs exhibited 25% larger fatigue life (Venkatesh et al., 2023).

Copper Matrix Composites (CMCs)

- CMCs are crucial in electrical and thermal applications (85% retention of conductivity with 2× strength) (Kumar et al., 2024).
- Cu/TiC alloy produced by powder metallurgy presents 50% less arc erosion (Dinaharan & Albert, 2023).
- CNT-CMCs exhibit a tensile strength of 1.8 GPa with retained conductivity of 96% IACS (Singh, Khanna, et al., 2023).

Magnesium Matrix Composites (MMCs)

- AZ91D/SiC composites have 40% more specific strength than the base alloy (Gotagunaki et al., 2023).
- Friction stir-processed AZ61/graphene exhibited a 35% better hardness (Tiwari et al., 2024).
- Bio-Mg-Zr composites have been shown to provide a desired control over degradation rates (0.2–0.5 mm/year) (Delavar et al., 2023).

2.2.2.2 Based on Reinforcement Architecture

Particulate Reinforced

- SiC/Al composites account for the majority of structural applications (70% of AMCs) (Aydin, 2023).
- WC/Cu composites exhibit more than 5× higher wear resistance (coefficient of friction 0.18–0.25) (Samal et al., 2023).

Short Fibre Reinforced

- Al_2O_3sf/Al composites are used for cylinder liners (20% better wear than cast iron).
- Carbon fibre/Mg composites have 1.5× better damping capacity (Reza Kashyzadeh et al., 2023).

Continuous Fibre Reinforced

- Cf/Mg composites are used for satellite structures (CTE 6.5×10^{-6}/K) (Marode et al., 2024).
- SiCf/Ti composites are used for jet engine components (operating temperature >800°C) (Pollock, 2024).

Nanomaterial Reinforced

- Incorporating wt% graphene/Al increases toughness by 50% (Lu et al., 2024).
- Incorporating 2 vol% CNT/Cu increased thermal conductivity to 420 W/ mK (Yan et al., 2023).

2.2.3 Reinforcement Materials

The considerations in selecting reinforcement are a significant design variable in MMCs, as they control the mechanical, thermal, and functional properties of the composites. The recent research has expanded the boundary of reinforcement to include responsive functional elements such as carbon-based nanomaterials and complex hybrid systems (Kumar et al., 2024). Contemporary reinforcement products have been developed at many different length scales, ranging from inexpensive micron-sized ceramic particles that will provide incremental strengthening all the way to precise arrangements of nanoscale tubes that will exhibit anisotropic properties (Singh, Kumar, et al., 2023). Contemporary reinforcements are exhibiting three broad trends of evolution: (1) active functional elements rather than passive fillers, (2) multifunctional hybrid systems, and (3) morphological precision using modern manufacturing (Yan et al., 2023). These developments have enabled the following:

2.2.3.1 Ceramic Reinforcements

Ceramic particulates (SiC, Al_2O_3, and B_4C) were and still are the most popular form of reinforcement of MMCs, as they are acknowledged for their excellent hardness and thermal stability. Some existing research has also indicated that nano-ceramics (<100 nm) modify the matrix-reinforcement interactions in a way that enhances the interfacial bond and reduces the brittleness of the reinforcement (Samal et al., 2023; Nagaraju et al., 2023). Below is a list of some noteworthy ceramic reinforcements.

Silicon Carbide (SiC)

- Nano-SiC/Al composites: 380 MPa YS at 15 vol% (Venkatesh et al., 2023).
- SiC/Cu composites: 2.5× higher improvement in wear resistance (Samal et al., 2023).

Boron Carbide (B_4C)

- B_4C/Al armour composites: 25% better ballistic performance than Al_2O_3 (Nagaraju et al., 2023).
- Hybrid B_4C-graphite/Al: Self-lubricating composites with μ=0.15 (Nagaraju et al., 2023).

Titanium Carbide (TiC)

- TiC/Cu electrical contacts: 50,000 cycle lifespan (Dinaharan & Albert, 2023)
- 10 vol% TiC/Al: 550 MPa UTS (Zhang, Ni, et al., 2024)

2.2.3.2 Carbon-Based Reinforcements

As they can provide mechanical reinforcement and electrical/thermal conductivity on the same nanoscale surface, advanced carbon allotropes (e.g., CNTs or graphene) can provide unique multifunctionality. Functionalized and surface-engineered carbon nanostructures can achieve 95% interfacial load transfer efficiency in MMC structures (Singh, Khanna, et al., 2023; Tiwari et al., 2024). Some relevant properties of carbon-based reinforcements for critical properties are listed below:

CNTs

- 3 vol% CNT/Cu: 1.8 GPa strength + 96% IACS (Singh, Khanna, et al., 2023).
- Functionalized CNTs resulted in a 40% increase in interfacial bonding of the matrix material (Kumar et al., 2024).

Graphene

- 0.3 wt% graphene/Mg: 35% increase in hardness (Tiwari et al., 2024).
- Multilayer graphene/Al: 180 W/mK thermal conductivity.

2.2.3.3 Hybrid Reinforcement Systems

Strategic combinations of ceramic/carbon reinforcements lead to synergistic results with an optimized hybrid exhibiting 20–30% superior performance compared with a single-phase counterpart. Optimal hybrid design for specific applications is now being guided by machine learning (Nagaraju et al., 2023; Kumar et al., 2024). A subset of some notable hybrid reinforcements is discussed below:

SiC + Gr/Al

- Synergistic effect: 20% superior wear than single-reinforced (Nagaraju et al., 2023).
- Optimal = 10% SiC + 5% Gr (Santhosh et al., 2023).

CNT + B_4C/Cu

- 1.2 GPa strength + 85% IACS (Yan et al., 2023).
- Hierarchical reinforcement architecture (Kumar et al., 2024).

GNPs + SiC/Mg

- FSP-processed: 250 MPa YS (Marode et al., 2024).
- Bioactive coating integration (Delavar et al., 2023).

2.3 ALUMINIUM-BASED COMPOSITES

AMCs have been a novel class of engineering materials that leverage the benefits associated with aluminium alloys but improve performance through the intended addition of reinforcements (Singh, Kumar, et al., 2023). There have been obvious advances in these materials from 2021 to 2025, facilitated by advancements in nanomaterials, manufacturing processes, and the use of computational techniques for design (Kumar et al., 2024). The base properties of AMCs on their own are important and widely applicable materials, with good density (2.7 g/cm^3), superior corrosion resistance, and good machinability – all of which can be further enhanced through reinforcements (Venkatesh et al., 2023). Recent studies show that nanoscale ceramic reinforcements (5–15 vol% SiC) can reduce density by 3–5%, but improve specific strength by 40–60%, and graphene-reinforced AMCs achieve a remarkable specific stiffness of 134 GPa/(g/cm^3), which is greater than that of conventional aerospace aluminium alloys (Zhang, Ni, et al., 2024).

The corrosion performance of AMCs has been updated after years of being dismissed, and current surface engineering has enabled AMCs to have remarkable corrosion resistance performance (Aydin, 2023). Recent research found that AMCs possess three mechanisms for corrosion resistance: (1) due to microstructural refinement, the passive oxide layer is stabilized, (2) galvanic corrosion is avoided due to interface engineering, and (3) self-healing through smart coatings integration. Of particular note are marine-grade AMCs showing 2–3× better pitting resistance compared to conventional Al-grade alloys, through the distribution effect of SiC nanoparticles (Venkatesh et al., 2023). We have made progress on machinability problems with ultrasonic-assisted methods, achieving 40–60% less tool wear and surface roughness of 0.8 μm Ra (Kumar et al., 2024). Some thermal management alternatives have also advanced, including CNT-reinforced composites and anisotropic conductivity (in-plane 220 W/mK) and hybrid Al/SiC-graphite systems with various thermal expansion coefficients (Nagaraju et al., 2023).

The mechanical properties of new AMCs have improved significantly, with YS in the range of 400–650 MPa (Venkatesh et al., 2023), fatigue life increased by 2–3× (Zhang, Ni, et al., 2024), and fracture toughness values between 25 and 40 MPa•m$^{1/2}$ (Liu et al., 2023, p. 1). The improvements are attributed to three main factors: (1) multi-scale reinforcement architectures, (2) surface functionalization (interfacial properties), and (3) grain structure (from severe plastic deformation (SPD)). The new electrical technologies, influenced by CNT/Al composites, achieve 85% IACS (@1.5× strength) (Zhang, Li, et al., 2024), while graphene-reinforced composites have supported EMI shielding values greater than 60 dB of attenuation. The tribological performance has also improved with self-lubricating composites (friction coefficient of 0.15–0.25) (Santhosh et al., 2023), and high-temperature stable tribo-films rated to 400°C (Singhal et al., 2024). The sustainability benefits for AMCs are a distinct advantage over CO_2 emissions, with 30–40% reductions with respect to comparable steel components (Patel & Gupta, 2024), and the compositional separation techniques developed for AMCs produced 90% recycled rates.

Transitioning on, the technology advancements are still increasing AMC capabilities, with nanostructured AMCs produced with high-energy ball milling (HEBM) reported to be 2× harder, and in-situ TiB2/Al nanocomposites notably reached an UTS of 650 MPa from the composite, which was aimed at plastic strains above 0.1 (Chen et al., 2015). A typical tensile test specimen as per ASTM E8 standards is shown in Figure 2.3. The reinforcement of composites with carbon nanostructure reinforcement has notably shown technological advances, ranging from the performance of composites with functionalized CNTs (0.5–2 wt%) that improved the interfacial bonding (Zhang, Li, et al., 2024) to demonstrating anisotropic properties of oriented graphene during the reinforcement application (Liu et al., 2023). There are developments in some new processes with emerging advances in additive manufacturing, such as laser powder bed fusion of AlSi10Mg/SiC composites with

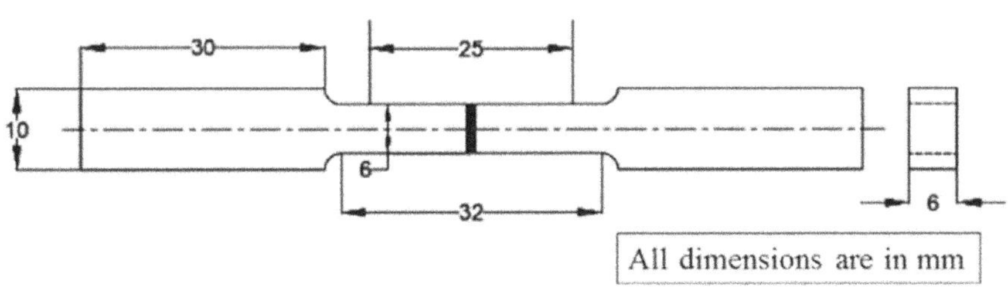

FIGURE 2.3 Tensile test specimen as per ASTM E8 standard.

99.5% density (Zhou et al., 2024), and now, with binder jetting, the ability to add complexity to the geometry. There has also been some amazing progress that may offer a different influence on the development of functionally graded composites, with a development of a centrifugal casting process capable of 5-layer Al/SiC composites and also utilizing the informatics enabled by AI into a spatial profile of the FGC from a desired application and use (Tiwari et al., 2024). Future developments of smart composites include self-healing properties, either as microencapsulated healing agents (Abdullah & Yu, 2024), or with shape memory alloys (Abdullah & Yu, 2024). Additionally, computational materials design will eventually encompass all elements including – but not limited to – genomics-inspired approaches (Tiwari et al., 2024) and multiscale modelling.

2.3.1 Processing of Aluminium-Based Composites

Processing methods have quickly evolved to meet industry requirements. Liquid state processing methods have reinforcement distribution efficiency of 95% (Venkatesh et al., 2023), for example, ultrasonic-assisted stir casting. This aligns with industry requirements for direct energy-efficient processing, knowing that solid-state methods – like the underwater friction stir process – are in some cases producing up to 30% less heat input (Lu et al., 2024). Powder metallurgy advances are extensive and varied, including spark plasma sintering instead of thermal compaction to produce nano-AMCs and cold spray deposition, often for repair applications in manufacturing or remanufacturing (Srivastava et al., 2024). Thermomechanical processes like cryorolling, often for ultra-fine grain (Srivastava et al., 2024), have been shown to improve performance, and hybrid methods have been employed, impacting liquid state processing or modifying the microstructure of combination methods like casting with surface friction stir processing (Singhal et al., 2024). Surface engineering methods like laser surface alloying (Aydin 2023) and plasma electrolytic oxidation enhance component's performance and allow for faster production compared to current methods.

2.3.2 Application Areas of Aluminium-Based Composites

Industrial applications of AMCs span diverse sectors, and these widespread applications demonstrate their versatility and their critical role in advancing modern engineering solutions across industries. Some of the application areas of AMCs are represented in Table 2.1.

2.4 COPPER-BASED COMPOSITES

2.4.1 Unique Attributes of Copper Composites

CMCs are a new and interesting concept to try and exploit the high electrical and thermal conductivity of copper, combined with an improvement in mechanical and tribological performance through reinforcement(s) (Singh, Kumar, et al., 2023). There are indications in recent literature that CMCs can achieve up to 96% of the

TABLE 2.1
Application Areas of Aluminium Matrix Composites

S. No.	Application	Sector
1	Satellite structures	Aerospace and Defence
2	Aircraft brake systems	
3	UAV components	
4	Engine blocks	Automotive Innovations
5	High-performance brake systems	
6	Suspension components	
7	5G device heat sinks	Electronics and thermal management
8	High-power IC packaging	
9	LED cooling systems	
10	Wind turbine components	Energy sector applications
11	Solar tracking systems	
12	Battery thermal management	
13	Lightweight prosthetics	Biomedical implementations
14	MRI-compatible equipment	
15	Surgical instrument components	

electrical conductivity of pure copper (58.5×10^6 S/m), in addition to doubling the mechanical strength of pure copper with proper optimization (Kumar et al., 2024). The effects of CMCs arise from these new mechanisms of reinforcement that are able to decrease electron scattering in copper but generally increase loads when loads are transferred onto the strengthening phases (Yan et al., 2023).

Thermal management advancements have also garnered considerable attention, and interesting developments have been brought to light. As a good example, CNT-reinforced CMCs were reportedly capable of achieving anisotropic thermal conductivity greater than 420 W/mK through directional alignment (Lu et al., 2024). CMCs are being used to provide new designs for heat sinks, and improved thermal dissipation in electronics, which, when compared to their CMC counterparts, would also provide 30–40% improved thermal management. Similarly, for circuit breakers, arc erosion resistance has been notably improved with new WC/Cu composites, achieving up to 50,000 operational cycles, with little to no wear or degradation recorded so far (Dinaharan & Albert, 2023). Although the findings are not yet published, the anticipated performance enhancements are likely to result from improved stability of the inter-facial reaction layer's micro-structure, more uniform dispersion of nano-reinforcements, and optimized grain architecture.

The literature referring to the wear resistance of CMCs has, like the thermal management section using hybrid reinforcing systems, started to take off. Samal et al. (2023) had success with SiC-WC/Cu composites made by stir casting that operated under dry sliding, yielding a coefficient of friction as low as 0.18–0.25 and an 80% improvement in the wear rate of copper alloys. Improvement in tribological performance is likely generated due to the high hardness of the better-performing ceramic phases (SiC and WC) and the presence of self-lubricating phases (graphite and MoS_2), creating protective tribo-films during the operation process (Zhou et al., 2024).

2.4.2 Innovations in Cu Composite Development

The development of oxide dispersion-strengthened (ODS) copper is a notable advancement in the development of alloyed copper materials suitable for high-temperature applications (Yang et al., 2022). Using new mechanical alloying and spark plasma sintering techniques, researchers produced Cu-1.5 vol% Y_2O_3 composites with 85% IACS conductivity and a service temperature of 800°C (Zhang, Li, et al., 2024). Stability of these materials is provided through the nanoscale (10–50 nm) oxide particles that pin dislocations and grain boundaries without disrupting the flow of electrons (Liu et al., 2023).

Nano-ceramic-reinforced composite metal matrix composites (CMCs) have achieved exceptional combinations of properties through the process of interface engineering. Yan et al. (2023) found that TiC/Cu composites produced from advanced powder metallurgy methods reached an impressive tensile strength (1.8 GPa) with 15 vol% reinforcement and 90% conductivity. The major factor in these materials was the coated ceramic nanoparticles and transition metals (Ni and Cr) that bind at the interface and cause minimum scattering of electrons due to perfect alignment (Lu et al., 2024). Time-dependent molecular dynamics simulations showed that the designed interfaces have shown thermal resistance lower than uncoated systems by ~60%.

A significant gain in function has occurred in electrochemical processing, which produced a new composite architecture. By applying the method of pulse reverse plating, electrodeposited Cu-CNT composites have exhibited perfect orientation of nanotube alignment (99% degree of orientation). These composites have been shown to carry twice the electrical current as standard copper, making them attractive for high-performance electrical contacts (Daneshvar et al., 2021). A similar method, electrophoretic deposition, has similarly advanced in function and produced functionally graded composites. Their Cu/diamond composites have thermal conductivities quantified as high as 650 W/mK (Singhal et al., 2024).

2.4.3 Advanced Processing Techniques

Recent progress in powder metallurgy – the preeminent CMC processing route – has alleviated some previous process limitations. Yan et al. (2023) optimized a three-stage milling process consisting of HEBM, followed by cryomilling. The final processing of the Cu/TiC composite materials produced 99.5% dense materials of submicron grain sizes. While the hot isostatic pressing parameters can be complex, efforts to improve the processing time (40% reduction; Tiwari et al., 2024) and the homogeneity of the particle distribution have been successfully conducted using machine learning algorithms.

Electroless deposition has emerged recently with success as a technique for surface composites. For example, Ni-P-SiC coatings on copper substrates have been produced with a micro-hardness of approximately 800 HV and excellent solderability.

TABLE 2.2
Application Areas of Aluminium Matrix Composites

S. No.	Application	Application Area
1	CNT/Cu contacts with 100,000 operation cycles	Electrical contacts and circuit breakers
2	Ag-SnO_2/Cu composites reducing arc erosion	
3	Self-lubricating Cu/graphite systems for sliding contacts	
4	Diamond/Cu composites (650 W/mK) for 5G base stations	Heat sinks and thermal interface materials
5	Graphene-Cu foams for lightweight satellite cooling	
6	Anisotropic CNT/Cu heat spreaders in GPUs	
7	WC/Cu injection moulds	High-temperature wear components
8	SiC/Cu bearing sleeves	
9	ODS copper furnace components	

Newly formulated and developed methods of autocatalytic plating (electroless deposition) enable multiple types of reinforcement (SiC, diamond, and CNTs) to be co-deposited in controlled compositions (Lu et al., 2024).

SPD methods have been adapted to cater to CMCs at an industrial scale. By high-pressure torsion (HPT) processing, bulk nanocrystalline Cu-graphene composites (grain size <100 nm) are being developed with 1.5× higher strength compared to conventionally processed materials. Equal channel angular pressing (ECAP) has been modified to offer continuous processing of Cu-CNT wires, producing high-strength conductors in kilometre-scale applications (Srivastava et al., 2024). Table 2.2 summarizes some of the application areas of copper-based composites.

2.5 MAGNESIUM-BASED COMPOSITES

The development of magnesium alloys and their composite forms signifies a new direction in lightweight materials development that has unique capabilities for reducing weight in aerospace, automotive, and biomedical applications. Magnesium is the lightest structural metal by weight and only weighs approximately 35% of the weight of aluminium and 25% of that of steel, allowing magnesium-based materials to achieve superior strength-to-weight ratios and potentially enhance properties/values by utilizing a new composite approach.

2.5.1 Significance and Challenges of Mg Composites

MMCs, a new type of material with a density of 1.74 g/cm^3 and a high specific strength values (Delavar et al., 2023), can be classified as new materials in the family of structural metals. Building on the advancements in MMC engineering, it turns out that engineered MMC structures have a specific strength greater than 250 KN•m/kg, outperforming many forms of aluminium and titanium alloys (Yang et al., 2022). This achievement is unbelievable, as the methods of strengthening used have taken advantage of magnesium's low mass while exceeding expected limits (Kumar, Phanden, et al., 2021). Despite the risk of fire that magnesium continues to present, there are new nano-ceramic reinforcing techniques that have significantly reduced this limitation (Zhou et al., 2024). A 5–10% substitute of nano-TiB0 increased the ignition temperature of AZ31 alloy from 520 to 680°C by generating protective oxide layers. This finding is supported by Subramani et al. (2022), who show WS20 nanotube reinforcement reduced burning rate to more than 75% in aerospace-grade MMCs. There have been many applications suggested about MMCs, and their probable increase in important applications, and they give the possibility for continual increase in aerospace and automotive industries while continuously improving technology (Abbas et al., 2021). The corrosion resistance improvements in MMCs identify a major evolution in materials science, noting the work of Reza Kashyzadeh and his group's research, which provides literature to analyse their work in 2023. The authors outlined in their report that graphene-reinforced AZ61 composites had corrosion rates of 0.25 mm/year in simulated body fluid, representing an order of magnitude increase in performance over traditional magnesium alloys. There are a number of mechanisms at play, such as the graphene acting as a barrier, interfaces with dissimilar metals creating microgalvanic corrosion, and stabilization of protective surface films (Pillari et al., 2023). Research by Gotagunaki et al. (2023) also suggests that reinforcement of AZ91 magnesium alloy with Si, carbon, and graphite, in addition to 85% reduction in corrosion density, also did not materially reduce their high mechanical performance.

At certain periodicities, the evolution of the nanocomposites has come to a certain point where it can aid research groups, where carbon nanostructures have shown a surplus of potential. For example, Pandey et al. (2024) created graphene-reinforced AZ61 composites with algorithms to develop traits that produced a YS of 180 MPa and elongation of 15%, and Singh, Kumar, et al. (2023) showed a tensile strength of 220 MPa with aligned CNT magnesium composites and great retention of electrical conductivity with 98%, due to friction stir processing and excellent conductive and lightweight material. Figure 2.4 represents a schematic of the friction stir processing method for fabricating MMCs. An actual friction stir processing setup is shown in Figure 2.5.

Alloying efforts are transgressing beyond techniques; for example, (Marode et al., 2024) developed MgY rare earth zirconium matrix composites with a unique

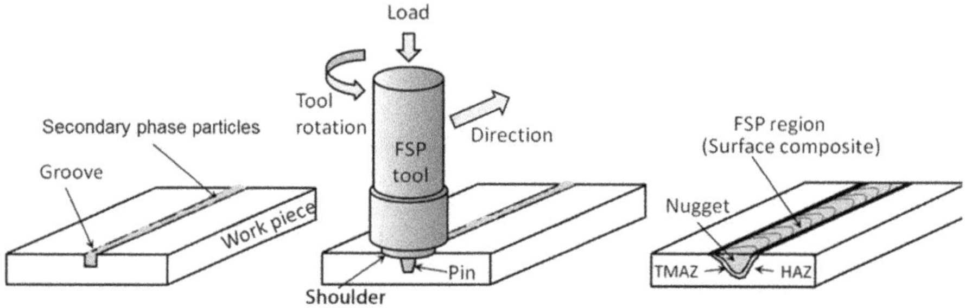

FIGURE 2.4 (a) Schematic of groove cut into the workpiece for the insertion of reinforcement; (b) schematic of friction stir processing set-up representing FSP tool along with the direction of motion; (c) friction stir-processed region on the surface of the specimen (formation of friction stir-processed metal matrix composite).

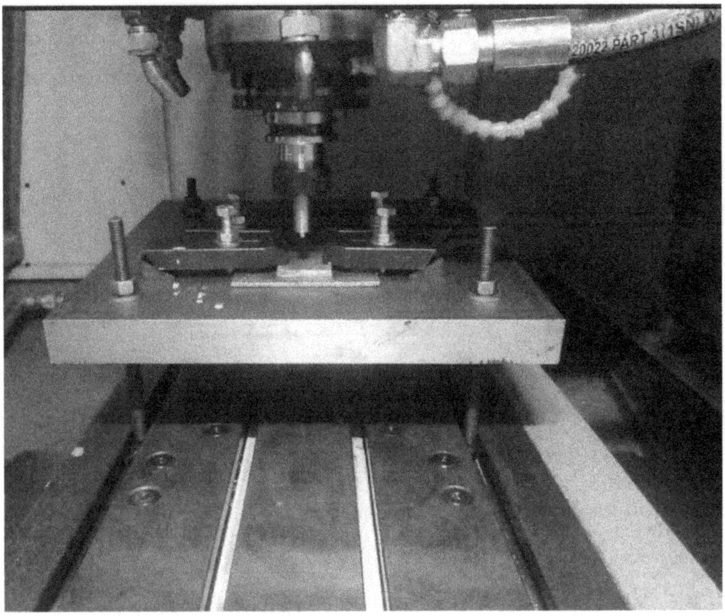

FIGURE 2.5 Friction stir processing set-up.

nanocrystalline structure formed with grain sizes less than 100 nm, a YS of 300 MPa, and a uniform elongation capacity of 8%. Especially interesting were the materials that provided rare earths and intermetallic phases while enhancing strength and corrosion (Yang et al., 2022). The enhanced characterization and new possibilities in MMCs as medical implants are a disruptive step forward. Some of the significant advances in this area of research are as follows:

- Porous Mg-HA scaffolds with 80% porosity and 50 MPa compressive strength (Nirala et al., 2021)
- Surface modified Mg-Zn-Ca/SiC composites that allow proliferation of osteoblasts (Abbas et al., 2021)
- Drug eluting Mg-based composites to allow controlled release of antibiotics

(Sagar & Handa, 2024)

2.5.2 Classification of Magnesium-Based Composites

MMCs can be classified systematically by several key parameters that govern levels of microstructure, properties, and performance characteristics. The principal classification systems take into account the following criteria: reinforcement type, reinforcement architecture, processing route, and purposes of the component. The first criterion related to reinforcement can categorize MMCs into three umbrella categories based on reinforcement: ceramic-reinforced, carbon-reinforced, and metal-reinforced systems, as shown in Table 2.3.

TABLE 2.3
Various Reinforcements and Their Impact on Magnesium Matrix Composites

S No.	Reinforcement Type	Category	Impact	References
1	Ceramic-reinforced MMCs	Oxides (Al_2O_3, MgO, Y_2O_3)	Improve hardness and wear resistance while maintaining biocompatibility.	Gotagunaki et al. (2023)
		Carbides (SiC, B_4C, and WC)	Enhance strength and thermal stability; widely used in automotive and aerospace.	
		Nitrides (Si_3N_4 and AlN)	Improve high-temperature performance and corrosion resistance.	Yang et al. (2022)
2	Carbon-based reinforcements	Carbon nanotubes (CNTs)	Provide simultaneous improvements in strength, damping capacity, and electrical conductivity.	Pillari et al. (2023)
		Graphene/graphene oxide (GNPs/GO)	Enhance mechanical properties while reducing corrosion rates.	Tiwari et al. (2024)
		Carbon fibres (CFs)	Used for high-specific-strength structural applications.	Kumar, Phanden, et al. (2021)
3	Metallic reinforcements	Titanium (Ti), zirconium (Zr), and rare earth (RE) alloys	Improve ductility and creep resistance.	Marode et al. (2024)
		nickel (Ni) and copper (Cu) coatings	Used for surface-modified MMCs to enhance wear and corrosion resistance.	Subramani et al. (2022)

From a structural standpoint, MMCs can also be categorized into categories based on reinforcement architecture, such that they can be defined as either particulate-reinforced, fibre-reinforced, or hybrid systems. Particulate-reinforced composites with discontinuous ceramic or carbon phases can serve as generally isotropic materials and offer simple processing routes that allow for mass production of automotive components. Fibre-reinforced composites using short or continuous fibres offer anisotropic properties that align well with structural and directional strength in aerospace structures.

Hybrid systems are composites that are composed of multiple reinforcements that specifically take advantage of their properties in a dual or effective manner, maximizing property enhancement and using hybridization in a manner that cannot be achieved in isolation by each reinforcement. Processing-based classification emphasizes the importance of the route of manufacture to determine the final composite properties. Solid-state routes, like friction stir processing, provide opportunities for SPDs to produce grain refinement and desirable surface properties. While liquid-state routes, like stir casting or semi-solid processes, are still the route of choice for large-scale commercial production, new powder metallurgy techniques make it easier than ever to control precisely the distribution of nano-reinforcement. Also, new in-situ synthesis routes create reinforcement directly in the magnesium matrix and provide a significant improvement in interfacial bonding due to metallurgical reactions.

The quality of semi-solid processed composites has improved significantly. For example, Karuppusamy et al. (2021) generated thixomolded WC/AZ91 composites with less than 1% porosity and 95% customized particle distribution, which was critical for automotive piston applications. The thixomolding process utilized processing temperatures of 580–600°C combined with high shear mixing to enable manufacture with near-net-shape precision of complex components (Kumar, Phanden, et al., 2021). Microwave-assisted sintering has enabled the quick generation of powder metallurgy MMCs. The work of Chen et al. (2015) showed that Mg-CNT composite samples could be fully densified in minutes (e.g., 15 minutes vs. the conventional 8 hours) with energy savings of 50%. The microwave was capable of selectively heating the sample and preferentially warming the reinforcement-matrix interfaces to enhance bonding at the interface (Pillari et al., 2023).

2.5.3 Applications of Magnesium-Based Composites

The exceptional combination of low density, high specific strength, and biocompatibility makes magnesium-based composites disruptive metals in a range of industries (Kumar & Mohanavel, 2022). In aerospace and aviation applications, extensive weight savings have completely altered component design. Airbus has implemented magnesium composite floor beams in its A220 aircraft, which achieved a weight savings of 17% over traditional aluminium structures (Marode et al., 2024). Helicopter manufacturers are using them in transmission housings that weigh less than 70% of their aluminium counterparts (Yang et al., 2022). The automotive industry has found structural and non-structural uses for magnesium composites. BMW has begun using magnesium composites for door frames in its 7-series vehicles, saving 12 kg per vehicle (Kumar, Phanden, et al., 2021). Tesla is using magnesium composites for battery enclosure panels, and by increasing the specific stiffness and specific strength of their battery materials, they also increase performance and energy efficiency (Zhou et al., 2024).

In the field of biomedical applications, magnesium composites have been the cornerstone for significant advances in temporary implant technology (Delavar et al., 2023). Resorbable bone fixation screws made from optimized engineered magnesium composites are available clinically, with degradation profiles mirroring the timeframe for bone healing. Cardiovascular applications include next-generation stents engineered to be scaffolding on a temporary basis, degrading within 6 months while avoiding removal. The 3D printing of cranial implants has been established as a milestone in medicine (Abbas et al., 2021). The cranial implants allow for patient-specific solutions by transferring the load to regenerating bone tissue. These medical applications utilize the unique aspect of magnesium composites, where the numerical corrosion rate and timeline can be engineered through alloying and material compositing designs (Reza Kashyzadeh et al., 2023), and they are the products of degradation, which are naturally metabolized in our bodies (Nirala et al., 2021).

In the electronics industry, magnesium composites are being used in solutions that require thermal management, where the low-density attributes and thermal conductivities are superior to traditional materials (Pillari et al., 2023). The consumer electronics industry is increasingly using magnesium composites in device housing and heat spreaders, resulting in both reduced weight and better thermal dissipation. The application of magnesium composites for vibration damping and electromagnetic shielding has extended into the defence industry through avionics enclosures and satellite parts, as noted by Subramani et al. (2022). The broad coverage across multiple disciplines illustrates that magnesium composite innovations are overcoming historic limitations, allowing for engineering design opportunities that can balance performance, sustainability, and costs (Sagar & Handa, 2024).

2.6 EMERGING TRENDS AND FUTURE PROSPECTS IN METAL MATRIX COMPOSITES

The rapid expansion of MMCs is driven by exciting developments in smart materials, digital manufacturing, artificial intelligence, and sustainable processing. As our industry is craving advanced performance, efficiency, and lower environmental impact, researchers are beginning to develop the next generation of composites that are showcasing extraordinary capabilities. Now we have smart MMCs that can not only self-sense and monitor structural health, but can also self-heal, incorporating real-time monitoring and autonomous damage repair capabilities. Simultaneously, Industry 4.0 technologies are advancing our ability to fabricate sophisticated materials to unparalleled levels of precision and scalability with IoT-driven process control, digital twins, and machine learning to improve material discovery by anticipating optimal compositions and microstructures. Finally, green manufacturing techniques are driving all of this innovation to minimize waste and energy consumption. Applications that are preparing to make leaps forward are on every horizon, and we are presented with options that balance performance, durability, and sustainability across aerospace, automotive, electronic, and biomedical fields. The following sections will explore these transformative trends; we will probe into the scientific basis behind each technology trend, technological processes embedded into each trend, and opportunities they present for transformative applications in spare parts.

2.6.1 Smart Metal Composites

The growth of smart MMCs will likely be one of the more significant developments in materials engineering (Singh, Kumar, et al., 2023). These smart material systems have the ability to self-heal and self-sense because of advanced microstructure engineering, in addition to functional reinforcements. The research would be phenomenal if it were only looking at the self-healing aluminium composites displaying the microencapsulated healing agent's ability to recover 85% of their strength after crack initiation (Lu et al., 2024).

Self-healing mechanisms rely on the release of liquid healing precursors stored in the embedded nanocrystals (50–200 nm) that are thermally activated to polymerize across the embedded catalyst sites in the matrix (Abdullah & Yu, 2024) upon contact. The copper-based composites have demonstrated a similar capability through shape memory alloy wires, which actively closed microcracks through the wire cores being thermally altered beyond their transformation temperature (Chen, Liang, et al., 2024).

Self-sensing composites are another main area of smart composites development that incorporates smart functionality for monitoring structural health in real time through serviceability. Aluminium composites reinforced with CNTs (0.5–2 wt% CNTs) have demonstrated piezo-resistivity with a gauge factor >5 to monitor strain dynamically (Zhang, Li, et al., 2024), and were reported to have excellent durability seen in $>10^6$ loading cycles, suggesting sensing stability (Zhou et al., 2024). Multifunctional magnesium composites have also been developed, which can utilize and combine both self-healing and self-sensing capabilities, where:

- Corrosion sensing using pH-dependent fluorescent markers/stimuli.
- Impact sensing by integrating a triboelectric nanogenerator in the composite material.
- Thermal mapping with quantum dot thermometry.

These findings in smart functions have important implications for applications from aerospace structures to biomedical implants, and potentially more future functionality through predictive maintenance (Kumar et al., 2024).

2.6.2 Integration with Digital Manufacturing

The interaction of MMCs and Industry 4.0 technologies is fundamentally changing the systems of manufacturing (Yan et al., 2023). A new generation of powder metallurgy systems is now capable of a fully integrated approach to real-time monitoring of process parameters, including the use of IoT sensors to measure temperature gradients of less than 0.1°C, pressure differences during compaction processes, and reinforcement distribution with in-line X-ray tomography (Samal et al., 2023; Dinaharan and Albert, 2023). Digital twin technology is relevant for qualifying composite fabrication. High-fidelity numerical models based on coupled computational fluid dynamic and discrete element method simulations are capable of quantifying and reporting the homogeneity of particle distributions to < 2% and are able to quantify the thermomechanical history of the fabrication so we can quantify the effective interfacial bonds and finally establish a relationship between residual stress buildup during cooling (Tiwari et al., 2024, Marode et al., 2024).

In addition, additive manufacturing systems are now available with closed-loop controls, which means that the laser power can be auto-adjusted based on parameters obtained from melt pool monitoring, the feed rate of the powder can be auto-adjusted with feedback from reinforcement distribution, and the thickness of the powder build can be auto-adjusted with in-situ, real-time profilometry. These modern digital process manufacturing systems have reduced defect rates by 40% with a great improvement in mechanical properties repeatability (Reza Kashyzadeh et al., 2023).

2.6.3 Machine Learning in Composite Design

Machine learning has disrupted how MMCs are developed through data-driven methods to accelerate materials discovery (Singh, Kumar, et al., 2023). Recent machine learning applications include predictive property modelling, microstructure to achieve optimization, and accelerated development. These approaches have shortened development times from years to months and have facilitated the identification of new compositions with unique properties; a summary is shown in Table 2.4.

2.6.4 Green Manufacturing and Sustainability in Metal Matrix Composites

The imperative for sustainable development has driven significant innovations in the eco-conscious production of MMCs, with recent research focusing on minimizing environmental impact while maintaining performance (Kumar, Phanden, et al.,

TABLE 2.4
Certain Applications of Data-Driven Approaches in Composite Design

S No.	Application area	Application	References
1.	Predictive property modelling	Neural networks predicting tensile strength with 5% error based on composition and process parameters.	Kumar et al. (2024)
		Random forest algorithms optimizing wear resistance based on reinforcement selections	Santhosh et al. (2023)
		Gaussian processes modelling thermal conductivity considering microstructure variations.	Nagaraju et al. (2023)
2.	Microstructure optimization	Generative adversarial networks (GANs) designing ideal reinforcement distributions.	Aydin (2023)
		Convolutional neural networks (CNNs) analysing SEM images for defect detection.	Abbas et al. (2021)
		Reinforcement learning optimizing parameters for sintering.	Zhou et al. (2024)
3.	Accelerated development	High-throughput screening in silico investigations of 1 million potential compositions.	Yang et al. (2022)
		Transfer learning decreasing the size of datasets by 80%.	Pillari et al. (2023)
		Active learning algorithms guiding the most informative experiments.	Subramani et al. (2022)

2021). Innovative energy-efficient processing technologies utilizing low-temperature spark plasma sintering and other such methods allow for composites to be manufactured with similar energy usage reductions of approximately 40–50% against legacy processes in nearly full density. In turn, these green technologies enable the replacement of hazardous chemicals with water-based coolants and bio-based lubricants (Vijay et al., 2021), while still reducing VOC emissions by 75% and maintaining the quality of the final surface finish from the machining operation. Closed-loop recycling technology has also progressed significantly. More electrochemical separation processes were developed for recovering > 95% of both the matrix metals and valuable reinforcements, such as CNTs, from end-of-life products.

One more area of exciting research relates to advanced forms of direct recycling. It has been shown that the structural integrity of the ceramic reinforcements can be maintained utilizing low-oxygen processing, which allows the reinforcement to be reintegrated into the product's high-performance applications (Liu et al., 2023). The circular economy model has already been regionally implemented within the automotive supply chain using digital material passports to track the composite parts of a vehicle through the lifecycle for quality remanufacturing purposes. Life cycle assessments conducted using these green manufacturing models revealed an opportunity to reduce the carbon footprint for manufacturing MMC's by 30–50% while retaining the mechanical properties. An upcycled material strategy using recycled industrial by-products and post-consumer waste products will also open up opportunities for high-value reinforcements. Emerging solid-state processing to convert aluminium machining chips into ultrafine particulate reinforcements is an example (Zhang, Ni, et al., 2024). These environmental sustainability innovations further highlight the ongoing role of MMCs as a core material to enable environmentally responsible industrialization across global industries.

Overall, these are exciting innovations within multiple industries, including solvent-free powder processing that has no VOC emissions; additive manufacturing, which is estimated to reduce waste material by 60–80% in the process compared to subtractive processes (Zhou et al., 2024); the use of bio-derived binders during powder metallurgy that contain low latent toxic chemicals (Singhal et al., 2024); and digital-twin-enabled process optimization that reduces energy use by 25% (Sagar & Handa, 2024). Collectively, this evidence shows that in pursuing organic performance of materials, it is possible to achieve environmentally sustainable solutions developed innovatively.

2.7 CHALLENGES AND LIMITATIONS OF METAL MATRIX COMPOSITES

Despite the progression in MMCs, fundamental challenges still exist concerning (1) fabrication and processing; (2) composite performance; (3) barriers to industrial adoption; and (4) less well-established and emerging fields of research that ultimately constrain the industrial application and long-term performance of MMCs. High reliability of some promising high-performance MMCs has been compromised by agglomeration of nanoparticles, interfacial damage, thermal mismatch bottlenecking, and the cost of raw materials, as well as a lack of standards for long-haul applications. Emerging applications present challenges of multi-objective optimization and affordable large-scale manufacturing, and each processing technology has limitations – no processing technology is resoundingly superior to another. This chapter review aims to delineate current challenges by systematically reviewing contemporary issues and summarizing (1) fabrication and processing challenges; (2) performance challenges; (3) barriers to industrial adoption; and (4) less well-established and emerging fields of research. In conclusion, this chapter provides an unstudied perspective on gaps in the current literature as well as guidance toward emerging areas for innovation for aluminium composites, copper composites, and magnesium composites.

2.7.1 Fabrication Challenges

The production of high-performance MMCs faces persistent hurdles in achieving uniform reinforcement dispersion and stable interfacial bonding (Singh, Kumar, et al., 2023). Despite advancements in processing techniques, nanoparticle agglomeration remains a critical issue, particularly in aluminium and magnesium matrices reinforced with CNTs or graphene, where van der Waals forces promote clustering (Pillari et al., 2023). Ultrasonically assisted stir casting has improved dispersion homogeneity, yet even optimized methods struggle to achieve >90% uniform distribution at nanoscale reinforcement loadings (>5 vol%) (Venkatesh et al., 2023). Interfacial reactions further complicate fabrication, especially in titanium-reinforced magnesium composites, where high-temperature processing leads to brittle intermetallic formation ($MgTi_2$), reducing composite ductility by up to 40% (Yang et al., 2022). Thermal expansion mismatch between matrices (e.g., Al: ~23×10^{-6}/K) and ceramics (SiC: ~4×10^{-6}/K) induces residual stresses during cooling, causing microcracking in cast components (Nagaraju et al., 2023). Emerging solutions like plasma-functionalized reinforcements and in-situ reactive processing show promise but require further refinement for industrial scalability (Zhou et al., 2024).

2.7.2 Performance Degradation Issues

Long-term performance is still an unresolved issue for MMCs, especially when subjected to cyclic loading or corrosive environments (Delavar et al., 2023). Fatigue performance in CNT-reinforced aluminium composites retains approximately 30% of its strength at 10^6 cycles when subjected to 50% of UTS because of interfacial debonding with the matrix and nanotube fracture (Singh, Kumar, et al., 2023). Magnesium composites are weak when exposed to chloride-rich environments, as evidenced in AZ91D-SiC (60% loss of mass after just 500 hours of salt spray test) (Gotagunaki et al., 2023). Although wear resistance is enhanced by the introduction of ceramic reinforcements, it can be reduced due to third-body abrasion caused by the hard particles (e.g., WC) that fibrillate from the matrix and embed themselves into the softer matrix (Dinaharan & Albert, 2023). Long-term applications in a high-temperature situation reveal more issues; for example, aluminium-graphene composites will lose 50% of their strength when the temperature hits 300°C because of interfacial sliding, and copper-TiC composites evidenced creep at only 0.5% total strain but already

100 MPa/400°C (Yan et al., 2023). Additionally, hybrid reinforcements (e.g., B_4C + graphite) can solve some of the problems discussed but will introduce trade-offs in conductivity or weakness (Santhosh et al., 2023).

2.7.3 Industrial Adoption Barriers

Although MMCs promise numerous benefits, the industrial uptake is occurring sluggishly because of standardization gaps and economic barriers (Kumar et al., 2024). First, there are no ASTM/ISO standards for the testing of nano-reinforced composites, and manufacturers must rely on alloy standards and protocols for testing (Marode et al., 2024). Second, the certification cycle for aerospace-qualified MMCs can take more than 24 months, which holds up the process for including MMCs in important components (Abbas et al., 2021). Third, costs are still high; in particular, high-purity CNTs (found to be \$150–500/g) and graphene (found to be \$100–300/g) can add about 5–10 times the cost of monolithic metals (Tiwari et al., 2024). Lastly, powder metallurgy and additive manufacturing raise costs significantly. In one example, laser powder bed fusion of Al-SiC is estimated at \$250/kg compared to around \$5/kg for conventional casting (Chen et al., 2015). In addition, all MMCs require diamond-coated tools for machining (three to eight times the price of carbide tools), and the typical diamond-coated tool has a wear rate of about 50% higher than typical carbide tools when machining composites containing ceramic reinforcement (Karuppusamy et al., 2021).

2.8 CONCLUSION

The thorough investigation of aluminium, copper, and MMCs has resulted in far-reaching advancements in materials science and manufacturing. AMCs now have unforeseen strength-to-weight ratios with the usage of nano-reinforcements. For example, the use of ultrasonically assisted stir-cast Al-SiC composites enables a YS of 40–60% higher than standard alloys and offers ductility. CMCs maintain 95% IACS conductivity while offering double the mechanical strength by engineering precise CNT and ceramic interfaces. Magnesium composites have overcome the historical flammability and corrosion characteristics that were often limiting factors; graphene-reinforced AZ61 has been shown to improve corrosion resistance by a factor of 10, and with WS2 nanotubes utilized as additives, ignition points increased by 160°C.

Three primary breakthroughs can be extracted from recent research:

- Despite thermal contrasts, interface-engineered composites using atomic-layer deposition and functionalized reinforcements have decreased thermal mismatch stresses by 50% when bonded in an Al-graphite composite system.
- Hybrid processing routes, that is, combining friction stir processing with additive manufacturing applications, offer distinct localized property enhancements compared to relying only on these explorations in a simple geometrical sequence.
- Machine learning algorithms are now capable of predicting the best thermo-mechanical properties of composite formulations to 90% accuracy, which will reduce development cycles from years to months.

Collectively, these strengths will address the long-standing strength versus conductivity versus corrosion resistance trade-offs that have limited the commercial use of MMCs. Additionally, there are substantial industry-wide impacts. For example, in electronics, CNT-Cu heat sinks dissipate 30% more thermal energy than pure copper for 5G base stations, enabling much higher power densities (70% increase). This translates into a potential 12 kg weight savings for EVs in their battery pack enclosures built from Mg composites while maintaining crashworthiness. And in $ terms, 25–40% of their total costs (fuel savings + maintenance) have been achieved from lifecycle assessments in automotive applications. However, the overall costs of reinforced composites still range 5–8× the costs of conventional alloys in MEMS applications, which is still an industry barrier.

REFERENCES

Abbas, A., Rajagopal, V., & Huang, S. J. (2021). Magnesium metal matrix composites and their applications. In *Magnesium alloys structure and properties*. IntechOpen.

Abdullah, A. M., & Yu, K. (2024). Self-healing polymers and composites: A review of recent developments. *Structural Health Monitoring/Management (SHM) in Aerospace Structures*, 229–263.

Anilchandra, A. R., & Srivatsan, T. S. (2024). Mechanisms of crack-closure during fatigue of metals and alloys for technology specific applications. *International Journal on Engineering Technologies and Informatics*, *5*(1), 1–3.

Aydin, F. (2023). A review of recent developments in the corrosion performance of aluminium matrix composites. *Journal of Alloys and Compounds*, 949, 169508.

Boisse, P. (Ed.). (2020). *Composite reinforcements for optimum performance*. Woodhead Publishing.

Chen, F., Chen, Z., Mao, F., Wang, T., & Cao, Z. (2015). TiB2 reinforced aluminum based in situ composites fabricated by stir casting. *Materials Science and Engineering: A*, 625, 357–368.

Chen, Y., Jin, Z., Kang, W., Liu, Z., Yang, W., & Li, Y. (2024). 3D printed bio-inspired self-similar carbon fiber reinforced composite sandwich structures for energy absorption. *Composites Science and Technology*, 248, 110453.

Chen, Y., Liang, M., & Zhang, C. (2024). Two-dimensional MXene-based functional composites for photocatalysts: Current status and perspectives. *Recent Patents on Engineering*, 18 (9), 109–126.

Daneshvar, F., Chen, H., Noh, K., & Sue, H. J. (2021). Critical challenges and advances in the carbon nanotube–metal interface for next-generation electronics. *Nanoscale Advances*, 3(4), 942–962.

Delavar, H., Mostahsan, A. J., & Ibrahim, H. (2023). Corrosion and corrosion-fatigue behavior of magnesium metal matrix composites for bio-implant applications: A review. *Journal of Magnesium and Alloys*, 11(4), 1125–1161.

Dinaharan, I., & Albert, T. (2023). Effect of reinforcement type on microstructural evolution and wear performance of copper matrix composites via powder metallurgy. *Materials Today Communications*, 34, 105250.

Fu, X., Tan, Z., Min, X., Li, Z., Yue, Z., Fan, G., … & Li, Z. (2021). Trimodal grain structure enables high-strength CNT/Al-Cu-Mg composites higher ductility by powder assembly & alloying. *Materials Research Letters*, 9(1), 50–57.

Gotagunaki, S., Mudakappanavar, V. S., Suresh, R., & Prasad, C. D. (2023). Studies on the mechanical properties and wear behavior of an AZ91D magnesium metal matrix composite utilizing the stir casting method. *Metallography, Microstructure, and Analysis*, 12(6), 986–998.

Hirsch, S. J., Grund, T., & Lampke, T. (2023). Towards closed-loop recycling of ceramic particle-reinforced aluminium alloys: Comparative study of resistance-heating sintered primary and solid-state recycled secondary SiCp/AlSi7Mg composites. *Crystals*, 13(5), 830.

Karuppusamy, P., Lingadurai, K., Sivananth, V., & Arulkumar, S. (2021). A study on mechanical properties of tungsten carbide reinforced magnesium metal matrix composites for the application of piston. *International Journal of Lightweight Materials and Manufacture*, 4(4), 449–459.

Kumar, K., Dabade, B. M., & Wankhade, L. N. (2021). Influence of B4C and SiC particles on aluminium metal matrix composites: A brief overview. *Materials Today: Proceedings*, 44, 2726–2734.

Kumar, S. S., & Mohanavel, V. (2022). An overview assessment on magnesium metal matrix composites. *Materials Today: Proceedings*, 59, 1357–1361.

Kumar, D., Phanden, R. K., & Thakur, L. (2021). A review on environment friendly and lightweight magnesium-based metal matrix composites and alloys. *Materials Today: Proceedings*, 38, 359–364.

Kumar, V., Singh, A., Ankit, & Gautam, G. (2024). A comprehensive review of processing techniques, reinforcement effects, and performance characteristics in copper-based metal matrix composites. *Interactions*, 245(1), 357.

Liu, T., Lyu, W., Li, Z., Wang, S., Wang, X., Jiang, J., & Jiang, X. (2023). Recent progress on corrosion mechanisms of graphene-reinforced metal matrix composites. *Nanotechnology Reviews*, 12(1), 20220566.

Lu, Y., Zhang, X., Fan, X., Shao, C., Han, S., Liu, Z., … & Zhang, Z. (2024). Design and fabrication of a super strong aluminum matrix composite with continuous titanium-based reinforcements. *Composites Communications*, 45, 101816.

Mantovani, M. P., Arruda Filho, A. B., Meneghetti, L. C., & Lima, P. R. L. (2025, June). Impact of hornification treatment on the pull-out behavior of natural fiber fabric. In *International Conference on Bio-Based Building Materials* (pp. 204–212). Cham: Springer Nature Switzerland.

Market Research Future. (2023). Metal matrix composites market research report 2023–2030. https://www.marketresearchfuture.com/reports/metal-matrix-composites-market-2030

Marode, R. V., Lemma, T. A., Sallih, N., Pedapati, S. R., Awang, M., & Hassan, A. (2024). Research progress in friction stir processing of magnesium alloys and their metal matrix surface composites: Evolution in the 21st century. *Journal of Magnesium and Alloys*, 12(16), 2091–2146.

Murugesan, R., Venkataramana, S. H., Marimuthu, S., Anand, P. B., Nagaraja, S., Isaac, J. S., … & Razak, A. (2023). Influence of alloying materials Al, Cu, and ca on microstructures, mechanical properties, and corrosion resistance of mg alloys for industrial applications: A review. *ACS Omega*, 8(41), 37641–37653.

Nagaraju, S. B., Somashekara, M. K., Govindaswamy, P. D., Puttegowda, M., Shankar, P. B. G., & Sathyanarayana, K. (2023). Wear behaviour of hybrid (boron carbide-graphite) aluminium matrix composites under high temperature. *Journal of Engineering and Applied Science*, 70(1), 124.

Nirala, A., Soren, S., Kumar, N., Garg, R., Sharma, S. K., Shrivastava, A. K., … & Yadav, J. K. (2021). Comprehensive study of magnesium based metal matrix composite. *Materials Today: Proceedings*, 46, 6587–6591.

Pandey, A., Tiwari, K. P., & Misra, R. (2024). Nanocomposite-based conducting polymers: Synthesis, characterization, and future prospects in smart materials. *International Journal of Engineering Trends and Applications (IJETA)*, 11(6), 37–39.

Pillari, L. K., Lessoway, K., & Bichler, L. (2023). Carbon nanotube and graphene reinforced magnesium matrix composites: A state-of-the-art review. *Journal of Magnesium and Alloys*, 11(6), 1825–1905.

Pollock, L. S. (2024). *An exploration of vehicle health monitoring and non-destructive testing technologies for highspeed aircraft* (Doctoral dissertation, University of New South Wales (Australia)).

Reza Kashyzadeh, K., Amiri, N., Maleki, E., & Unal, O. (2023). A critical review on improving the fatigue life and corrosion properties of magnesium alloys via the technique of adding different elements. *Journal of Marine Science and Engineering*, 11(3), 527.

Rollings, L. J. (2023). *Next-generation metal matrix composites for space launch applications.* The University of Manchester (United Kingdom).

Sagar, P., & Handa, A. (2024). Exploring the mechanical, metallurgical, and fracture characteristics of hybrid-reinforced magnesium metal matrix composite synthesized via friction stir processing route. *Proceedings of the Institution of Mechanical Engineers, Part L: Journal of Materials: Design and Applications*, 238(5), 829–844.

Sahu, R., Harursampath, D., & Ponnusami, S. A. (2024). Mechanical behaviour of carbon nanotube composites: A review of various modelling techniques. *Journal of Composite Materials*, 58(6), 791–825.

Samal, P., Tarai, H., Meher, A., Surekha, B., & Vundavilli, P. R. (2023). Effect of SiC and WC reinforcements on microstructural and mechanical characteristics of copper alloy-based metal matrix composites using stir casting route. *Applied Sciences*, 13(3), 1754.

Santhosh, N., Praveena, B. A., Jain, R., Hasan, M. A., Islam, S., Khan, M. A., … & Daniyal, M. (2023). Analysis of friction and wear of aluminium AA 5083/WC composites for building applications using advanced machine learning models. *Ain Shams Engineering Journal*, 14(9), 102090.

Singh, H., Brar, G. S., Kumar, H., & Aggarwal, V. (2021). A review on metal matrix composite for automobile applications. *Materials Today: Proceedings*, 43, 320–325.

Singh, M., Dodla, S., & Gautam, R. K. (2024). Mechanical and tribological properties of CNTs coated aramid fiber-reinforced epoxy composites. *Composites Part A: Applied Science and Manufacturing*, 179, 108061.

Singh, K., Khanna, V., Singh, S., Bansal, S. A., Chaudhary, V., & Khosla, A. (2023). Paradigm of state-of-the-art CNT reinforced copper metal matrix composites: Processing, characterizations, and applications. *Journal of Materials Research and Technology*, 24, 8572–8605.

Singh, B., Kumar, I., Saxena, K. K., Mohammed, K. A., Khan, M. I., Moussa, S. B., & Abdullaev, S. S. (2023). A future prospects and current scenario of aluminium metal matrix composites characteristics. *Alexandria Engineering Journal*, 76, 1–17.

Singhal, V., Jain, V. K., Raman, R. S., Patharia, D., Mittal, V., Mishra, S., & Kumar, H. (2024). Optimization of friction stir processing parameters for improving structural and mechanical properties in in situ AA5083-H111/Al–Fe composites. *Proceedings of the Institution of Mechanical Engineers, Part C: Journal of Mechanical Engineering Science*, 238(10), 4477–4490.

Singla, S., Sagar, P., Handa, A., & Kang, A. S. (2023). Recent advances in magnesium-based metal matrix surface composites developed via friction stir processing route:: An overview. *Metallography, Microstructure, and Analysis*, 12(3), 385–400.

Srivastava, N., Singh, L. K., Yadav, M. K., & Kassa, M. K. (2024). Numerical and experimental investigation of the mechanical properties of MWCNT/RHA reinforced AlP0507-based hybrid aluminum metal matrix composites. *Cogent Engineering*, 11(1), 2297470.

Subramani, M., Huang, S. J., & Borodianskiy, K. (2022). Effect of WS2 nanotubes on the mechanical and wear behaviors of AZ31 stir casted magnesium metal matrix composites. *Journal of Composites Science*, 6(7), 182.

Tiwari, A., Kumar, N., & Banerjee, M. K. (2024). Applications of genetic algorithm in prediction of the best achievable combination of hardness and tensile strength for graphene reinforced magnesium alloy (AZ61) matrix composite. *Results in Control and Optimization*, 14, 100334.

Venkatesh, V. S. S., Rao, G. P., Patnaik, L., Gupta, N., Kumar, S., Saxena, K. K., … & Al-Kafaji, F. H. K. (2023). Processing and evaluation of nano SiC reinforced aluminium composite synthesized through ultrasonically assisted stir casting process. *Journal of Materials Research and Technology*, 24, 7394–7408.

Vijay, P., Raju, K. B., Ramji, K., & Kamaluddin, S. (2021). Effect of tungsten carbide on Al6061/SiC hybrid metal matrix composites. *Composites Theory and Practice*, 21(4), 169–180.

Wu, Q., & Pei, N. (2024). Effect of SiC concentration on the mechanical properties and fracture modes of SiC/Al composites via cluster analysis of acoustic emission signals. *Journal of Mechanical Science and Technology*, 38(2), 519–526.

Yan, Y. F., Kou, S. Q., Yang, H. Y., Shu, S. L., Qiu, F., Jiang, Q. C., & Zhang, L. C. (2023). Ceramic particles reinforced copper matrix composites manufactured by advanced powder metallurgy: Preparation, performance, and mechanisms. *International Journal of Extreme Manufacturing*, 5(3), 032006.

Yang, H., Chen, X., Huang, G., Song, J., She, J., Tan, J., … & Pan, F. (2022). Microstructures and mechanical properties of titanium-reinforced magnesium matrix composites: Review and perspective. *Journal of Magnesium and Alloys*, 10(9), 2311–2333.

Yuan, X., Zhang, Z., Mu, X., Shan, C., Gao, X., & Zhu, B. (2024). Recent progress on interface characterization methods of carbon fiber reinforced polymer composites. *Chemical Engineering Journal*, 499, 156220.

Zhang, X., Li, X., Wang, J., Li, B., Liu, L., Hou, X., … & Li, S. (2024). Stimulating adequate strengthening effect through alloying regulation for powder metallurgy CNTs reinforced aluminum matrix composites. *Journal of Materials Research and Technology*, 28, 1725–1733.

Zhang, H., Ni, D. R., Xiao, B. L., Liu, F. C., & Ma, Z. Y. (2024). Recent progress of aluminum alloys and aluminum matrix composites produced via laser powder bed fusion: A review. *Journal of Materials Science*, 59(22), 9857–9891.

Zhang, H. F., Wang, L., Zhang, S., Wu, C. L., Zhang, C. H., & Sun, X. Y. (2023). Design, fabrication, microstructure and properties of in-situ synthesized TiC reinforced stainless steel matrix composite coating by laser cladding. *Materials Characterization*, 204, 113177.

Zhou, Y., Wang, L., Kong, D., Zhang, B., Liu, T., Yan, Y., … & Dong, C. (2024). Ultra-high strength metal matrix composites (MMCs) with extended ductility manufactured by size-controlled powder and spherical cast tungsten carbide. *Composites Part A: Applied Science and Manufacturing*, 182, 108194.

3 Reinforcements for Metal and Hybrid Metal Matrix Composites and Current Market Opportunities

Gaurav Arora, Kanwarjeet Singh, Pankaj Kumar Singh, Vaibhav Chandra, Praveen Kumar, and Kunal Arora

3.1 REINFORCEMENT

A suitable material and manufacturing method are the crucial elements for the development of any engineering product. The field of material science consists of pure materials, alloys, composites, and smart materials. The present era of material science is continuously working on the research and development of composite and smart materials. A composite can be defined as the combination of two materials, and the resulting material has its own distinctive properties (Kelly, 1967). A composite is a combination of matrix and reinforcement suitably combined to have a superior performance material. Magnesium (Mg), aluminium (Al), zinc (Zn), copper (Cu), and titanium (Ti) alloys are a few popular choices as matrix, whereas nitrides, oxides, and carbides of various elements are a few examples of reinforcements suitably selected and combined to fabricate composite materials. Based on the type of matrix, composites are polymer matrix composites (PMC), metal matrix composites (MMCs), and ceramic matrix composites (CMC) (Clyne and Withers, 1993). A composite can be monolithic (having one type of reinforcement) and hybrid (having two or more than two types of reinforcement), depending upon the applications for which it is developed. Literature has proved that the hybrid composites are superior to the monolithic composites and alloys (Arora and Sharma, 2017). It is a true fact that composite materials are suitably replacing the various products in sports, automotive, household appliances, aerospace, construction, defence, and medical sectors (Pilato and Michno, 1994).

A discontinuous phase suitably combined with the matrix alloy to fabricate a composite material is known as reinforcement. It can also be defined as the element embedded in the matrix for the improvement of the characteristic properties of the developed

 DOI: 10.1201/9781003620570-3

composites. The other important fact to be noticed about the reinforcement is that it remains physically and chemically distinct even after the formation of the composite. Generally, the reinforcing phase is rigid and stiffer as compared to the matrix phase and is responsible for imparting strength to the manufactured composite (Campbell 2010). A reinforcement plays the following functions in composite materials:

a. The matrix phase grabbed the reinforcing phase and transferred the applied load to the reinforcement, resulting in the composite bearing stresses.
b. Imparting grain refinement.
c. Enhancement of physical and mechanical properties, namely, impact resistance, strength, and toughness of the fabricated composites.
d. Allow fabrication of tailor-made or customized composites as per the required characteristic properties by altering the type, weight/volume fraction, morphology, and orientation of reinforcement.

3.2 TYPES OF REINFORCEMENT

Generally, the quantity or fraction of reinforcement is less than the quantity or fraction of matrix in a composite material. The selection of matrix and reinforcement is done suitably for the required or specific applications. Figure 3.1 represents the classification of reinforcements utilized to fabricate various types of composite materials.

3.2.1 Particulate Reinforcement

Particulate reinforcements are the small, distinct particles embedded in the matrix material to improve the performance of composite materials. Basically, the aspect ratio, that is, l/d of reinforcement, is a key factor for deciding the characteristic properties of the developed composite materials. The aspect ratio of particulate reinforcement is approximately one. Figure 3.2 illustrates a schematic representation of a particulate-reinforced composite material. This type of reinforcement is induced to improve the strength, thermal conductivity, and wear resistance of the composite. Availability, competitive cost, and ease of manufacturing have made the particulate reinforcements the most popular amongst the researchers (Ibrahim et al., 1991).

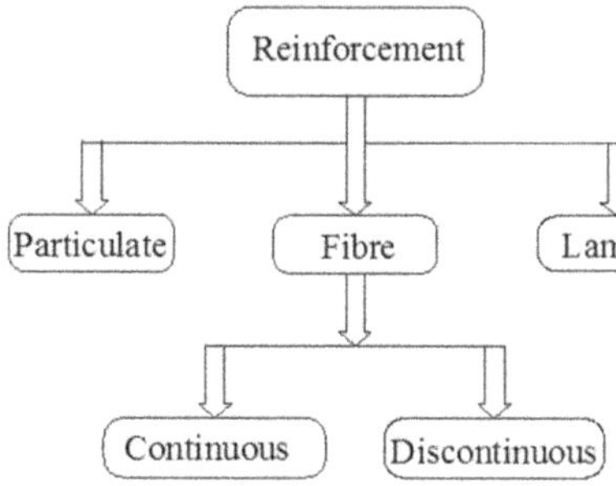

FIGURE 3.1 Classification of reinforcement.

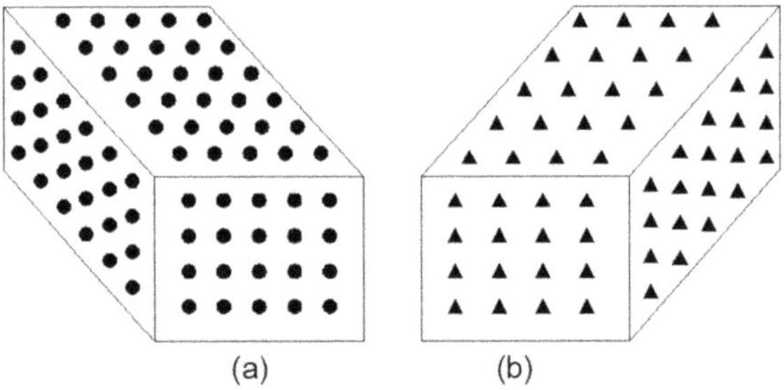

FIGURE 3.2 Particulate (a) spherical and (b) triangular reinforced composite material.

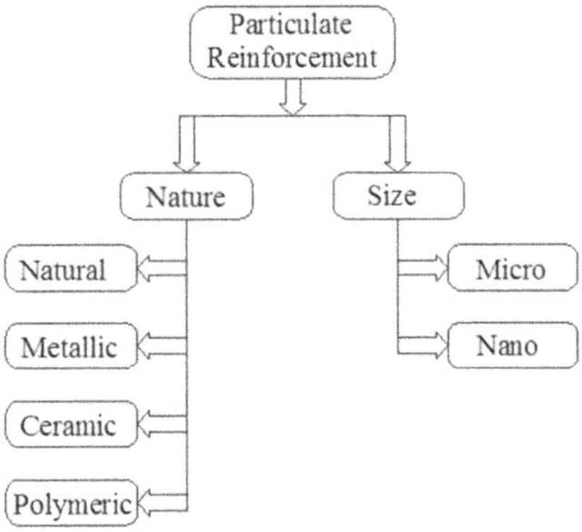

FIGURE 3.3 Classification of particulate reinforcement.

A particulate reinforcement can be classified based on the type and size of the particles employed for the development of composite materials. The classification of particulate reinforcements is presented in Figure 3.3. As per the nature of the reinforcement, it is classified as natural, that is, shells of nuts and particles of bio-waste. The particles of various metals such as copper, aluminium, zinc, and titanium are a few popular types of metallic particulate reinforcements. The other classification is ceramic particulates, which are oxides, carbides, and nitrides, for example, silicon carbide (SiC), titanium oxide (TiO2), and boron nitride (B4N). Further, polymeric particles such as rubber, Kevlar (aramid), nylon, and polyvinyl chloride (PVC) are also utilized to develop composite materials for suitable applications.

Based on the size of particles, particulate reinforcement can be of micro size, having particle sizes ranging below 100 micrometres (μm), and nano size, scaling smaller than 100 nanometres (nm).

3.2.1.1 Metal Matrix Composites with Particulate Reinforcement

Literature has indicated extensive study on the particulate-reinforced composites. Numerous researchers have also highlighted enhancements in the characteristic properties of developed MMCs with the inclusion of particulate reinforcements. Analysis of the microstructure and mechanical properties of Al 2014 matrix composites reinforced with Al_2O_3 particulate reinforcement using casting was conducted by Srivatsan (1996). An enhancement of strength and elastic modulus with the inclusion of reinforcement was reported by the author. The microstructural analysis exhibited anisotropy due to the uneven distribution of the reinforcement. A study on the TiC, Al_2O_3, TiN, and Cr_3C_2 particulate reinforced steel matrix composites was performed by Pagounis and Lindroos (1998). The composites were fabricated utilizing a hot isostatic process, and a comparative analysis was conducted with the magnesium- and aluminium-based composites. The results revealed an increment in the wear resistance, whereas a decrement in the ductility and tensile strength of the steel matrix composites was reported.

The evaluation of wear behaviour of particulate-reinforced Al matrix composites by varying the temperature and load values was carried out by Korkut (2004). The particulates of SiC, M_7C_3, Al_2O_3, and SiFe were used as reinforcement for the fabrication of composites. The outcome of research indicated that the wear resistance of the fabricated composites improved with the inclusion of particulate reinforcement. The study also emphasized that wear properties were impacted by the volume fraction of the added reinforcements. The mechanical and microstructural characteristics of AL-6Cu-0.4Mn matrix composites reinforced with SiC particulates via powder metallurgy technique were evaluated by Slipenyuk et al. (2006). The research highlighted the effect of particle size ratio of matrix to reinforcement, particle size, and fraction of volume on the properties of manufactured composites. The results indicated that the addition of a high volume fraction was feasible for the uniform distribution of particulate reinforcements. The results also revealed that a high tensile strength and yield stress were obtained utilizing small-size reinforcements with uniform distribution. Another important outcome was that a large particle size reinforcement needs to be included where a high Young's modulus of the developed material has to be achieved.

In another research finding, the researchers investigated the wear characteristic and crack density of Al/SiCp composites developed utilizing the direct metal laser sintering method (Ghosh and Saha, 2011). The fraction of volume and particle size of SiC was varied to evaluate the crack density and wear performance of the fabricated composites. An increment of crack density was observed for the addition of SiC particles beyond 15% volume fraction. The results also exhibited that the wear resistance of composites improved with the decrease in particle size.

The prediction of tensile strength for friction stir-welded joints of TiC particulate-reinforced AA6061 was carried out by Gopalakrishnan and Murugan (2011). A stir casting approach involving bottom pouring of reinforcement was employed to develop Al6061/TiCp composites with varying weight percentages of TiCp. The majority of the weld trials resulted in joint efficiencies exceeding 90%. The findings of research also indicated that amongst the available parameters, the tool pin profile had the greatest influence on the tensile strength of welded joints, followed by the speed of welding. In another research work, the evaluation of the mechanical characteristics of Al-Ca intermetallic particulate-reinforced aluminium matrix composites manufactured utilizing the powder metallurgy technique was carried out (Chaubey et al., 2012). The reinforcing powders were blended manually and ball-milled for the development of composites. An 112 MPa tensile strength was obtained for the pure aluminium alloy during testing. The results demonstrated that the value of tensile strength improved to 250 MPa and 280 MPa for the ball-milled 20% and 40% volume fraction composites, whereas 140 MPa and 165 MPa tensile strength was observed for the manually blended composites.

Myriounis et al. analysed how heat treatment affects the fatigue characteristics of A359/SiCp composites. T6 and modified T6 types of heat treatments were performed to study the precipitation hardening mechanisms (Myriounis et al., 2012). The endurance limits of fabricated composites were between 70% and 85% of ultimate tensile strength (UTS). The composites with T6 heat treatment have shown an improvement in fatigue characteristics as compared to composites with modified T6 treatment. This improved behaviour was observed due to the fine interface between the matrix and reinforcement. In another investigation, Sozhamannan et al. (2012) focused on the effect of parameters such as time of holding and temperature of processing on the resulting mechanical behaviour of Al-Si-Mg/SiCp composites developed by the stir casting route. The temperature of processing varied from 700°C to 900°C, whereas the time of holding ranged from 10 minutes to 30 minutes. A fine distribution of SiC particulates was observed during microstructural characterization at temperatures of 750°C and 800°C. The outcome of the study revealed that the UTS of fabricated composites was enhanced up to a temperature of 800°C, and after that, it was decreased. In contrast, the authors observed a decrease in the UTS of developed composites with increasing holding time.

The impact of TiC particulate reinforcement on the corrosion resistance of Al6061 matrix composites manufactured using the stir casting method was explored by investigators (Murthy and Singh, 2015). The polarization tests indicated that the corrosion resistance of composites was enhanced in comparison to the matrix alloy. The EIS (impedance measurement) analysis demonstrated that the polarization resistance of developed composites improved with the increased content of TiC particulate reinforcement. Uzun and Usca (2018) conducted a study on the influence of the addition of Cr particulates (varying weight ratios) on the wear resistance and hardness of Cu/Crp composites developed via powder metallurgy technique. The results exhibited an improved hardness and wear resistance of composites with the increased ratio of Cr particulates. The author stated that this improved wear resistance and hardness of composites was achieved due to the even distribution of Cr particulates observed during microstructural characterization.

The effect of the inclusion of $FeTiO_3$ particulates on the corrosion and tribological properties of AA6061 matrix composites was examined by experts (Morampudi et al., 2022). The AA6061 matrix composites were fabricated by varying the percentage of FeTiO3 reinforcement via the stir casting method. During the microstructural characterization, a fine distribution of $FeTiO_3$ was observed. An increase in $FeTiO_3$ reinforcement content enhanced the hardness and wear resistance of the composites due to the strengthening mechanism of particulates. The literature has also proved that the performance of hybrid composites is superior in comparison to the monolithic composites.

Microstructural and mechanical characterization of $Al/SiC+Al_2O_3$ hybrid composites synthesized using the squeeze infiltration technique was evaluated by Kim et al. (1994). The contribution of varying particle size of SiC, volumetric content of reinforcements, and applied pressure to the microstructure of the developed hybrid composite was analysed. An even distribution of reinforcement, along with the refining of matrix material, was observed. Additionally, the wear resistance and hardness of the hybrid composites are enhanced by increasing the content of reinforcement. A study by researchers has shown an improvement in the elastic modulus and tensile strength of Al hybrid composites with alumina particulate and short fibre reinforcements (Jung et al., 2000). The results exhibited that due to the formation of cluster structure, an extra strengthening is reflected in the elastic modulus and tensile strength of fabricated hybrid composites. Further, results also show that the increase in particulate content, along with a reduction in cluster size and fibre length, contributes to an enhanced strengthening effect attributed to the clusters.

The damping capacity and mechanical behaviour of magnesium matrix hybrid composites with particles of SiC and short fibres of mullite ($Al_2O_3.SiO_2$) were examined by Gu et al. (2004). An increase in the strength of the hybrid composites as compared to pure magnesium was observed. Further, results revealed a lower damping capacity of Mg hybrid composites in comparison to pure Mg. Chandrakanth et al. (2010) investigated the mechanical and physical characteristic properties of graphite- and TiC-reinforced copper hybrid composites manufactured by the microwave sintering route. The hybrid composites were developed by varying the volumetric content of graphite and TiC. The results exhibited that the hardness and sintered density of microwave-processed composites were superior to those of the composites sintered conventionally. The evaluation of wear and friction characteristics of Cu/SiCp/Grp hybrid composites developed by the powder metallurgy approach was attempted by Zhan and Zhan (2004). Slide wear testing under a dry environment at varying loads (20 N, 50 N, and 110 N) and a 0.42 m/s fixed sliding velocity was performed by the authors. The research findings showed that fabricated hybrid composites exhibited an improved resistance to wear under low loads due to the formation of a graphite layer on the surface. The delamination wear mechanism was dominant at the high values of loading conditions, as it resulted in elevated friction coefficient and wear due to the formation of a three-body tribosystem.

The wear and hardness behaviour of AZ91/TiC-TiB_2 hybrid composites was investigated by Ma et al. (2005). Both the used reinforcements were synthesized in situ from Al-Ti-B4C. The composite samples were fabricated utilizing a stir casting approach. The wear testing was performed at the load values of 5N and 35N and constant velocity on a pin-on-disc setup. The results of the investigations revealed an enhanced wear resistance and hardness of the developed hybrid composite in comparison to the AZ91 alloy due to the presence and even distribution of TiC and TiB_2 reinforcements.

In another research work, authors studied the impact of the speed of sliding on the deformation of the surface and wear characteristics of Al2219 matrix hybrid composite reinforced with SiC and graphite particulates (Basavarajappa et al., 2007). The hybrid composites were manufactured via the liquid metallurgy method. The slide wear tests were carried out utilizing pin-on-disc equipment at a load of 40 N. The testing was performed at various sliding speeds of 1.53, 3.0, 4.6, and 6.1 m/s and with a sliding distance of 5000 m under controlled conditions. The inclusion of graphite and SiC as reinforcements enhanced the wear resistance of Al2219 hybrid composites for all values of sliding speeds. The author divided the subsurface deformation stretching from the contact surface down to the depth where the composite's bulk hardness was attained into three regions. For region one, a subsurface deformation of 60 μm was observed for SiCp-reinforced Al2219 and SiC-Gr particulate-reinforced Al2219 hybrid composites. For the second region, a subsurface deformation ranging from 60 to 180 μm was recorded for Al2219/SiCp composites, and 60–180 μm was recorded for Al2219/SiCp/Grp hybrid composites. The third region exhibited no surface deformation of the composites.

The impact of turning machining parameters on the quality of the surface of Al356/SiC/B4C particulate hybrid composites utilizing Taguchi's orthogonal array approach was analysed by Babu et al. (2012). As per the results obtained by the authors, the most dominant parameter was feed rate, followed by the cutting speed, whereas a little impact of depth of cut was noted on surface roughness value. Further, the optimum turning machining parameters obtained through ANOVA for the minimum value of surface roughness are a cutting speed of 90 m/min, a feed rate of 0.1 mm/rev, and a depth of cut of 0.5 mm. James et al. (2018) investigated the microstructural, mechanical, wear, and corrosion characteristic properties of Al6061/ZrO_2/Al_2O_3 particle-reinforced hybrid composites, which were manufactured using the stir casting technique. A fair distribution and presence of ZrO_2 and Al_2O_3 reinforcements were observed during microstructural characterization. An enhancement in microhardness, tensile strength, and wear resistance of the fabricated composites as compared to the matrix alloy was recorded by the author. Further, a reduction in the corrosion rate of hybrid composites with the increased content of ZrO_2 and Al_2O_3 reinforcements was noted during corrosion testing.

Mechanical characterization of Al matrix hybrid composites with varying percentages of SiC and B4C, along with the addition of 2% graphite, fabricated via the stir casting route, was examined by Raj et al. (2022). The results of optical micrographs revealed an even distribution of included reinforcements. An improved hardness, yield, compressive and tensile strength, and elongation percentage were observed due to the addition of hard B4C reinforcement. The author stated that due to the formation of the Mg2Si phase, the mechanical characteristics of the hybrid composites were enhanced.

The literature has proved that MMCs with particulate reinforcements are engineered materials that can be customized by selecting the matrix and reinforcement for suitable applications. The particulate-reinforced MMCs aim to combine the superior mechanical, tribological, and corrosion-resistant characteristics, making them attractive for the development of various industry and engineering components. Previous available research has also demonstrated that the composites reinforced with fibre reinforcements also exhibited improved physical, mechanical, microstructural, and wear characteristic properties.

3.2.2 Fibre Reinforcement

A thin, long, and elongated filament of material is known as a fibre. Fibres can be derived from natural or man-made sources and characterized by their capacity to be woven into fabrics or twisted into threads. They play a crucial role in various applications in the textile industry, composite materials, and especially as reinforcement. A broad classification of fibre reinforcement is represented in Figure 3.4.

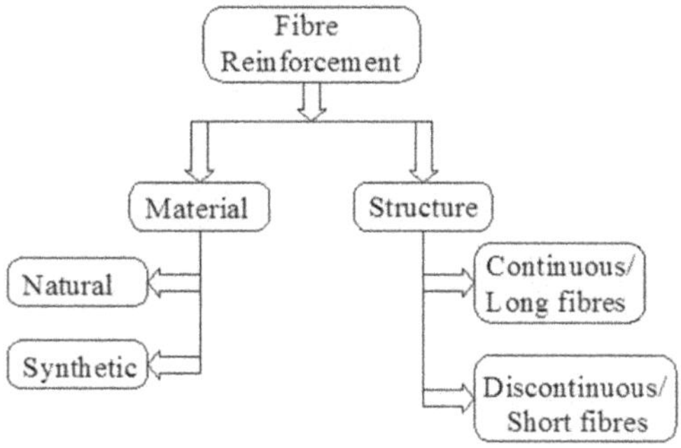

FIGURE 3.4 Classification of fibre reinforcement.

The fibres developed from natural sources such as animals, plants, and minerals are called natural fibres. Flax, cotton, jute, sisal, coconut husk fibres, and hemp are a few examples of natural fibres derived from plants, whereas silk, rabbit hair, cashmere, and wool are examples of animal-based fibres. Asbestos and glass fibre are kinds of mineral fibres. All of these fibres are utilized as reinforcement to strengthen the fabrication of composite materials. Continuous or long fibres can also be specified as filaments. These types of fibres are typically characterized by a significant aspect ratio, that is, length-to-diameter ratio. They are very thin and light in weight but provide prominent mechanical properties if used as a reinforcement.

Glass fibres, aramid fibres, and carbon fibres represent some of the most popular continuous fibres extensively utilized in composite production for a range of engineering and industrial purposes. Glass fibres are characterized by low cost and possess fair tensile strength with excellent corrosion resistance, making them suitable for marine and structural composite applications. Aramid fibres are well known for their superior thermal stability, toughness, and impact resistance, making them fit for aerospace and ballistic protection products. On the other hand, carbon fibres are significantly stiffer and stronger than the glass fibres, with the added advantage of superior weight-to-strength ratios. These fibres also exhibited great creep and fatigue resistance, making them ideal for high-end sporting goods, aerospace, and motorsports components.

On the other hand, discontinuous or short fibres are specified as chopped or short fibres and are generally randomly oriented. They are represented by a low aspect ratio, that is, length-to-diameter ratio. These fibres can be derived naturally, namely, from wool and cotton, or synthetically, namely, from carbon and polyester. In addition to being cost-effective, they are valued for their excellent processability and formability, making them suitable for various composite applications. The application of discontinuous fibre composites spans various fields, notably in aerospace and automotive, as well as in sports equipment and the textile industry. The representation of continuous and discontinuous fibre composites is presented in Figure 3.5.

3.2.2.1 Metal Matrix Composites with Fibre Reinforcement

The composites developed with fibre reinforcements have been widely researched and suitably applied due to their versatility and superior mechanical properties. As per the literature review, the inclusion of fibres for the fabrication of composites is

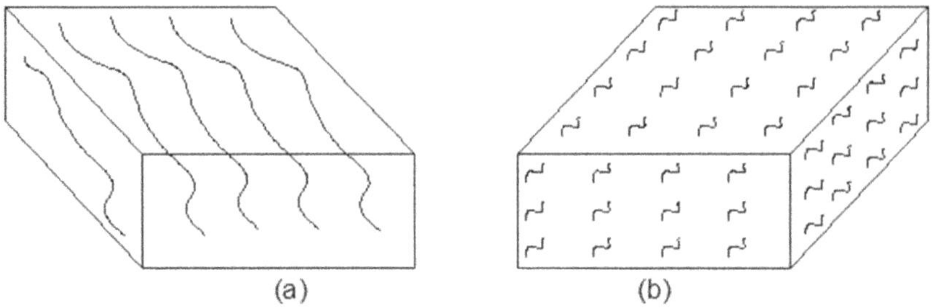

FIGURE 3.5 Fibre (a) continuous and (b) discontinuous reinforced composite material.

very unchallenging. Additionally, this inclusion also contributes to the enhancement of the durability, strength, and wear resistance of developed composite materials. Further, aspect ratio, orientation, and bonding between the matrix and fibre contribute to the final performance of the fabricated composite (Seshan et al., 1996). The continuous fibre-reinforced composites offer prominent capacity of load bearing; in contrast, discontinuous reinforced composites are chosen for their cost efficiency and ease of processing (Bahl, 2021).

The tensile and compression behaviour of Al/SiC fibres, Al/α-alumina fibres, and Al/carbon fibres was evaluated for unidirectional continuous fibres in 0° and 90° by Schulte and Minoshima (1993). The results revealed an improvement in the value of compression characteristics and Young's modulus with the increase in the content of fibres. The fatigue property of the fibre-reinforced composites was also enhanced in comparison to the pure Al alloy. In other research, the impact of the inclusion of Nextel 440 and 610 continuous fibre-reinforced aluminium matrix composites (Al 6061 and high-purity aluminium (HPAL)) on the mechanical behaviour and acoustic emission is evaluated by Pacheco et al. (1998). The HPAL/610 composite exhibited the highest value of strength, which can be attributed to the inclusion of high-strength fibres. On the other hand, Al6061/610 composite demonstrated the lowest strength, which was likely due to the elevated residual stresses. A few bursts were noticed due to the continued dislocation and breaking of fibres.

A comparative study of the mechanical characteristics of aluminium matrix and titanium matrix composites reinforced with SiC, carbon, and Al_2O_3 continuous fibres was executed by Vassel (1999). The Ti MMCs were fabricated using solid-state technique; in contrast, the Al-MMCs were manufactured utilizing liquid methods. At the temperature below 300°C, a high longitudinal strength was possessed by Al-MMCs in comparison to TiMMCs. The superior tensile characteristics were exhibited by Al-MMCs reinforced with carbon fibres for the temperature range of 20°C–300°C. Lee et al. evaluated the wear characteristics of AA6061 matrix composites with whiskers of SiC and fibres of alumina. The composites were developed using the direct squeeze infiltration technique (Lee et al., 1992). The fabricated MMCs have shown an improved wear resistance due to the inclusion of reinforcements. The results of microstructural analysis exhibited a uniform distribution and fair interfacial bonding. The results also revealed that MMCs with normally oriented reinforcements demonstrated superior wear resistance than those with randomly oriented reinforcements.

The contribution of interfaces in evaluating physical and mechanical characteristics of discontinuous fibre-reinforced MMCs has been discussed by Mitra and Mahajan (1995). The author emphasized that the interface is primarily responsible for facilitating the transfer of load and particle-dislocation interactions, which ultimately enhances the strength of the fabricated composites. The author presented a wide range of literature on different combinations of discontinuous fibre and metal matrices. The outcome exhibited that the interfacial bonding between reinforcement and matrix depends upon the composition of matrix and reinforcement, alloying elements, fabrication technique, pre-treatment of reinforcement, ageing of matrix, etc. Additionally, the existence of closely packed planes and charged defects, stoichiometric-deficient composition, and reduced heat of formation of reinforcements are the contributing factors for the enhancement of interfacial bonding.

The hardness and microstructural characterization of AA336/Al_2SiO_5 discontinuous fibre composites were examined by Nath and Singh (1999). The composites were fabricated by varying the percentage of Al_2SiO_5 utilizing the squeeze infiltration technique. The unreinforced alloy and developed MMCs were aged and analysed. An even distribution of reinforcements was observed in the optical micrographs. The results also revealed an improved hardness of aged MMCs in comparison to the unreinforced alloy. The outcome also exhibited that the reinforced matrix age hardened more rapidly with the increase in the fraction of reinforcement. In another study, Sklenicka et al. (2015) examined the creep behaviour of AZ91 and QE22 composites fabricated by the squeeze casting route. The MMCs were prepared by varying the percentage of Al_2O_3 short fibres as reinforcement. The outcome of research revealed an improved creep resistance of MMCs due to the direct and indirect strengthening phenomena.

The mechanical and wear characteristics of Al-Si/Cf/SiCp and SiCw hybrid composites fabricated via the pressure casting technique were examined by Cheng et al. (1992). The author concluded that with the hybridization of a small amount of SiC whiskers and particulates, it was possible to fabricate multi-filament aluminium matrix composites reinforced with carbon fibres. An improved tensile strength, hardness, and wear resistance of the developed hybrid composites were recorded by the authors. In another research work, the impact of alumina and aluminosilicate fibres on the tensile strength of Al-Si alloy hybrid composites was investigated by Jiang et al. (1994). All the hybrid composite samples were manufactured by the pressure infiltration technique. A comparison of tensile strength at a temperature of 300°C and room temperature of the developed samples was carried out by the authors. The research findings revealed that at 300°C, the hybrid composites indicated an improved UTS compared to the Al-Si alloy. In contrast, the UTS of the Al-Si alloy is higher than that of the fabricated hybrid composites at room temperature.

Mechanical and wear characterization of Al-Si-Cu hybrid composites reinforced with varying volumetric percentages of carbon and alumina fibres, manufactured via a squeeze infiltration approach, were analysed by Song and Han (1997). The outcome of the study exhibited an even distribution of reinforcements along with a fair interfacial bonding between reinforcement and matrix. A decrease in the elastic modulus and tensile strength of Al/C/Al_2O_3 hybrid composites was noted with an increase in carbon fibre content during testing. In contrast, an improved wear resistance of

$Al/C/Al_2O_3$ hybrid composites in comparison to Al/Al_2O_3 composites was recorded by the authors. Fu et al. (2004) investigated the wear characteristics of Al/SiC/Saffilf and Al/Al_2O_3/Saffilf manufactured via the squeeze casting route. The wear testing of fabricated hybrid composites was performed on a pin-on-disc setup under dry and lubricated environments. A high wear resistance was shown by Al/SiC/Saffilf hybrid composites at high load and temperature for a dry sliding environment, along with adhesive wear as the dominant wear mechanism. On the other hand, under the lubricated sliding environment, Al/Al_2O_3/Saffilf hybrid composites have shown the superior wear-resistant characteristics at room temperature along with microploughing as the dominant wear mechanism.

The influence of heat treatment on AlSiCuMgNi matrix hybrid composites with short carbon and Al_2O_3 fibres developed via a squeeze infiltration approach was studied by Du et al. (2007). The heat treatment of hybrid composites involved solution treatment, quenching, and precipitation hardening. Microstructural characterization findings revealed an even distribution of reinforcements in the matrix. An improvement in the friction coefficient and wear resistance of heat-treated hybrid composites was observed during testing. The results of statistical average values also exhibited that the wear rate and coefficient of friction were higher for the as-cast composites in comparison to heat-treated hybrid composites. The wear study also revealed that the dominant wear mechanism was slight ploughing with delamination and ploughing grooves for heat-treated and as-cast composites, respectively.

The effect of the addition of carbon fibres and graphene on the mechanical and electrical performance of hot-press sintered Cu/Cf/G hybrid composites was examined by Zhang et al. (2019). An enhancement in the hardness and tensile strength was noted with the inclusion of reinforcements. A decreasing trend in the electrical conductivity with the increasing content of graphene was observed during testing experiments, but it still exceeded 80% of the IACS value.

The available research findings have shown that the fibre-reinforced MMCs are high-performance composites due to their low weight, improved strength, and durability. These composites can be suitably applied for the development of products for various industries, namely, aerospace, automotive, marine, and sports. The ability of customization via material selection and fibre orientation has made these composites ideal for various functional and structural applications. The literature has also exhibited that the characteristic properties of MMCs can be enhanced by utilizing laminate-type reinforcements.

3.2.3 Laminate Reinforcement

Laminate reinforcement is a crucial aspect of composite design, especially for durable, strong, and lightweight structures. The art of strengthening a composite material through the inclusion of layers in a particular orientation or sequence is laminate reinforcing. Various types of reinforcements, such as glass, Kevlar, and carbon, can be used to develop laminar reinforced MMCs. This laminar reinforcement contributes to the improvement of stiffness, fatigue behaviour, corrosion resistance, weight reduction, and strength of the developed composites. Composites with laminar reinforcements are specially crafted engineering materials manufactured by layering sheets of reinforcement held together in a matrix. These reinforcements in each layer can be oriented in a particular direction to achieve specific physical and mechanical characteristics. The orientation of reinforcement and the sequence of stacking play a vital role in the design of the laminate. Based on the orientation, these laminates can be classified as follows:

a. Unidirectional laminates: All the reinforcements are aligned in the same direction. It offers an improved strength in the aligned direction, but is poor in the opposite or other direction.
b. Angle ply laminates: The reinforcement layers are oriented or aligned at ±45°, resulting in an enhanced torsional strength and shear resistance.
c. Cross-ply laminates: The layers of reinforcement are placed at 0° and 90° alternatively to provide strength in both transverse and longitudinal directions.
d. Quasi-isotropic laminates: The reinforcement layers are arranged in various orientations (0°, +45°, -45°, and 90°) to simulate isotropic behaviour in a plane.

The performance of laminated composites is a function of the type of reinforcement, its orientation, and its stacking sequence. Basically, it is the reinforcement part that bears the applied load in laminated composites. The researchers can optimize the orientation and sequence of stacking in laminates to achieve the specific requirements. To evaluate the mechanical performance of laminated composite structures, classical laminate theory (CLT) is generally applied by researchers. This framework allows for predicting the overall stiffness matrix of the designed laminate depending on the material's characteristics and orientation. These laminar reinforced composites are suitably applied for the development of products in the aerospace, automotive, sports, construction, marine, and many engineering industries. Figure 3.6 presents the schematic representation of single-layer and multilayer laminate-reinforced composites.

3.2.3.1 Metal Matrix Composites with Laminar Reinforcement

It is widely recognized that, in addition to particulate and fibre-reinforced MMCs, researchers have also explored the potential of laminar reinforcement-based MMCs. The impact behaviour of zinc matrix laminate composites reinforced with steel sheets

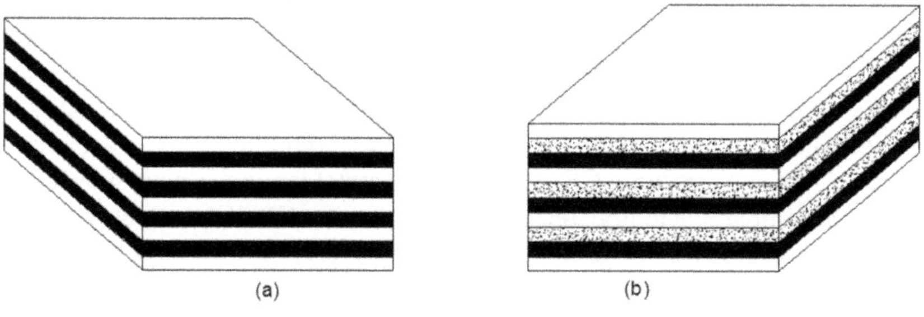

FIGURE 3.6 Laminate (a) single and (b) multi-reinforced composite material.

was examined by Vazquez and Damborenea (1995). The author concluded that an enhancement in the impact resistance can be achieved with the increased layer of reinforcement sheets. This improvement is caused by the cracking, sliding, and delamination phenomena that occurred between the reinforcement and matrix during investigations.

The investigation of deformation and failure behaviour of AA2024/Al_2O_3 laminated composites was performed by Karamis et al. (2009). The MMCs were fabricated in the two types of laminar fashion. The one type "A" of MMCs was made with the three layers of reinforced AA2024 and two layers of AA2024, whereas the other type of MMCs "B" was fabricated with two layers of reinforced AA2024 and three layers of AA2024 via the squeeze casting technique. The ballistic test was performed to assess the behaviour of the fabricated composites. The outcome of the study exhibited that for the A type of MMCs, big craters were seen at the projectile holes for both entrance and exit, whereas for the B type of MMCs, petal-like structures were formed at the entrance, and the exit of the hole appeared swollen.

Yu et al. (2015) evaluated the electrical and mechanical characteristic properties of laminar copper matrix composites reinforced with carbon nanotubes, which were fabricated via the electrodeposition route. Enhanced mechanical properties and a small impact on the conductivity of MMCs were recorded with the increased content of reinforcement films. The outcome of the investigation also revealed that the conductivity and tensile strength of MMCs were superior for CNTs in the longitudinal direction in comparison to the CNTs in the lateral direction.

The mechanical and microstructural behaviour of laminar boron carbide-reinforced Al6061 composites was analysed by Chen et al. (2017). The MMCs were fabricated utilizing sintering, followed by the extrusion process and hot rolling. An even distribution of B4C particles without any porosity was observed during microstructural characterization. The results also exhibited that an enhanced UTS and yield strength were shown by the rolled laminar composites as compared to the extruded laminar composites.

Hou et al. (2019) compared the mechanical characteristics of Ni/CNTs with those of the unreinforced Ni laminar composites. The composites were successfully manufactured via the electrodeposition technique. The author recorded an improvement in tensile strength and a decrease in the elongation of fabricated CNT-reinforced laminar composites with the inclusion of more CNTs. The nanohardness of the Ni/CNTs laminar composites was also enhanced with the addition of CNTs. These improvements in the mechanical characteristics of laminar composites were observed due to the even distribution and fine interfacial bonding of reinforcement in the matrix.

The impact of carbon fibre-reinforced polyether ether ketone and glass fibre-reinforced polyether imide on the low- and high-impact performance of fibre metal laminates was investigated by Cortes and Cantwell (2007). The thick Ti alloy foils were used to develop fibre metal laminates. The research findings imply that for the cross-ply composite structure, CF PEEK/Ti fibre metal laminates have shown higher resistance in comparison to fibre metal laminates having GF/PEI. In contrast, the GF/PEI fibre metal laminates indicated the enhanced perforation energy for high-velocity impact environment testing.

Le Bourlegat et al. (2010) examined the effect of titanium-based carbon fibre epoxy on the mechanical performance of Al/Ti/CF-E hybrid laminates produced by an autoclave system. A comparative analysis of the characteristics of the fabricated hybrid laminates with glass laminate aluminium reinforced epoxy (GLARE) and carbon fibre reinforced aluminium laminate (CARALL) was performed by the authors. The results revealed that developed hybrid laminates exhibited a high tensile stress and Young's modulus as compared to GLARE and CARALL. Further, hybrid laminates also demonstrated a high interlaminar shear strength (ILSS) compared to GLARE and CARALL.

The influence of the lay-up framework and strain rate on the flexural and tensile behaviour of carbon and glass fibre hybrid Al6061 laminates was researched by Rajkumar et al. (2014). An enhancement in the tensile strength and a decrease in the flexural strength of the developed hybrid laminates were noted with the increase in strain rate. For the layup configuration having a carbon fibre layer at the outer side rather than the glass fibre layer, the mechanical performance of hybrid laminates was enhanced. A comparative analysis of the mechanical properties of carbon fibre-reinforced plastic/aluminium hybrid laminates with laminates of aluminium and laminates of carbon fibre-reinforced plastic was carried out by Zhou et al. (2021). The results revealed that the value of tensile strength of hybrid laminates was enhanced, whereas the value of tensile modulus dropped with the increase of CFRP layers in hybrid laminates. On the other hand, the highest tensile strength was observed for plain CFRP laminates, and the highest tensile modulus was recorded for plain aluminium laminates. The flexural performance of the developed hybrid laminates was noted to fall between that of plain aluminium and CFRP laminates.

The mechanical characteristics of Al/glass/PP hybrid laminates fabricated by ECM were evaluated by Yeole et al. (2021). Aluminium sheets of 1.5 mm thickness and 12 mm long glass fibres were used to develop the hybrid laminates. A fine interfacial bonding between glass fibre-reinforced PP was observed during microstructural characterization. An improvement in tensile and flexural strength of hybrid laminates was observed during examination. In another research work, the impact of the addition of nanoplatelets of graphene on the mechanical performance of Al7075/Cf epoxy/graphene nanoplatelets hybrid laminates was studied by Gurbanov and Babanli (2021). A fine interfacial bonding between the reinforcement and matrix was observed during microstructural investigations. The results also exhibited an improved tensile and bending strength of hybrid laminates with the inclusion of graphene nanoplatelets.

Based on the above-proved investigations, it can be concluded that incorporating laminate reinforcements or developing laminar composites is an evolved and adaptable approach for the design of novel materials. In view of increasing industrial demands for materials that are stronger, possess high efficiency, and are lightweight, the importance of laminated composites is expected to grow substantially. Literature has shown that the utilization of agro- and industrial wastes as reinforcement for the MMCs is quite popular amongst the researchers nowadays. Studies have clearly indicated that these agro- and industrial wastes contributed a remarkable improvement in the characteristic properties of the developed MMCs. Additionally, the usage of these wastes as reinforcement decreased the overall cost of reinforcement, the disposal cost of these wastes, and environmental pollution.

3.3 AGRO AND INDUSTRIAL WASTE REINFORCEMENT

The integration of agro- and industrial wastes as reinforcements in composite materials is an emerging and sustainable approach aimed at addressing both the material needs and environmental concerns. The SiC, Al_2O_3, B4C, Si_3N_4, nylon, aramid, and acrylic fibres are a few examples of synthetic or man-made reinforcements suitably employed as reinforcement for the development of MMCs. These synthetic reinforcements are quite expensive and complicated to manufacture; in contrast, the wastes produced from agriculture and industries can be suitably applied as reinforcement. Red mud, palm oil clinker, fly ash, coconut shell ash (CSA), rice husk ash (RHA), and sugarcane bagasse ash (SBA) are a few illustrations of these wastes which have been utilized for the development of MMC materials. The MMCs fabricated with these wastes as reinforcement exhibited superior wear and mechanical performance and have suitable applications in the fields of automotive, aerospace, marine, sports, and various engineering industries (Arora and Sharma, 2017; Lancaster et al., 2013).

3.3.1 Metal Matrix Composites with Agro Waste Reinforcement

The utilization of agricultural waste as reinforcement offers an eco-friendly and sustainable alternative to synthetic reinforcements. Due to the increasing concerns of waste management and environmental pollution, the use of these agro wastes as reinforcement represents a value-added resource. Generally, these residues of agriculture are burnt, contributing to waste disposal costs and pollution. It is a true fact that incorporating agro waste as reinforcement helps in reducing overall cost and promotes a sustainable environment. A few popular agrowastes are coconut shell, bamboo leaf, rice husk, corn cobs, banana fibres, breadfruit seed hull, sugarcane bagasse, etc., which are utilized as reinforcement by the researchers for the development of MMCs.

The physical and mechanical characteristics of Al1100 reinforced with CSA composites fabricated by the compo casting method were investigated by Kumar et al. (2016). A decrease in the density of Al1100/CSA composites in comparison to Al100 alloy was recorded by the authors. Further, the hardness of the developed composites was increased with the addition of CSA due to the presence of MgO and SiO_2 in CSA. In another research finding, the influence of the addition of CSA on microstructural, mechanical, and tribological performance of AA7075/Al_2O_3/CSA hybrid composites was examined by Kumar et al. (2022). The hybrid composite samples were fabricated using a stir casting approach. A fine distribution of CSA and Al_2O_3 was observed in the optical micrographs. The value of UTS and the hardness of the developed composites increased with the inclusion of reinforcements. In comparison to the base Al alloy, the hybrid composites exhibited a lower friction coefficient, wear rate, and corrosion rate.

Bannaravuri and Birru (2018) studied the characteristic properties of Al-Cu/BLA (bamboo leaf ash) composites manufactured via the stir casting method. The composite samples were developed by varying the weight percentage of BLA in the Al-Cu matrix material. An even distribution of BLA was noticed during SEM examination. Due to the addition of BLA, a decrease in the density of fabricated composites was observed. On the other hand, an improvement in hardness, tensile strength, and wear resistance of Al/BLA composites was recorded by the author. The impact of the inclusion of bamboo leaf ash on the hot deformation behaviour of Al6063/SiC/BLA hybrid composites was investigated by Alaneme et al. (2020). The deformation behaviour was examined at varying strain rates and temperature values. The results revealed that an enhancement in flow stress with the increment in strain rate and a decrease in deformation temperature was observed.

The material performance of AA6351 reinforced with varying weight percentages of RHA and SiC as reinforcement was studied by Arora and Sharma (2018). A decrease in density was noted with the addition of RHA in comparison to the density of the matrix alloy. An enhancement in the microhardness, tensile strength, and wear resistance of AA6351/SiC and AA6351/RHA composites was recorded by the authors. The influence of the inclusion of one synthetic reinforcement, B4C, and one agrowaste reinforcement, RHA, on the corrosion characteristics of Al6061 hybrid composites was examined by Nithyanandhan et al. (2020). The hybrid composite samples were immersed in aluminium chloride solution to evaluate the corrosion behaviour of the developed composites. The research findings indicated that with the increased proportions of B4C and RHA, the resistance to corrosion of AA6351/B4C/RHA hybrid composites improved.

Mechanical and physical characteristics of Al6063/corn cob ash (CCA) composites manufactured via a stir casting approach were analysed by Odoni et al. (2020). The composite samples were prepared by varying the weight percentage of CCA. The fabricated Al6063/CCA composites demonstrated a reduced density and acceptable porosity level in comparison to the Al6063 matrix alloy. The optical micrographs revealed a fine distribution of CCA in matrix material. The results exhibited an improved hardness and wear resistance of Al6063/CCA composites with the increased content of CCA. In another research finding, the impact of the inclusion of CCA on the wear and mechanical performance of Al6082/boron nitride/B4C/CCA hybrid composites was examined by Manu et al. (2022). The analysis of microstructure exhibited a fine distribution of reinforcement and a refined grain structure during microstructural characterization. The fabricated hybrid composites indicated elevated impact strength and hardness in contrast to the matrix alloy. The results of the applied Taguchi method have shown the lowest rate of wear at the load value of 15 N, sliding distance of 750 m, and sliding velocity of 1 m/s.

The outcome of the inclusion of breadfruit seed hull ash (BSHA) on the mechanical and microstructural characteristics of Al-Si-Fe/BSHA composites developed by the double stir casting route was studied by Atuanya et al. (2012). All the composite samples were fabricated by varying weight fractions of BSHA and incorporating it into the alloy matrix. A reduced grain size of the matrix was observed during microstructural examination. The results revealed an enhanced hardness and tensile strength and a reduction in impact energy of manufactured composites with the inclusion of BSHA reinforcement. In continuation of this, Kaliyannan et al. (2024) investigated the impact of the addition of breadfruit seed husk ash (SHA) on the performance of LM25/SHA/Gr hybrid composites fabricated using the squeeze casting technique. Graphite nanoparticles with varying weight percentages were introduced for the development of hybrid composite samples. The results exhibited a decrease in the density of hybrid composites with the inclusion of SHA. The developed hybrid composites exhibited superior abrasion resistance, UTS, and corrosion resistance properties across the various characterization techniques performed by the authors.

The tribological and mechanical properties of SBA-reinforced AlSi10Mg alloy matrix were examined by Shankar et al. (2018). The stir casting technique was used for the development of composite samples. The results of mechanical characterization indicated an improvement in hardness, tensile strength, and impact strength, whereas a reduction in ductility was observed with the inclusion of SBA contents. Additionally, the wear resistance of the fabricated composite is enhanced with the increasing weight percentage of SBA particulates. Moreover, Chandla et al. (2020) investigated the mechanical performance of Al6061 hybrid composites having alumina and SBA as reinforcements. The testing composite samples were prepared via a stir casting approach. A fine distribution of reinforcement was observed in microstructural images. The results revealed an enhancement in microhardness, impact, and tensile strength of hybrid composites in comparison to the alloy matrix. In contrast, these properties have shown a decreasing trend with the incorporation of SBA content of large particle size. Further, existing research studies have indicated that industrial wastes also hold potential as reinforcement for the fabrication of MMCs.

3.3.2 Metal Matrix Composites with Industrial Waste Reinforcement

The utilization of industrial wastes as reinforcement is one of the emerging ways for the development of composites and materials for various engineering and other applications. This utilization also reduces the landfill dependency and disposal costs of these industrial wastes. The incorporation of industrial waste into the development of composites contributes to enhancing their performance.

The mechanical behaviour of aluminium composites reinforced with red mud fabricated by the powder metallurgy approach was examined by Chinta et al. (2018). The red mud powder was ball-milled, and varying weight percentages of milled powder were included in the matrix alloy for the manufacturing of composites. The testing results exhibited an improved hardness and compression strength of the developed composites with the increased content of red mud. Additionally, the effect of incorporating red mud on thermal and mechanical characteristics of LM25/red mud/sillimanite hybrid composites manufactured via the stir casting route was analysed by Yadav and Kumar (2023). Both reinforcements were added in the same proportion. The results of microstructural analysis demonstrated a fair and even distribution of reinforcements. Further, the hardness and tensile strength of developed composites increased with the inclusion of red mud and sillimanite reinforcements. In contrast, a decrement in thermal conductivity of the hybrid composites was observed during testing experiments.

The tribological performance of Al/fly ash composites with different weight percentages of fly ash particulates developed through a stir casting approach was analysed by Sharma et al. (2017). The wear rate of manufactured composites has shown a declining trend with the increase of fly ash particulates. The least value of the friction coefficient of the composite was observed for the Al/fly ash composite, having 4% weight of fly ash. Moreover, Reddy et al. (2020) investigated the mechanical and wear performance of Al7075/B4C/fly ash hybrid composites fabricated via liquid metallurgy technique. A better interfacial bonding between the fly ash and B4C with the alloy matrix was recorded in the SEM images. Additionally, an enhancement in microhardness, percentage elongation, tensile strength, and wear resistance of the hybrid composites was noted by the authors.

In addition to the above research findings, the impact of the inclusion of wet grinder stone dust (WSD) particulates on the wear performance of Al6063/WSD composites was studied by Xavier and Suresh (2016). The composite samples were fabricated using a two-step stir casting approach. An improvement in hardness and a decrease in the coefficient of friction, wear rate, and ductility of developed composites were recorded with the increased weight proportion of WSD.

In view of the above literature facts and results, it can be concluded that the utilization of industrial wastes as reinforcement is a sustainable and practical approach for the development of composite materials and the minimization of waste in the surrounding environment. Additionally, it is a well-known fact that metal and hybrid composites are the most popular materials across various engineering sectors due to their superior material performance. As the engineering industry continues to seek lightweight, environmentally friendly, and superior performance alternatives to existing materials, the demand for advanced and hybrid composite materials continues to grow in the near future.

3.4 INDUSTRIAL APPLICATIONS AND DEMAND FOR METAL AND HYBRID COMPOSITES

It is a well-known fact that the hybrid composites offer superior performance in comparison to the monolithic composites and pure alloy materials. Further, literature has also proved that the metal matrix and hybrid composites are gaining significant traction in modern industries as they demonstrate a combination of lightweight, high-strength, and wear-resistant characteristic properties. These composites are suitably tailored, namely, various combinations of available matrices and reinforcements to achieve the required properties and meet the industrial demands.

Metal matrix and hybrid composites are widely used in industries where enhanced thermal stability, high strength-to-weight ratio, and improved wear resistance are very much required. The various industrial applications of these metal and hybrid composites are as follows (Kempaiah et al., 2015; Srinivasan et al., 2021; Rohatgi 1993; Krishna et al. 2015; Elanchezhian et al. 2018; Vijayakumar et al. 2020; Nturanabo et al. 2019; Sharma et al. 2020; Srivyas and Charoo, 2019; Ramesh and Nataraj, 2020):

a. Aerospace sectors: The ability to withstand mechanical stresses and high temperatures of these composites makes them a promising candidate for the aerospace industry. Turbine engines, aircraft structural parts or frames, and satellite components are a few products developed from these metal and hybrid composites.
b. Electronics industry: The dimensional stability and perfect thermal conductivity have made the metal and hybrid composites a suitable choice for electronic industry products, namely, packaging materials and heat sinks.

c. Marine applications: The offshore structures and ship hulls are the products made from metal and hybrid composites due to their corrosion-resistant property.
d. Sports and leisure industry: The metal matrix and hybrid composites are lightweight and durable in nature; therefore, these are utilized to produce bicycle frames, tennis rackets, and golf clubs.
e. Automotive industry: Since HMMSCs possess a high strength-to-weight ratio, they are employed in manufacturing brake systems, engine components, and transmission parts, ultimately improving fuel efficiency while maintaining durability.

The global demand for metal and hybrid composites is growing steadily, driven by the high performance, lightweight, and durable materials across various engineering and other industries. The push for fuel efficiency and environmental regulations is propelling the aerospace and automotive sectors to adopt these composites extensively. Further, the ability to customize the required composite performance for the specific industrial applications makes these metal and hybrid composites highly attractive for future technological developments.

REFERENCES

Alaneme, K. K., S. A. Babalola, L. H. Chown, and M. O. Bodunrin. "Hot Deformation Behaviour of Bamboo Leaf Ash–Silicon Carbide Hybrid Reinforced Aluminium Based Composite." *Manufacturing Review* 7 (2020): 17.

Arora, G., and S. Sharma. "A Review on Monolithic and Hybrid Metal–Matrix Composites Reinforced with Industrial-Agro Wastes." *Journal of the Brazilian Society of Mechanical Sciences and Engineering* 39, no. 11 (2017): 4819–35.

Arora, G., and S. Sharma. "A Comparative Study of AA6351 Mono-Composites Reinforced with Synthetic and Agro Waste Reinforcement." *International Journal of Precision Engineering and Manufacturing* 19, no. 4 (2018): 631–38.

Atuanya, C. U., A. O. A. Ibhadode, and I. M. Dagwa. "Effects of Breadfruit Seed Hull Ash on the Microstructures and Properties of Al–Si–Fe Alloy/Breadfruit Seed Hull Ash Particulate Composites." *Results in Physics* 2 (2012): 142–49.

Babu, T. M., M. A. Sugin, and N. Muthukrishnan. "Investigation on the Characteristics of Surface Quality on Machining of Hybrid Metal Matrix Composite (Al-SiC-B_4C)." *Procedia Engineering* 38 (2012): 2617–24.

Bahl, S. "Fiber Reinforced Metal Matrix Composites: A Review." *Materials Today: Proceedings* 39 (2021): 317–23.

Bannaravuri, P. K., and A. K. Birru. "Strengthening of Mechanical and Tribological Properties of Al-4.5% Cu Matrix Alloy with the Addition of Bamboo Leaf Ash." *Results in Physics* 10 (2018): 360–73.

Basavarajappa, S., G. Chandramohan, A. Mahadevan, M. Thangavelu, R. Subramanian, and P. Gopalakrishnan. "Influence of Sliding Speed on the Dry Sliding Wear Behaviour and the Subsurface Deformation on Hybrid Metal Matrix Composite." *Wear* 262, no. 7–8 (2007): 1007–12.

Campbell, F. C. "Introduction to Composite Materials." In *Manufacturing Processes for Advanced Composites*, 30. Amsterdam: Elsevier, 2010.

Chandla, N. K., Yashpal, S. Kant, M. M., and C. S. Jawalkar. "Experimental Analysis and Mechanical Characterization of Al 6061/Alumina/Bagasse Ash Hybrid Reinforced Metal Matrix Composite Using Vacuum-Assisted Stir Casting Method." *Journal of Composite Materials* 54, no. 27 (2020): 4283–97.

Chandrakanth, R. G., K. Rajkumar, and S. Aravindan. "Fabrication of Copper–TiC–Graphite Hybrid Metal Matrix Composites through Microwave Processing." *The International Journal of Advanced Manufacturing Technology* 48 (2010): 645–53.

Chaubey, A. K., S. Scudino, N. K. Mukhopadhyay, M. S. Khoshkhoo, B. K. Mishra, and J. Eckert. "Effect of Particle Dispersion on the Mechanical Behavior of Al-Based Metal Matrix Composites Reinforced with Nanocrystalline Al–Ca Intermetallics." *Journal of Alloys and Compounds* 536 (2012): S134–7.

Chen, H. S., W. X. Wang, H. H. Nie, J. Zhou, Y. L. Li, and P. Zhang. "Microstructure and Mechanical Properties of B_4C/6061Al Laminar Composites Fabricated by Power Metallurgy." *Vacuum* 143 (2017): 363–70.

Cheng, H. M., A. Kitahara, S. Akiyama, K. Kobayashi, and B. L. Zhou. "Fabrication of Carbon Fibre-Reinforced Aluminium Composites with Hybridization of a Small Amount of Particulates or Whiskers of Silicon Carbide by Pressure Casting." *Journal of Materials Science* 27 (1992): 3617–23.

Chinta, N. D., N. Selvaraj, and V. Mahesh. "Mechanical Characterization of Aluminium–Red Mud Metal Matrix Composites." *Materials Today: Proceedings* 5, no. 13 (2018): 26911–17.

Clyne, T. W., and P. J. Withers. *An Introduction to Metal Matrix Composites*. Cambridge: Cambridge University Press, 1993.

Cortes, P., and W. J. Cantwell. "The Impact Properties of High-Temperature Fiber-Metal Laminates." *Journal of Composite Materials* 41, no. 5 (2007): 613–32.

Du, J., Y. Liu, S. Yu, and W. Li. "Effect of Heat-Treatment on Friction and Wear Properties of Al_2O_3 and Carbon Short Fibres Reinforced AlSi12CuMgNi Hybrid Composites." *Wear* 262, no. 11–12 (2007): 1289–95.

Elanchezhian, C., B. V. Ramnath, G. Ramakrishnan, K. S. Raghavendra, M. Muralidharan, and V. Kishore. "Review on Metal Matrix Composites for Marine Applications." *Materials Today: Proceedings* 5, no. 1 (2018): 1211–18.

Fu, H. H., K. S. Han, and J. I. Song. "Wear Properties of Saffil/Al, Saffil/Al_2O_3/Al and Saffil/SiC/Al Hybrid Metal Matrix Composites." *Wear* 256, no. 7–8 (2004): 705–13.

Ghosh, S. K., and P. Saha. "Crack and Wear Behavior of SiC Particulate Reinforced Aluminium Based Metal Matrix Composite Fabricated by Direct Metal Laser Sintering Process." *Materials & Design* 32, no. 1 (2011): 139–45.

Gopalakrishnan, S., and N. Murugan. "Prediction of Tensile Strength of Friction Stir Welded Aluminium Matrix TiCp Particulate Reinforced Composite." *Materials & Design* 32, no. 1 (2011): 462–67.

Gu, J., X. Zhang, and M. Gu. "Mechanical Properties and Damping Capacity of (SiCp + $Al_2O_3 \cdot SiO_2$ f)/Mg Hybrid Metal Matrix Composite." *Journal of Alloys and Compounds* 385, no. 1–2 (2004): 104–8.

Gurbanov, N. A., and M. B. Babanli. "Investigation of Effects of Graphene Nanoplatelets Addition on Mechanical Properties of 7075-T6 Aluminium Matrix Hybrid Fibre Metal Laminates." *Metallophysics & Advanced Technologies/Metallofizika i Novejsie Tehnologii* 43, no. 12 (2021): 1589–99.

Hou, Z. C., L. Q. Xiong, Y. F. Liu, L. Zhu, and W. Z. Li. "Preparation of Super-Aligned Carbon Nanotube-Reinforced Nickel-Matrix Laminar Composites with Excellent Mechanical Properties." *International Journal of Minerals, Metallurgy, and Materials* 26 (2019): 133–41.

Ibrahim, I. A., F. A. Mohamed, and E. J. Lavernia. "Particulate Reinforced Metal Matrix Composites: A Review." *Journal of Materials Science* 26 (1991): 1137–56.

James, S. J., M. Ganesan, P. Santhamoorthy, and P. Kuppan. "Development of Hybrid Aluminium Metal Matrix Composite and Study of Property." *Materials Today: Proceedings* 5, no. 5 (2018): 13048–54.

Jiang, J. Q., H. N. Liu, A. P. Ma, and R. S. Tan. "The Structure and Tensile Properties of Al-Si Alloy Hybrid Reinforced with Alumina-Aluminosilicate Short Fibre." *Journal of Materials Science* 29 (1994): 3767–73.

Jung, S. W., J. H. Lee, J. B. Nam, H. W. Nam, and K. S. Han. "Analysis of Strengthening Mechanism in Hybrid Short Fiber/Particle Reinforced Metal Matrix Composites." *Key Engineering Materials* 183 (2000): 1297–302.

Kaliyannan, P., M. Perumal, J. Manivannan, and R. Venkatesh. "Enhancing Mechanical and Functional Properties of LM25 Alloy through Squeeze Cast Hybrid Nanocomposite Incorporating Breadfruit Seed Husk Ash and Graphite Nanoparticles." *International Journal of Cast Metals Research* 37, no. 1 (2024): 12–21.

Karamis, M. B., F. Nair, and A. A. Cerit. "The Metallurgical and Deformation Behaviours of Laminar Metal Matrix Composites after Ballistic Impact." *Journal of Materials Processing Technology* 209, no. 10 (2009): 4880–89.

Kelly, A. "The Nature of Composite Materials." *Scientific American* 217, no. 3 (1967): 160–79.

Kempaiah, U. N., N. Santhosh, D. H. Kumar, A. H. Chelgeri, P. Vishnu, and S. Raorane. "Characterization of High Performance Al 5083/SiCP/Fly Ash Hybrid Metal Matrix Composite for Advanced Aerospace Applications." *International Journal of Advanced Innovation Research* 3 (2015): 334–41.

Kim, M. S., K. M. Cho, and I. M. Park. "Microstructure of Squeeze Cast AC4A Al/Al_2O_3 + SiCp Hybrid Metal Matrix Composite." *Journal of Korea Foundry Society* 14, no. 3 (1994): 258–66.

Korkut, M. H. "Effect of Particulate Reinforcement on Wear Behaviour of Aluminium Matrix Composites." *Materials Science and Technology* 20, no. 1 (2004): 73–81.

Krishna, S. M., T. N. Shridhar, and L. Krishnamurthy. "Research Significance, Applications and Fabrication of Hybrid Metal Matrix Composites." *International Journal of Innovative Science Engineering and Technology* 2 (2015): 227–37.

Kumar, A., K. Kumar, S. Saurav, and S. S. Raju. "Study of Physical, Mechanical and Machinability Properties of Aluminium Metal Matrix Composite Reinforced with Coconut Shell Ash Particulates." *Imperial Journal of Interdisciplinary Research* 2, no. 2454 (2016): 1362.

Kumar, A., R. C. Singh, and R. Chaudhary. "Investigation of Nano-Al_2O_3 and Micro-Coconut Shell Ash (CSA) Reinforced AA7075 Hybrid Metal–Matrix Composite Using Two-Stage Stir Casting." *Arabian Journal for Science and Engineering* 47, no. 12 (2022): 15559–73.

Lancaster, L., M. H. Lung, and D. Sujan. "Utilization of Agro-Industrial Waste in Metal Matrix Composites: Towards Sustainability." *Proceedings of World Academy of Science, Engineering and Technology* 7, no. 73 (2013): 1136. World Academy of Science, Engineering and Technology (WASET).

Le Bourlegat, L. R., C. A. Damato, D. F. Da Silva, E. C. Botelho, and L. C. Pardini. "Processing and Mechanical Characterization of Titanium-Graphite Hybrid Laminates." *Journal of Reinforced Plastics and Composites* 29, no. 22 (2010): 3392–400.

Lee, C. S., Y. H. Kim, K. S. Han, and T. Lim. "Wear Behaviour of Aluminium Matrix Composite Materials." *Journal of Materials Science* 27 (1992): 793–800.

Ma, B. X., H. Y. Wang, Y. Wang, and Q. C. Jiang. "Fabrication of (TiB_2–TiC)p/AZ91 Magnesium Matrix Hybrid Composite." *Journal of Materials Science* 40 (2005): 4501–4.

Manu, S., N. Radhika, V. Sidvilash, and T. Mohanraj. "Investigation on the Mechanical and Wear Behaviour of Al-6082-BN-B4C-Corn Cob Ash Hybrid Composites." *Tribology in Industry* 44, no. 2 (2022): 294.

Mitra, R., and Y. R. Mahajan. "Interfaces in Discontinuously Reinforced Metal Matrix Composites: An Overview." *Bulletin of Materials Science* 18 (1995): 405–34.

Morampudi, P., V. V. Ramana, K. Bhavani, C. K. Reddy, and K. S. R. Vikas. "Wear and Corrosion Behavior of AA6061 Metal Matrix Composites with Ilmenite as Reinforcement." *Materials Today: Proceedings* 52 (2022): 1515–20.

Murthy, H. A., and S. K. Singh. "Influence of TiC Particulate Reinforcement on the Corrosion Behaviour of Al 6061 Metal Matrix Composites." *Advanced Materials Letters* 6, no. 7 (2015): 633–40.

Myriounis, D. P., T. E. Matikas, and S. T. Hasan. "Fatigue Behaviour of SiC Particulate-Reinforced A359 Aluminium Matrix Composites." *Strain* 48, no. 4 (2012): 333–41.

Nath, D., and V. Singh. "Ageing Characteristics of Aluminium Alloy Aluminosilicate Discontinuous Fibre Reinforced Composites." *Scripta Materialia* 40, no. 7 (1999): 791–94.

Nithyanandhan, T., P. Sivaraman, R. Ramamoorthi, P. S. Kumar, R. Kannakumar, and A. N. Kumar. "Enhancement of Corrosion Behaviour of AL6061-B4C-RHA Reinforced Hybrid Composite." *Materials Today: Proceedings* 33 (2020): 372–77.

Nturanabo, F., L. Masu, and J. B. Kirabira. "Novel Applications of Aluminium Metal Matrix Composites." *In Aluminium Alloys and Composites*, 1, (2019): 71–94.

Odoni, B. U., F. O. Edoziuno, C. C. Nwaeju, and R. O. Akaluzia. "Experimental Analysis, Predictive Modelling and Optimization of Some Physical and Mechanical Properties of Aluminium 6063 Alloy Based Composites Reinforced with Corn Cob Ash." *Journal of Materials and Engineering Structures «JMES»* 7, no. 3 (2020): 451–65.

Pacheco, T., H. Nayeb-Hashemi, and H. E. M. Sallam. "The Effects of Matrix and Fiber Properties on the Mechanical Behavior and Acoustic Emission in Continuous Fiber Reinforced Metal Matrix Composites." *Materials Science and Engineering: A* 247, no. 1–2 (1998): 88–96.

Pagounis, E., and V. K. Lindroos. "Processing and Properties of Particulate Reinforced Steel Matrix Composites." *Materials Science and Engineering: A* 246, nos. 1–2 (1998): 221–34.

Pilato, L. A., and M. J. Michno. *Advanced Composite Materials*. New York: Springer Science & Business Media, 1994.

Raj, P., B. Deepanraj, N. Senthilkumar, and T. Tamizharasan. "A Study on Effect of Primary and Secondary Reinforcements in Hybrid Metal Matrix Composite." In *AIP Conference Proceedings*, Vol. 2393, No. 1. AIP Publishing, 2022.

Rajkumar, G. R., M. Krishna, H. N. Narasimhamurthy, Y. C. Keshavamurthy, and J. R. Nataraj. "Investigation of Tensile and Bending Behavior of Aluminum Based Hybrid Fiber Metal Laminates." *Procedia Materials Science* 5 (2014): 60–68.

Ramesh, P., and M. Nataraj. "Automotive Industry Application of Aluminium-Based Hybrid Metal Matrix Composite." *International Journal of Heavy Vehicle Systems* 27, no. 1-2 (2020): 18–32.

Reddy, T. P., S. J. Kishore, P. C. Theja, and P. P. Rao. "Development and Wear Behavior Investigation on Aluminum-7075/B4C/Fly Ash Metal Matrix Composites." *Advanced Composites and Hybrid Materials* 3, no. 2 (2020): 255–65.

Rohatgi, P. K. "Metal Matrix Composites." *Defence Science Journal* 43, no. 4 (1993): 323.

Schulte, K., and K. Minoshima. "Damage Mechanisms under Tensile and Fatigue Loading of Continuous Fibre-Reinforced Metal-Matrix Composites." *Composites* 24, no. 3 (1993): 197–208.

Seshan, S., A. Guruprasad, M. Prabha, and A. Sudhakar. "Fibre-Reinforced Metal Matrix Composites—A Review." *Journal of the Indian Institute of Science* 76, no. 1 (1996): 1.

Shankar, S., A. Balaji, and N. Kawin. "Investigations on Mechanical and Tribological Properties of Al-Si10-Mg Alloy/Sugarcane Bagasse Ash Particulate Composites." *Particulate Science and Technology* 36, no. 6 (2018): 762–70.

Sharma, A. K., R. Bhandari, A. Aherwar, R. Rimašauskienė, and C. Pinca-Bretotean. "A Study of Advancement in Application Opportunities of Aluminum Metal Matrix Composites." *Materials Today: Proceedings* 26 (2020): 2419–24.

Sharma, V. K., R. C. Singh, and R. Chaudhary. "Effect of Fly Ash Particles with Aluminium Melt on the Wear of Aluminium Metal Matrix Composites." *Engineering Science and Technology, an International Journal* 20, no. 4 (2017): 1318–23.

Sklenicka, V., K. Kucharova, M. Kvapilova, and M. Svoboda. "Factors Influencing Creep Resistance in Discontinuously Reinforced Magnesium Metal Matrix Composites." *Metallurgical and Materials Transactions A* 53, no. 4 (2015): 221–29.

Slipenyuk, A., V. Kuprin, Y. Milman, V. Goncharuk, and J. Eckert. "Properties of P/M Processed Particle Reinforced Metal Matrix Composites Specified by Reinforcement Concentration and Matrix-to-Reinforcement Particle Size Ratio." *Acta Materialia* 54, no. 1 (2006): 157–66.

Song, J. I., and K. S. Han. "Mechanical Properties and Solid Lubricant Wear Behavior of Al/ Al_2O_3/C Hybrid Metal Matrix Composites Fabricated by Squeeze Casting Method." *Journal of Composite Materials* 31, no. 4 (1997): 316–44.

Sozhamannan, G. G., S. B. Prabu and V. S. K. Venkatagalapathy. "Effect of processing paramters on metal matrix composites: stir casting process." *Journal of Surface Engineered Materials and Advanced Technology* 2, (2012): 11–15.

Srinivasan, V., S. Kunjiappan, and P. Palanisamy. "A Brief Review of Carbon Nanotube Reinforced Metal Matrix Composites for Aerospace and Defense Applications." *International Nano Letters* 11, no. 4 (2021): 321–45.

Srivatsan, T. S. "Microstructure, Tensile Properties and Fracture Behaviour of Al_2O_3 Particulate-Reinforced Aluminium Alloy Metal Matrix Composites." *Journal of Materials Science* 31 (1996): 1375–88.

Srivyas, P. D., and M. S. Charoo. "Application of Hybrid Aluminum Matrix Composite in Automotive Industry." *Materials Today: Proceedings* 18 (2019): 3189–200.

Uzun, M., and U. A. Usca. "Effect of Cr Particulate Reinforcements in Different Ratios on Wear Performance and Mechanical Properties of Cu Matrix Composites." *Journal of the Brazilian Society of Mechanical Sciences and Engineering* 40, no. 4 (2018): 197.

Vassel, A. "Continuous Fibre Reinforced Titanium and Aluminium Composites: A Comparison." *Materials Science and Engineering: A* 263, no. 2 (1999): 305–13.

Vazquez, A. J., and J. D. Damborenea. "Laminar Composites of Steel and Zinc Matrix: Charpy Tests." *International Journal of Materials and Product Technology* 10, no. 1–2 (1995): 143–51.

Vijayakumar, K., L. Prabhu, B. S. Subin, S. Satheen, and K. Vaishnav. "Development of Hybrid Aluminium Metal Matrix Composites for Marine Applications." In *IOP Conference Series: Materials Science and Engineering*, Vol. 993, No. 1, 012016. IOP Publishing, 2020.

Xavier, L. F., and P. Suresh. "Wear Behavior of Aluminium Metal Matrix Composite Prepared from Industrial Waste." *The Scientific World Journal* 2016, no. 1 (2016): 6538345.

Yadav, A. K., and N. Kumar. "Mechanical and Thermal Behaviour of Stir Cast Sillimanite/ Red Mud/LM25 Al-Alloy Metal Matrix Composite." *Material Science and Engineering Technology* 54, no. 12 (2023): 1648–58.

Yeole, P., H. Ning, and A. A. Hassen. "Development and Characterization of a Polypropylene Matrix Composite and Aluminum Hybrid Material." *Journal of Thermoplastic Composite Materials* 34, no. 3 (2021): 364–81.

Yu, J. I. N., Z. H. U. Lin, X. U. E. Wei-Dong, and L. I. Wen-Zhen. "Fabrication of Superaligned Carbon Nanotubes Reinforced Copper Matrix Laminar Composite by Electrodeposition." *Transactions of Nonferrous Metals Society of China* 25, no. 9 (2015): 2994–3001.

Zhan, Y., and G. Zhang. "Friction and Wear Behavior of Copper Matrix Composites Reinforced with SiC and Graphite Particles." *Tribology Letters* 17 (2004): 91–8.

Zhang, X., W. Yang, J. Zhang, X. Ge, X. Liu, and Y. Zhan. "Multiscale Graphene/Carbon Fiber Reinforced Copper Matrix Hybrid Composites: Microstructure and Properties." *Materials Science and Engineering: A* 743 (2019): 512–19.

Zhou, C., Y. Li, G. Zhu, G. Luo, X. Zhao, and Q. Yu. "Tensile and Flexural Behavior of Metal/ CFRP Hybrid Laminated Plates." *Polymer Composites* 42, no. 6 (2021): 2882–97.

4 Existing Development Techniques and Machining in MMCs for Automotive and Aeronautical Industries

Narender Singh, Rajesh Kumar Saluja, Neeraj Kumar Gahlot, and Gurpreet Singh Saini

4.1 INTRODUCTION

The advancement of engineering materials from traditional monolithic metals to meticulously engineered composites has been driven by consumer requirements for improved mechanical properties, reduced weight, and enhanced durability (Simões 2024). One important family of materials that fills the performance gap between traditional metals and cutting-edge ceramics is metal matrix composites (MMCs) (Pooja, Tarannum, and Chaudhary 2025). In the middle of the 20th century, aircraft engineers started developing materials that were lightweight, exceptionally durable, and able to withstand demanding service conditions (Yavuz and Ozkol 2024). The development of early MMCs produced a suitable material for use in space and military aircraft by strengthening aluminium matrices with carbon and boron fibres. Then, the advancement in fabrication techniques and better knowledge of matrix-reinforcement interactions increased the use of MMCs in the automotive industry until the 1980s and 1990s (Nagaraju et al. 2023). The amazing wear resistance, dimensional stability, and heat dissipation of MMCs led to their inclusion by manufacturers in the drive shaft, brake rotor, and piston (Kumawat et al. 2023). Following this, when the emphasis shifted to energy efficiency and pollution reduction, MMCs were a crucial part of industry-wide initiatives (Moghayedi and Awuzie 2023). As a result, hybrid MMCs (HMMCs) developed with a variety of reinforcement materials, including carbon-based nanometrics and ceramics, have subsequently expanded their scope of applications.

Furthermore, modern MMCs started to be modified for specific applications where carefully and appropriately chosen matrices (titanium, magnesium, and aluminium) and metal reinforcement (SiC, Al2O3, B4C, and CNTs) have been used (Sambasivam et al. 2023). In addition to this flexibility of material design, modern

DOI: 10.1201/9781003620570-4

production techniques, including powder metallurgy (PM), adaptive manufacturing, and squeeze casting, have been utilized (Baliarsingh et al. 2024). These processes offer improved microstructural control, with optimal strength-to-weight ratio, fatigue resistance, and thermal properties. In addition, adequate progress has been made in machining MMCs, despite the limitations of traditional methods caused by tool wear and poor surface finish. Consequently, the use of advanced unconventional machining techniques such as electrical discharge machining (EDM), laser-assisted machining (LAM), and abrasive water jet machining (AWJM) has seen a significant rise in recent years (Tengjiao 2024; Grzesik and Ruszaj 2021). These techniques were all created to reduce the difficulties associated with shaping and finishing MMCs.

Recently, MMC's development has been changing the way these materials are imagined, increasingly involving artificial intelligence, machine learning, and computational resources (Sarmah and Gupta 2024). Also, predictive modelling and real-time feedback systems allow rapid design cycles and customized characteristics. Currently, the focus is on recycling stability, not only to increase research performance but also to use recycled resources and green production techniques. As a result of the initiative to reduce environmental impact and increase efficiency of automotive and aerospace industries, MMCs are being used as an important material for the future (Gopal et al. 2025).

In view of this, the chapter presents a complete overview of MMCs' machining and manufacturing processes based on the current and anticipated future research, along with their background, techniques, and review of applications.

4.2 ADVANCED FABRICATION TECHNIQUES FOR METAL MATRIX COMPOSITES

This chapter presents a detailed classification of advanced fabrication techniques employed in the development of modern composite and engineered materials. The techniques are systematically grouped into four major categories: liquid-state processing, solid-state processing, vapour deposition methods, and additive manufacturing (AM). This foundational knowledge is essential for selecting the most appropriate process for advanced material design and manufacturing. Figure 4.1 depicts the systematic classification of advanced machining methods utilized for MMCs.

4.2.1 Liquid-State Processing

The most widely used method for producing MMCs is liquid-state processing because of its adaptability and simplicity. This process involves adding reinforcement to a matrix of molten metal, which then solidifies (Biswas et al. 2025). Squeeze casting, stir casting, and infiltration are three methods, each offering unique benefits for strengthening of element complexity, bonding quality, and dispersion. Despite its widespread use, issues such as porosity, poor wettability, and uneven particle spread are being investigated and addressed. The following section explains the various methods in detail.

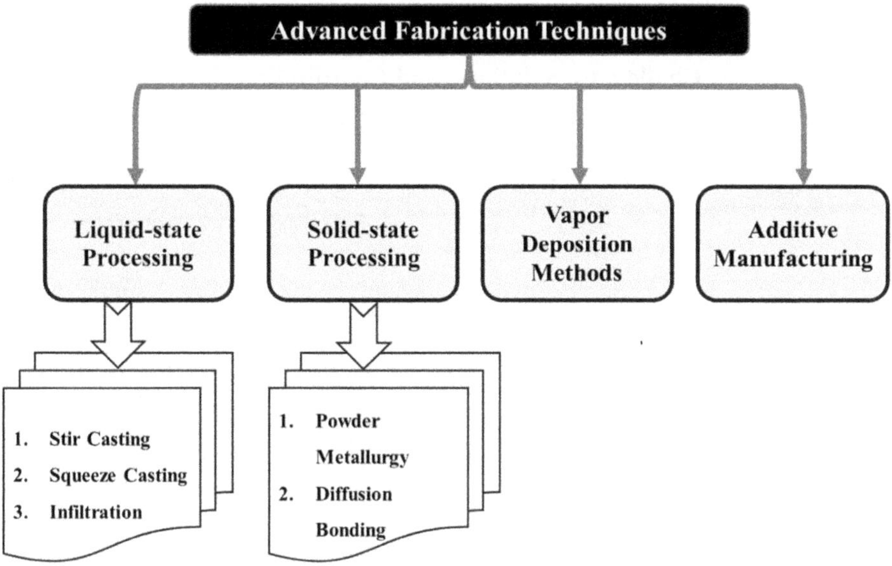

FIGURE 4.1 Classification of advanced fabrication methods.

4.2.1.1 Stir Casting

One of the most popular and affordable liquid-state building methods for MMCs is stir casting (S. K. Sharma, Saxena, Salem, et al. 2024b). The metal matrix (such as magnesium, aluminium, or their alloys) is melted, and the molten metal is physically stirred to form a vortex into which reinforcing particles (such as graphite, SiC, or Al2O3) are added. Before being inserted into moulds for solidification, continuous stirring guarantees that the reinforcement is distributed evenly throughout the melted metal. In this process, the following steps are carried out: melting the material, stirring it up, adding particles, stirring again, and then casting to prevent agglomeration. Stir casting offers benefits such as simplicity, scalability, low production cost, and the ability to produce composites with a wide range of reinforcement types and volume fractions (Sankhla and Patel 2022). However, it faces challenges such as poor wettability between matrix and reinforcements, particle agglomeration and sedimentation, gas entrapment leading to porosity, and non-uniform reinforcement distribution. Research between 2023 and 2025 has focused on improving particle dispersion using ultrasonic stirring, coating reinforcements to enhance wettability, and optimized process parameters to reduce defects and enhance mechanical properties. Table 4.1 presents recent research focused on overcoming the challenges inherent to stir casting processes.

4.2.1.2 Squeeze Casting

Squeeze casting is a semi-solid processing technique widely used for fabricating MMCs that combines the advantages of casting and forging (Nehra et al. 2025). The process involves pouring molten metal into a preheated mould containing a reinforcement preform or particles, followed by the application of high pressure during solidification. This pressure helps to force the molten metal into the reinforcement structure,

TABLE 4.1
Summary of Key Challenges, Solutions, and Contributions in Stir Casting

Challenge	Solution	Key Contribution	References
Poor wettability between the metal and particles	Surface coating of reinforcements with metals (e.g., Ni and Cu) or carbides	Ni-coated SiC particles improved wettability and composite strength	X. Sun et al. (2025b)
Particle agglomeration and sedimentation	Ultrasonic-assisted stirring to break clusters and maintain dispersion	Ultrasonic stirring enhanced uniform particle distribution	Mehdi et al. (2025)
Gas entrapment and porosity	Vacuum-assisted casting or inert gas shielding	Vacuum stirring reduced porosity by 30% in Al/ Al_2O_3 MMCs	Oke et al. (2025)
Non-uniform reinforcement distribution	Optimization of stirring speed, time, and temperature	Process parameter optimization resulted in homogeneous composites	Ashok Raj et al. (2024)

reducing porosity and improving the metal-reinforcement interface bonding. The key steps include melting the matrix metal, placing the reinforcement in the mould, pouring the molten metal, applying high pressure (typically 50–150 MPa) until the metal solidifies, and finally ejecting the solidified composite. The technique offers benefits such as improved mechanical properties due to reduced defects, near-net shape production, refined microstructure, and enhanced wettability of reinforcements (Ali et al. 2024). Mismatch and challenges include high equipment cost, difficulty in uniformly distributing reinforcements, residual stresses due to thermal mismatch, and limited complex shape fabrication. Table 4.2 highlights recent studies aimed at addressing the challenges associated with squeeze casting processes.

TABLE 4.2
Summary of Key Challenges, Solutions, and Contributions in Squeeze Casting

Challenge	Solution	Key Contribution	References
High equipment and operational costs	Development of low-cost hydraulic systems and modular moulds	Modular moulds reduced setup cost by 20%	Sinchai, Boonyang, and Simmala (2024)
Non-uniform reinforcement distribution	Ultrasonic vibration during casting to improve particle dispersion	Ultrasonic-assisted squeeze casting enhanced the uniformity of SiC in Al	Bharat and Bose (2024)
Residual stresses due to thermal mismatch	Use of graded reinforcement layers and controlled cooling rates	Graded B_4C reinforcements reduced crack formation in the Mg matrix	Demir et al. (2023)
Limited complexity in shapes	Integration with additive manufacturing for mould design	3D-printed moulds enabled complex MMC shapes with squeeze casting	Jeevankumar et al. (2024)

4.2.1.3 Infiltration Technique

Infiltration is an effective method to produce MMCs, especially useful for high volume fractions of reinforcement and good mechanical bonding (Yunlei Wang and Monetta 2023). This method involves making a mould that is filled with ceramic reinforcement (such as SiC, Al2O3, and B4C), and usually binders or sintering are used to keep particles in proper shape. After that, the melted metal (matrix), such as magnesium or aluminium, is pushed into the porous structure of the mould, which is performed by squeezed castings such as capillary action, gas pressure, vacuum aid, or methods of external pressure. The composite is allowed to harden after infiltration, producing a thick MMC with favourable load transmission properties. The ability to construct big and complicated shapes, high reinforcement loading, near-net-shape fabrication, and efficient matrix-reinforcement bonding are some advantages of this technology (Baliarsingh et al. 2024). However, it also presents partial infiltration issues due to such porosity formation, chemical interaction at the contact point, thermal mismatched stress, and poor wettability. Scientists have developed advanced wetting agents, pressure-heroic infiltration methods, and surface treatments to solve these problems and improve the reliability of the process. Table 4.3 summarizes current research efforts focused on overcoming difficulties encountered in infiltration processes.

4.2.2 Solid-State Processing

Solid-state processing methods provide more control over microstructure and reduce interface interactions by fabricating MMCs without melting the metal matrix (Mohammed et al. 2024). It is possible to produce close-net-shaped components

TABLE 4.3
Summary of Challenges, Solutions, and Contributions in Infiltration Technique

Challenge	Solution	Key Contribution	References
Poor wettability between the matrix and the ceramic	Use of alloying elements like Ti, Mg, or reactive wetting agents	Mg addition improved Al wetting on SiC by 35%	Ge, Guo, and Shen (2025)
Incomplete infiltration and porosity	Pressure-assisted (squeeze) infiltration under controlled atmosphere	Achieved >98% densification in Al/Al_2O_3 composites using squeeze casting	Gupta et al. (2024)
Chemical reactions at the interface	Application of barrier coatings (e.g., carbon and BN) on reinforcement	BN coating prevented Al–SiC reaction and enhanced interface stability	Lee et al. (2025)
Residual stresses due to thermal mismatch	Controlled cooling and functionally graded reinforcements	Graded SiC preform reduced thermal stress cracks in Al matrix	Ram et al. (2024)
Difficulty in controlling the preform porosity	3D-printed or gel-cast preforms with uniform pore architecture	3D printing improved infiltration uniformity and mechanical strength	Arunprasand and Nallasamy (2024)

with equally distributed reinforcement and sophisticated grain architecture using techniques such as diffusion bonding and PM. These methods are especially useful for developing high-demonstration components with better dimensional stability and mechanical integrity. Although solid-state methods often require more expensive and sophisticated machinery, their ability to create high-strength, innocent composites makes them a suitable choice for important motor vehicle and aerospace applications (Liang et al. 2025).

4.2.2.1 Powder Metallurgy

Another popular way to produce MMCs is PM, which combines metal powder with ceramic reinforcement elements (e.g., SiC, Al_2O_3, and B_4C) to improve mechanical properties. Typically, the PM process consists of sintering, compaction, and powder blending (Balasubramanian, Vignesh, and Padmanaban 2024). The first step is to achieve homogeneity by evenly mixing the metal and reinforcing powders. After that, a green compact is made by compaction under high pressure. Then, to promote diffusion bonding and density, the compact is sintered at high temperatures below the melting point of the matrix metal. The PM offers many advantages, including compatibility with high-melting-point reinforcements, homogeneous reinforcement distribution, near-net-shape fabrication, and low material waste (Goh et al. 2024). Nevertheless, it encounters obstacles such as porosity, poor matrix-reinforcement interfacial bonding, full densification difficulty, and reinforcement particle aggregation. Table 4.4 presents recent investigations targeting the challenges related to diffusion bonding techniques.

TABLE 4.4
Summary of Key Challenges, Solutions, and Contributions in Powder Metallurgy

Challenge	Solution	Key Contribution	References
Oxide layer formation at the interface	Use of vacuum/inert atmosphere and surface activation via ion cleaning	Ion beam cleaning improved Al/SiC joint integrity by 22%	H. Sun et al. (2025a)
Long bonding time	Application of transient liquid-phase (TLP) bonding for faster diffusion	Reduced bonding time by 40% in Ti/SiC MMCs using the TLP method	Zou et al. (2024)
Poor wettability between the matrix and the ceramic	Use of reactive interlayers (e.g., Ti, Cu, and Ni) to enhance bonding	The Ti interlayer promoted wetting and formed diffusion barrier layers	Mokhles et al. (2024)
Differential thermal expansion mismatch	Functionally graded interlayers to reduce stress concentration	FGM interlayer reduced CTE mismatch effects in Al-B_4C MMCs	Yin Wang et al. (2024)
Difficulty in scaling up for industrial use	Hybrid diffusion bonding with microwave or spark plasma assistance	Achieved uniform bonding in multi-layer MMCs with 30% shorter cycle time	

4.2.2.2 Diffusion Bonding

Diffusion bonding is also a solid-state connecting technique used to make MMCs when precise, high-strength connections are required without melting the base materials (J. Kumar, Singh, and S. Kalsi 2024b). To create metallurgical linkages, the process involves exposing two or more materials to high temperatures (often between 50% and 80% of the matrix metal's melting point) and moderate pressures for an extended period. This approach is particularly well-suited for fusing MMCs and dissimilar materials because it removes the heat loss and residual stresses that can be associated with welding via fusion (Rathinasamy et al. 2025). It has many advantages, including minimum deformation, smooth and defect-free joints, maintenance to strengthen integrity, and the ability to combine complex or multi-level systems. However, problems include accurate surface preparation, bonding durations, formation of oxide layer, limiting diffusion, coefficients of thermal expansion, and limited scalability for large-scale production. Table 4.5 outlines recent advancements addressing the issues inherent to PM processes.

4.2.3 Vapour Deposition Methods

Chemical vapour deposition (CVD) and physical vapour deposition (PVD) are two vapor deposition processes that are used to create high-purity, homogeneous, reinforced layers on metal substrates. The PVD process employs physical techniques such as evaporation or sputtering, while CVD involves creating solid layers by the chemical interaction of vapour-phase precursors on a heated substrate (Awan, Afsheen,

TABLE 4.5
Summary of Key Challenges, Solutions, and Contributions in Diffusion Bonding

Challenge	Solution	Key Contribution	References
Porosity in sintered composites	Use of hot isostatic pressing (HIP) post-sintering to reduce porosity	Demonstrated 15% improvement in density using HIP treatment	Dehghani, Afghahi, and Soleimani (2025)
Weak interfacial bonding	Application of surface treatment (e.g., coating reinforcement with Cu or Ni)	Ni-coated SiC improved bond strength and load transfer efficiency	J. Wang et al. (2025)
Non-uniform distribution of reinforcements	Ultrasonic-assisted powder mixing	Achieved uniform dispersion of nano Al_2O_3 in the Al matrix	Raja et al. (2024)
Agglomeration of nano-reinforcements	Use of surfactants and mechanical alloying	Prevented agglomeration in Al-TiB_2 MMCs	Sindhu, Goyal, and Sreekumar (2024)
Limited densification during sintering	Spark plasma sintering (SPS) for rapid and uniform heating	Reported 98% theoretical density using SPS in Al/SiC composites	S. K. Sharma, Saxena, Salem, et al. (2024b)

and Kausar 2025). These methods greatly increase the efficiency of surface heat restriction, corrosion resistance, and wear resistance. PVD is generally considered suitable for microelectronic devices and tool coatings, whereas CVD is particularly useful for depositing oxide, carbide, and nitride on complex geometries (Gyurika, Ali, and Jakab 2023). Regardless of their potential, both techniques are more favourable for coating applications than bulk composite production, due to their relative slowness and higher cost. Therefore, research continues to increase their utility in structural applications, increase adhesion power, and cut processing time.

4.2.4 Additive Manufacturing

In the production of MMCs, the AM provides unique versatility, as it enables the formation of layer-by-layer complex and customized components (S. K. Sharma, Saxena, Dixit, et al. 2024a). In this, metal powder and reinforcement can be deposited simultaneously. In addition, techniques such as direct energy deposition (DED), electron beam melting (EBM), and selective laser melting (SLM) can be used. This approach is helpful in creating topology-infectious structures with excellent mechanical and thermal properties as well as functionally classified materials (Huang, Shao, and Li 2023). This process enables complex internal features and localized reinforcement placements that cannot be done with traditional approaches. However, its widespread adoption is hindered by issues like porosity, anisotropic characteristics, thermal residual stress, and reinforcement agglomeration. It is anticipated that computational modelling, real-time process monitoring, and sophisticated hybrid manufacturing systems would resolve these problems and increase their capacity for high-performing MMC applications. Table 4.6 provides a summary of the various fabrication techniques used for MMCs.

TABLE 4.6
Summary of MMC Fabrication Techniques

Technique	Process Type	Key Advantages	Major Limitations	Applications
Stir Casting	Liquid	Cost-effective and scalable	Poor reinforcement dispersion	Automotive pistons and brake rotors
Squeeze Casting	Liquid	High density and good bonding	Equipment cost and complexity	Aerospace brackets and engine parts
Infiltration	Liquid	Suitable for complex shapes	Process sensitivity	Complex structural components
Powder Metallurgy	Solid	Precise control and near-net shape	High processing cost	Gears and electronic components
Diffusion Bonding	Solid	Strong metallurgical bond	Requires high temp/pressure	Laminate structures
Vapor Deposition	Coating	High surface quality	Limited thickness	Surface coatings and wear parts
Additive Manufacturing	Layered	Complex shapes and design freedom	High initial cost	Aerospace components and medical implants

4.3 ADVANCED MACHINING TECHNIQUES FOR MMCS

4.3.1 Conventional Machining Challenges

Traditional machining of MMCs includes many material removal processes, serving each specific manufacturing requirement (Parikh et al. 2023; Gutema and Lemu 2024). To produce cylindrical shapes, turning on a lathe is widely used to shape MMC rods and tubes. Milling is employed to create flat surfaces and complex profiles, in which multi-point cutting tools are used, with general operations, including end milling and face milling. Drilling is another important operation to produce holes in MMC components, especially for assembly and fastening purposes, usually using twist drills. Grinding is a final operation that is used to achieve tight-dimensional tolerance and better surface quality, especially in accurate parts. Sawing is applied to cut MMC sheets and profiles to the required sizes and is often used as a preparatory step before further machining. Figure 4.2 compares the machine tools used in conventional machining (e.g., turning, milling, and drilling) with those in non-conventional methods (e.g., AWJM, LAM, EDM, and USM) through schematic representations.

However, these processes produce many important challenges when used for MMC content. Traditional machining methods, like turning, milling, and drilling, struggle with high tool wear, chatter, and poor surface finish due to the abrasive nature of ceramic reinforcements (Rathinasamy et al. 2025). In order to address these issues, tool materials like PCD and CBN are often required. The increasing hardness and stiffness of MMCs produce high cutting and thrust forces, causing tool deflection and dimensional inaccuracy. The poor surface finish is another common issue, resulting in uneven cutting behaviour between the ductal matrix and brittle reinforcement. It also often leads to matrix smearing, particle pull-out, and surface damage. Apart from this, in drilling, chip clearance becomes difficult, and thermal issues can further

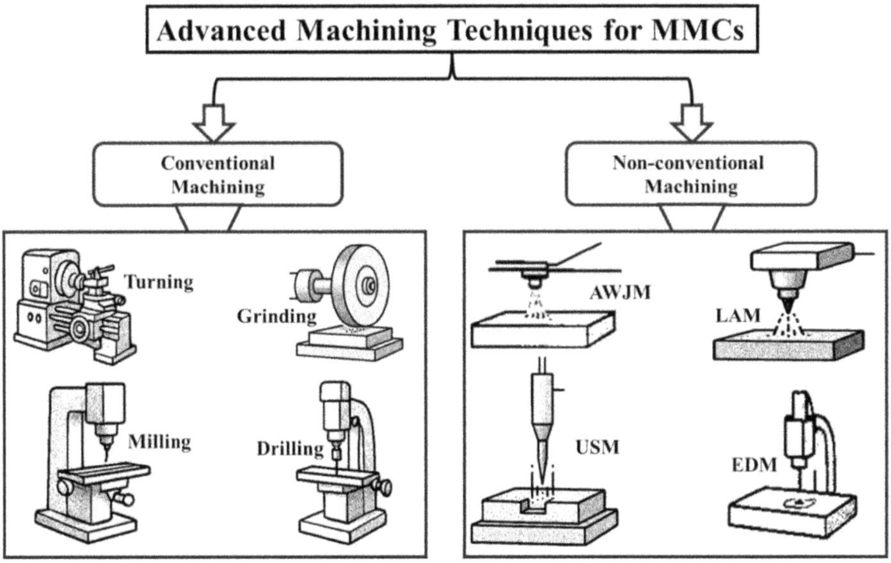

FIGURE 4.2 Conventional and non-conventional machining tools for MMCs.

reduce tool life due to low thermal conductivity. Grinding and sawing operations are also suffering from rapid wear of abrasive equipment, demanding diamonds and coatings to maintain productivity and accuracy (Gutema and Lemu 2024). Therefore, these challenges require the use of adapted machining parameters, effective cooling strategies, and optimized tool geometry to ensure efficient and cost-effective machining of MMCs.

4.3.2 Non-conventional Machining Methods

Non-conventional machining techniques have become rapidly significant when processing MMCs, as abrasive ceramic reinforcement, such as silicon carbide (SiC), alumina (AlO_3), or boron carbide (B_4C), offers specific challenges for typical machining processes. These reinforcements increase tool wear, machining forces, and surface damage when using traditional cutting tools. Non-traditional machining, in contrast, depends on non-contact energy-based mechanisms, including thermal, chemical, electrical, and mechanical energy to remove material rather than relying on direct equipment-content interaction (Ikumapayi, O.M. et al. Ikumapayi et al. 2023). This approach not only expands the equipment life but also improves surface integrity and dimensional accuracy. Techniques such as AWJM, LAM, EDMs, and ultrasonic machining (USM) are particularly useful for MMCs as they allow machining of complicated geometries with minimal mechanical stress (Tengjiao 2024; S. Sharma et al. 2022). These techniques are perfect for high precision automotive and aerospace components where accuracy and quality are crucial because they provide more control over heat distortion and microstructural damage.

4.3.2.1 Electrical Discharge Machining (EDM)

EDM is a non-contact thermal erosion method that removes the material from electro-conductive MMC through a sequence of high-intensity electrical discharges (sparks) between an electrode and workpiece; both are immersed in a dielectric fluid (such as kerosene or deionized water) (Equbal et al. 2024). The small amount of material melts and evaporates because each spark produces heavy localized heat (up to 12,000°C), while the dielectric fluid cools the area and cleans the residue. It has been observed that EDM is the only method that works well for high-precision machining of MMCs having complex shapes, particularly when conventional methods result in the rapid wear of tools, such as fine features, deep cavities, and micro-holes. One of the major benefits of EDM is that it can get extremely difficult and brittle MMCs without the need for physical contact, without breaking mechanical stress and tools (Pawar et al. 2025).

Additionally, it offers improved dimensional precision and surface finish in tight tolerance applications. Nevertheless, only electrically conductive MMCs may profit from it, and due to the poor material removal rates, the process is naturally slow. This heat-affected zone (HAZ), which may include surface microcracks and a recast layer, may also influence the mechanical properties of the machined product due to the repeated spark erosion. Electrode wear is an additional problem that might alter feature accuracy with prolonged use. Despite these challenges, EDM is a crucial process for high-hardness and complex-geometry MMC machining.

4.3.2.2 Laser-Assisted Machining (LAM)

LAM is a hybrid process. In this method, a focused laser beam is employed, and the material of a typical cutting tool is heated locally to improve the machining capacity of the hard and abrasive MMCs (Javed and Rashid 2024). The metal matrix phase is selectively softer, and the laser workpiece raises the surface temperature without compromising the robust ceramic reinforcement. The result makes machining smoother and improves surface smoothness by drastically lowering vibration, tool wear, and localized hot cutting forces. It is strengthened by stiff particles like SiC or Al2O3 in MMCs. The LAM is especially useful in applications like aerospace or biomedical components requiring excellent surface integrity and tight precision, as it enables increasingly higher material removal rates and better machining of complex materials in general (Wazeer et al. 2023). However, LAM demands accurate control of laser parameters such as power, spot size, and feed rate to prevent thermal damage. The equipment required for LAM is expensive, and integrating the laser system with CNC machines adds to the complication and cost (Sibisi et al. 2023). In addition, the effectiveness of laser heating can vary depending on the composition of MMCs, especially when the matrix has low thermal conductivity. Despite these limitations, the LAM provides a significant benefit by increasing mechanization and maintaining the quality and accuracy of the prepared portion.

4.3.2.3 Abrasive Water Jet Machining (AWJM)

AWJM is a specific cold cutting method that removes material by erosion using a high-pressure water jet. It usually works between 200 and 400 MPa in combination with abrasive particles such as silicon carbide, garnet, or alumina (Medibew et al. 2025). The water is passed through a small nozzle, and the abrasive is added just before the water is released. Finally, a high-velocity jet is produced that attacks the MMC surface and removes the material by mechanical erosion and micro-chipping. The unique aspect of this approach is cutting thick or stacked MMC sheets and profiles, especially heat-sensitive ones. As it avoids heat production, it prevents thermal deformation and the formation of heat-affected zones.

Nevertheless, AWJM has several drawbacks as well. The surface finish that is produced is typically thicker than that of other unconventional techniques like EDM or USM. As the material thickness increases, issues such as tapering of the cut and variation in kerf width may arise. The abrasive particles employed in the process produce significant nozzle wear, demanding regular servicing or replacement. Furthermore, controlling the disposal and recycling of abrasive solutions has both environmental and operational challenges (Arshad et al. 2025). Despite these disadvantages, AWJM remains a versatile and efficient technique for machining MMCs, particularly where accurate cuttings with low temperature effects are required.

4.3.2.4 Ultrasonic Machining

USM is a mechanical, non-thermal material removal method ideal for brittle, hard, and non-conductive MMCs, where conventional or thermal techniques may be ineffective or damaging (Peng et al. 2024). Unlike the intended geometry, a tool vibrates on ultrasonic frequencies (usually 20–40 kHz), which provides energy through an abrasive solution (containing particles such as boron carbide or silicon

TABLE 4.7
Summary of Machining Techniques for MMCs

Technique	Principle	Key Advantages	Limitations	Typical Use
EDM	Spark erosion	No contact, complex shapes	Limited to conductive materials	Aerospace nozzles
LAM	Laser preheating	Reduced tool wear	High energy cost	High-strength MMCs
AWJM	Abrasive jet cutting	No thermal effects	High abrasive cost	Thin MMC sheets
Ultrasonic	Vibratory abrasion	Precise, low wear	Limited material removal rate	Nozzle vane slotting in MMCs

carbide) that flows between the device and the workpiece. This technique involves high-frequency vibrations of a tool in abrasive slurry, making it suitable for brittle MMCs. It minimizes tool wear and offers good dimensional accuracy. Ultrasonic vibration enhances slurry particle movement, resulting in efficient material removal and reduced surface damage (Bharat and Bose 2024). Hybrid ultrasonic methods, such as rotary USM, further enhance machining rates and surface integrity. The technique produces no heat or electrical interference, preventing thermal distortion, residual stress, or microstructure changes. USM provides exceptional dimensional precision and surface integrity, particularly for composites with brittle reinforcements. But because of its poor material removal rate, it is not appropriate for mass production. Furthermore, constant vibration and abrasive impact cause tool wear, and the requirement for slurry management and disposal increases operating complexity (Ranjan et al. 2024). However, USM is still a useful technique for MMCs where high precision and material integrity are crucial and heat or conventional methods are insufficient. Table 4.7 provides an overview of the key machining techniques employed for MMCs.

4.4 APPLICATIONS IN AUTOMOTIVE AND AERONAUTICAL INDUSTRIES

4.4.1 Automotive Sector

MMCs are widely used in the automotive industry to enhance fuel efficiency, durability, and thermal stability (Khan et al. 2024). Components like brake rotors, piston heads, cylinder liners, and drive shafts leverage the wear resistance and lightweight characteristics of MMCs (A. Kumar et al. 2024a). HMMCs with graphite or CNTs improve lubrication and reduce friction, enhancing engine efficiency. Aluminium-based MMCs (e.g., Al-SiC and Al-B4C) are used in electric vehicle (EV) components due to their excellent thermal conductivity and low weight, critical for battery enclosures and heat sinks. PM has enabled the fabrication of complex gear and transmission components with high dimensional accuracy (Wu et al. 2024).

4.4.2 Aeronautical Sector

In aerospace applications, the high strength-to-weight ratio and thermal stability of MMCs contribute to improved flight performance and fuel economy (Chowdary, Kajavali, and Rao 2025). Titanium-based MMCs are used in turbine blades, fan components, and structural airframe parts where high-temperature resistance is essential (Bhat et al. 2025). AM is facilitating the production of topology-optimized MMC structures, reducing component weight while maintaining integrity. Moreover, MMCs are increasingly integrated into satellite and UAV systems where strength, stiffness, and thermal management are mission-critical (Martinez Tragacete and Gogu 2025). The use of MMCs in automotive, aeronautical, and aerospace sectors is summarized in Table 4.8.

TABLE 4.8
Applications of MMCs in Automotive, Aeronautical, and Aerospace Sectors

Sector	Application	MMC Material Used	Key Benefits
Automotive	Brake discs and rotors	Al-SiC, Al-Al_2O_3	High wear resistance, lightweight, and thermal conductivity
	Engine components (pistons and rods)	Al-SiC	Reduced weight, improved wear, and thermal resistance
	Cylinder liners	Al-SiC	Enhanced wear resistance and better heat dissipation
	Driveshafts and gears	Al-MMCs	Lower rotational mass and improved strength
	Suspension arms and control knuckles	Al-MMCs	High stiffness and weight reduction
	Heat sinks, thermal systems	Al-SiC and Cu-W	Excellent thermal conductivity and dimensional stability
Aeronautical	Aircraft structures (fuselage and wings)	Al-SiC	High strength-to-weight ratio and fatigue resistance
	Landing gear components	Ti-based MMCs (TiC and TiB)	High fatigue strength, lightweight
	Jet engine blades and vanes	Ti-based MMCs with ceramic	High temperature resistance, creep, and oxidation resistance
	Engine casings	Ni-based MMCs	Thermal and mechanical stability
	Nozzles and liners	Ni-Al MMCs	High temperature and corrosion resistance
Aerospace	Satellite frames and panels	Al-SiC	Dimensional stability and low weight
	Spacecraft radiators and thermal panels	Al-SiC	High thermal conductivity and low CTE
	Rocket nozzles and heat shields	Ti-MMCs and Ni-MMCs	Withstand extreme heat and thermal fatigue resistance
	Actuator housings and control surfaces	Lightweight MMCs	High stiffness and structural reliability
	Electronic packaging systems	Al-SiC	Strong, excellent thermal management

4.5 CONCLUSION

MMCs have emerged as vital materials in modern transportation technologies, particularly in sectors that demand strength, thermal stability, and lightweight construction. The ongoing evolution of fabrication methods – from traditional casting to modern AM – has significantly broadened the scope and performance of MMCs. Similarly, the emergence of non-conventional machining techniques has resolved many processing challenges, enabling the precision manufacture of complex MMC parts.

Research continues to steer towards smarter, greener, and more customizable composites. Innovations such as nano-reinforcements, AI-assisted material design, and digital twins are redefining the capabilities of MMCs. With the automotive and aeronautical industries pushing boundaries for performance and sustainability, MMCs and HMMCs stand poised to shape the next generation of high-performance components. This synergy of material science, digital innovation, and manufacturing technology will be the cornerstone of future breakthroughs in advanced composites.

4.6 SCOPE FOR FUTURE WORK

Despite the substantial advancements in the fabrication and machining of MMCs and HMMCs, several key research areas remain open for exploration.

4.6.1 Process Parameter Optimization through AI/ML

- Integration of artificial intelligence and machine learning can enable predictive modelling of fabrication and machining parameters.
- Data-driven optimization can reduce experimental iterations and improve material performance.

4.6.2 Development of Self-Healing and Smart MMCs

- Research into self-healing matrix composites can enhance material longevity, especially in high-stress environments.
- Incorporation of sensors and smart materials can allow real-time health monitoring of components.

4.6.3 Hybridization with Natural and Sustainable Reinforcements

- Use of bio-based and natural fibres as secondary reinforcements alongside traditional ceramics to develop eco-friendly HMMCs.
- Investigation into the long-term stability and compatibility of such hybrids.

4.6.4 Multi-scale and Multi-physics Modelling

- Advanced simulation approaches combining thermal, mechanical, and microstructural behaviour can offer holistic process insights.
- Bridging micro- and macro-scale effects will aid in the accurate prediction of composite performance under real-life conditions.

4.6.5 Advancements in Green Machining Technologies

- Exploration of cryogenic machining, minimum quantity lubrication (MQL), and dry machining for MMCs.
- Research into tool materials and coatings to enhance sustainability and reduce wear.

4.6.6 Customization via Additive Manufacturing

- Further development of AM techniques specific to MMCs, including functionally graded materials.
- Optimization of AM process parameters for defect-free, reinforced structures.

REFERENCES

Ali, Muhammad Asad, Nadeem Ahmad Mufti, Kashif Ishfaq, Rakhshanda Naveed, Muhammad Qaiser Saleem, and Asif Mahmood Qureshi. 2024. "Development and Characterization of Hybrid Aluminium Matrix Composites through Stir-Squeeze Casting Using Distinct Reinforcements for Structural Applications." *The International Journal of Advanced Manufacturing Technology* 133 (11): 5897–25. https://doi.org/10.1007/s00170-024-14102-9.

Arshad, Muhammad Hazak, Krishna Kumar Saxena, Xiaolei Chen, and Dominiek Reynaerts. 2025. "Introduction and Overview of Advanced Machining Processes." In *Advancements in Machining Processes Through Hybrid and Sequential Approaches*, 1–27. CRC Press

Arunprasand, T R, and P Nallasamy. 2024. "Advancements in Optimizing Mechanical Performance of 3d Printed Polymer Matrix Composites via Microstructural Refinement and Processing Enhancements: A Comprehensive Review." *Mechanics of Advanced Materials and Structures* 0 (0): 1–19. https://doi.org/10.1080/15376494.2024.2426776

Ashok Raj, R, S Sekar, K S Jayakumar, K Prakash, and D Antony Prabu. 2024. "Optimization on Stir Casting Process Parameters of Al7050 Based Composites Using Taguchi Technique." *Materials Today: Proceedings*. https://doi.org/10.1016/j.matpr.2024.03.035

Awan, Tahir Iqbal, Sumera Afsheen, and Sabah Kausar. 2025. "Physical Vapor Deposition Techniques." In *Thin Film Deposition Techniques: Thin Film Deposition Techniques and Its Applications in Different Fields*, 31–64. Singapore: Springer Nature Singapore. https://doi.org/10.1007/978-981-96-1364-9_2

Balasubramanian, K, R Vaira Vignesh, and R Padmanaban. 2024. "Numerical Modeling of Compaction Stresses in Powder Metallurgy Processing of Fly-Ash Based Ceramic Tiles." In *Challenges and Opportunities in Industrial and Mechanical Engineering: A Progressive Research Outlook*, 173–84. CRC Press.

Baliarsingh, S Soumyaranjan, Abhishek Gautam Tripathy, Barada Prasanna Sahoo, Diptikanta Das, Rasmi Ranjan Behera, Mantra Prasad Satpathy, and Ramanuj Kumar. 2024. "Comprehensive Review on Fabrication Methods of Metal Matrix Composites and a Case Study on Squeeze Casting." *Advances in Materials and Processing Technologies* 10 (4): 2862–85. https://doi.org/10.1080/2374068X.2023.2191457

Bharat, Nikhil, and P S C Bose. 2024. "Effect of Ultrasonic Aided Fabrication on Properties of Metal Matrix Composites: A Review." *Advances in Materials and Processing Technologies* 0 (0): 1–18. https://doi.org/10.1080/2374068X.2024.2341985

Bhat, Chinmai, Cho-Pei Jiang, Yulius Shan Romario, Sandeep Kumar Paral, and Ehsan Toyserkani. 2025. "Critical Review of Metal-Ceramic Composites Fabricated through Additive Manufacturing for Extreme Condition Applications." *Mechanics of Advanced Materials and Structures* 32 (10): 2153–80. https://doi.org/10.1080/15376494.2024.2376337

Biswas, Prasenjit, Jagadish Nayak, Arjun Kundu, Deepak Patel, Archana Mallik, and Sanjeev Das. 2025. "A Review on the Development of Processing Techniques for the Production and Casting of Al-Alloy and Metal Matrix Composite Material." *Iranian Journal of Science and Technology, Transactions of Mechanical Engineering* 49 (1): 35–66. https://doi.org/10.1007/s40997-024-00773-y

Chowdary, Jarubula Ranga Raya, Abdul Sattar Kajavali, and Veeravalli Ramakoteswara Rao. 2025. "Processing, Properties and Applications of Metal Matrix Composites." In *Fundamentals and Advances in Metal Matrix Composites*, 1–14. CRC Press.

Dehghani, Pouria, Seyed Salman Seyed Afghahi, and Farshad Soleimani. 2025. "Hot Isostatic Pressing (HIP) in Advanced Ceramics Production." In *Advanced Ceramic Materials*, edited by Amparo Borrell Tomás and Rut Benavente Martínez. Rijeka: IntechOpen. https://doi.org/10.5772/intechopen.1007176

Demir, Mehmet Emin, Yahya Hışman Çelik, Erol Kilickap, and Ali Kalkanli. 2023. "The Effect of B4C Reinforcements on the Microstructure, Mechanical Properties, and Wear Behavior of AA7075 Alloy Matrix Produced by Squeeze Casting." *Proceedings of the Institution of Mechanical Engineers, Part E* 237 (6): 2574–84. https://doi.org/10.1177/09544089221139095

Equbal, Azhar, Asif Equbal, Hussain VMS, Israr Equbal, Irfan Anjum Badruddin, Sarfaraz Kamangar, Abhilash Edacherian, and Abdul Azeem Khan. 2024. "A Critical Review on Electro-Discharge Machining of Non-Conductive Materials." *Machining Science and Technology* 28 (6): 1092–128. https://doi.org/10.1080/10910344.2024.2431027

Ge, Yu-Jie, Rui-Fen Guo, and Ping Shen. 2025. "Wettability and Interfacial Reactions in Al–Mg–Si/SiC and Al–Mg–Si/SiO2 Systems." *Materials Characterization* 219: 114599. https://doi.org/10.1016/j.matchar.2024.114599

Goh, Guo Dong, Kin Keong Wong, Nigel Tan, Hang Li Seet, and Mui Ling Sharon Nai. 2024. "Large-Format Additive Manufacturing of Polymers: A Review of Fabrication Processes, Materials, and Design." *Virtual and Physical Prototyping* 19 (1): e2336160. https://doi.org/10.1080/17452759.2024.2336160

Gopal, P M, V Kavimani, and S Sudhagar. 2025. "13 - Evolution and Recent Advancements of Composite Materials in Industrial Applications." In *Applications of Composite Materials in Engineering*, edited by Madhu Puttegowda, T G Yashas Gowda, J S Binoj, Sanjay Mavinkere Rangappa, and Suchart Siengchin, 317–34. Woodhead Series in Materials. Elsevier Science Ltd. https://doi.org/10.1016/B978-0-443-13989-5.00013-9

Grzesik, Wit, and Adam Ruszaj. 2021. "Physical Fundamentals of Conventional and Unconventional Machining Processes." In *Hybrid Manufacturing Processes: Physical Fundamentals, Modelling and Rational Applications*, 35–60. Cham: Springer International Publishing. https://doi.org/10.1007/978-3-030-77107-2_4

Gupta, P K, Alok Kumar Trivedi, M K Gupta, and Manish Dixit. 2024. "Metal Matrix Composites for Sustainable Products: A Review on Current Development." *Proceedings of the Institution of Mechanical Engineers, Part L* 238 (10): 1827–64. https://doi.org/10.1177/14644207241238197

Gutema, Endalkachew Mosisa, and Hirpa G Lemu. 2024. "Conventional Machining of Metal Matrix Composites towards Sustainable Manufacturing: Present Scenario and Future Prospects." *Journal of Composites Science* 8 (9). https://doi.org/10.3390/jcs8090356

Gyurika, István Gábor, Osamah Ihsan Ali, and Miklós Jakab. 2023. "Determination of Research Guidelines and Establishing of a Test Framework for the Development of New CVD Coating Formulations and New Approach Coating Equipment." In *Smart Technologies in Urban Engineering*, edited by Olga Arsenyeva, Tatiana Romanova, Maria Sukhonos, and Yevgen Tsegelnyk, 420–31. Cham: Springer International Publishing.

Huang, Zhiyuan, Guangbin Shao, and Longqiu Li. 2023. "Micro/Nano Functional Devices Fabricated by Additive Manufacturing." *Progress in Materials Science* 131: 101020. https://doi.org/10.1016/j.pmatsci.2022.101020

Ikumapayi, O M, B H Omietimi, P Onu, T S Ogedengbe, J R Oluwafemi, S A Afolalu, and E T Akinlabi. 2023. "Non-Traditional Machining Techniques in Manufacturing Industries: An Overview." *E3S Web Conferences* 430: 1213. https://doi.org/10.1051/e3sconf/202343001213.

Javed, Md Omar Al, and Adib Bin Rashid. 2024. "Laser-Assisted Micromachining Techniques: An Overview of Principles, Processes, and Applications." *Advances in Materials and Processing Technologies*, 11: 1583–626.

Jeevankumar, Pallagani, Poly Rose, Ammanabrolu Rajanikanth, Thankappan Pillai Rajasekharan, and Vummethala Seshu Bai. 2024. "Net-Shaping of Advanced Ceramic Composites by Gelcasting into Precision Molds Made by Rapid-Prototyping." *International Journal of Applied Ceramic Technology* 21 (2): 664–74. https://doi.org/10.1111/ijac.14568

Khan, Fardin, Nayem Hossain, Juhi Jannat Mim, S M Maksudur Rahman, M D Jayed Iqbal, Mostakim Billah, and Mohammad Asaduzzaman Chowdhury. 2024. "Advances of Composite Materials in Automobile Applications: A Review." *Journal of Engineering Research*. https://doi.org/10.1016/j.jer.2024.02.017

Kumar, Ashish, Virendra Pratap Singh, R C Singh, Rajiv Chaudhary, Deepak Kumar, and Abdel-Hamid I Mourad. 2024a. "A Review of Aluminum Metal Matrix Composites: Fabrication Route, Reinforcements, Microstructural, Mechanical, and Corrosion Properties." *Journal of Materials Science* 59 (7): 2644–711. https://doi.org/10.1007/s10853-024-09398-7

Kumar, Jatinder, Dilbag Singh, and Nirmal S. Kalsi. 2024b. "A Critical Review of Fabrication Techniques and Possible Interfacial Reactions of Silicon Carbide Reinforced Aluminium Metal Matrix Composites." In *Metal Matrix Composites: A Modern Approach to Manufacturing*, 42–90. Bentham Science Publishers. https://www.benthamdirect.com/content/books/9789815223439.chapter-3

Kumawat, Yogesh Kumar, Rishabh Sehgal, Irfan Ayoub, Rakesh Sehgal, and Vijay Kumar. 2023. "Recent Progress in the Development of Metallic Composite for Advanced Technologies." In *Nanoparticles Reinforced Metal Nanocomposites: Mechanical Performance and Durability*, edited by Santosh K Tiwari, Vijay Kumar, and Sabu Thomas, 53–87. Singapore: Springer Nature Singapore.

Lee, Seul Hee, Seyoung Kim, In-sub Han, Hyung Joon Bang, Soo-Hyun Kim, and Young-Hoon Seong. 2025. "Study on the Characteristics of SiC/SiC Composites with h-BN Fiber Interface Coating Fabricated by Liquid Silicon Infiltration Process with Various Coating Thickness." *Journal of the Korean Ceramic Society* 62 (2): 262–70. https://doi.org/10.1007/s43207-024-00474-x

Liang, Songfeng, Wu Xiaoyu, Wu Zhenzhen, and Wenting Liu. 2025. "Progress and Challenges on Materials Used to Optimize Flight Efficiency, Improve Safety and Reliability, and Reduce Cost of Electric Vertical Take-Off and Landing Aircraft." *Journal of Applied Polymer Science* 142 (24): e57029. https://doi.org/10.1002/app.57029

Martinez Tragacete, Marina, and Christian Gogu. 2025. "State of the Art Review of Levers in the Field of Structures and Materials for Low Carbon Footprint Aircraft." https://hal.science/hal-05056586

Medibew, Tesfaye Mengesha, Dawid Zieliński, Sisay Workineh Agebo, and Mariusz Deja. 2025. "Recent Research Progress in the Abrasive Machining and Finishing of Additively Manufactured Metal Parts." *Materials* 18 (6). https://doi.org/10.3390/ma18061249

Mehdi, Husain, Shallender Singh Bhati, Pradeep Kumar Mouria, Subhash Mishra, Mandeep Bhadana, and Lalit Batra. 2025. "Effect of Nanoparticles on Metal Matrix Composites Fabricated by Ultrasonic Stir Casting." *International Journal of Cast Metals Research* 0 (0): 1–17. https://doi.org/10.1080/13640461.2025.2496607

Moghayedi, Alireza, and Bankole Awuzie. 2023. "Towards a Net-Zero Carbon Economy: A Sustainability Performance Assessment of Innovative Prefabricated Construction Methods for Affordable Housing in Southern Africa." *Sustainable Cities and Society* 99:104907. https://doi.org/10.1016/j.scs.2023.104907

Mohammed, Sohail, Pranjal Nautiyal, Tanaji Paul, Cheng Zhang, and Arvind Agarwal. 2024. "A Critique on Boron Nitride Nanotube Reinforced Metal Matrix Composites." *Critical Reviews in Solid State and Materials Sciences* 49 (3): 482–519. https://doi.org/10.1080/10408436.2023.2229867

Mokhles, Mohammad, Seyed Mojtaba Zebarjad, Habib Danesh-Manesh, and Tushar R Dandekar. 2024. "Enhancing Ni–Ti Shape Memory Alloy Diffusion Bonding with Ti/Ni Reactive Multilayer Foils." *Journal of Materials Science: Materials in Engineering* 19 (1): 18. https://doi.org/10.1186/s40712-024-00162-y

Nagaraju, Sharath Ballupete, Hadlahalli Chandrashekar Priya, Yashas Gowda Thyavihalli Girijappa, and Madhu Puttegowda. 2023. "9 - Lightweight and Sustainable Materials for Aerospace Applications." In *Lightweight and Sustainable Composite Materials*, edited by Sanjay Mavinkere Rangappa, Sunita M Doddamani, Suchart Siengchin, and Mrityunjay Doddamani, 157–78. Woodhead Publishing. https://doi.org/10.1016/B978-0-323-95189-0.00007-X

Nehra, Ram Kishan, Manoj Sahni, Vishvesh Badheka, and Krunal Mehta. 2025. "Schematic Review of Manufacturing Methods, Process Parameters and Existing Challenges for Metal Matrix Composites." *International Journal of Cast Metals Research* 0 (0): 1–22. https://doi.org/10.1080/13640461.2025.2496603

Oke, Samuel Ranti, Nthabiseng Maledi, Ibrahim Adeyemi Okanlawon, Tijesunimi Ruth Bishi, and Oluwasegun Eso Falodun. 2025. "Tensile and Corrosion Properties of Copper and Alumina Hybrid Reinforcement on Aluminum Matrix Composite." *Journal of Composite Materials* 59 (13): 1595–607. https://doi.org/10.1177/00219983251315090

Parikh, V K, Vivek Patel, D P Pandya, and Joel Andersson. 2023. "Current Status on Manufacturing Routes to Produce Metal Matrix Composites: State-of-the-Art." *Heliyon* 9 (2). https://doi.org/10.1016/j.heliyon.2023.e13558

Pawar, Aniket B., Sachin S. Chavan, Pradeep V. Jadhav, and Dattatray B. Jadhav. 2025. "Emerging Trends in Hybrid ECM: A Review of Multi-Assisted Machining Techniques." *Engineering Research Express* 7 (1). https://doi.org/10.1088/2631-8695/adb53e

Peng, Zhenlong, Aowei Han, Chenlin Wang, Hongru Jin, and Xiangyu Zhang. 2024. "Ultrasonic Vibration Cutting of Advanced Aerospace Materials: A Critical Review of in-Service Functional Performance." *Journal of Intelligent Manufacturing and Special Equipment* 5 (1): 137–69. https://doi.org/10.1108/JIMSE-12-2023-0016

Pooja, Km., Nazia Tarannum, and Pallavi Chaudhary. 2025. "Metal Matrix Composites: Revolutionary Materials for Shaping the Future." *Discover Materials* 5 (1): 35. https://doi.org/10.1007/s43939-025-00226-6

Raja, D Elil, S Vijayan, Tushar Sonar, and S Prathap Singh. 2024. "Analyzing the Dry Sliding Wear Performance on Aluminum 6061 Alloy Reinforced with SIC And B_4C Hybrid Nanocomposite." *Advances in Materials Science* 24 (4): 42–56.

Ram, S C, Awani Bhushan, Mukesh Raushan Kumar, Anupam Gautam, Amit Moray, and Sunkulp Goel. 2024. "A Comparative Study of Processing Route, Microstructures, Fracture Analysis and Phase Formation of Al-Si/Mg2Si Functionally Graded Composites: An Overview." *Composite Interfaces* 31 (12): 1575–614. https://doi.org/10.1080/09276440.2024.2358568

Ranjan, Rahutosh, Dharminder Singh, Vikas Pandey, Ankit Sharma, A S K Sinha, and Naveen Mani Tripathi. 2024. "A Comprehensive Review on the Machining Process: Unconventional as an Alternative to Conventional Machining." *Advances in Pre-and Post-Additive Manufacturing Processes*, 163–95.

Rathinasamy, Saravanan, Omkumar Meenakshisundaram, Sathish Thanikodi, and Jothi Arunachalam Solairaju. 2025. "Development of Metal Matrix Composites Through Solid-State Techniques." *Fundamentals and Advances in Metal Matrix Composites*, no. 1, 51– 65.

Sambasivam, Sangaraju, Siddheshwar Shirbhate, Ahmed S Abed, Pravin P Patil, Irfan Khan, Rajesh Singh, Lavish Kansal, and Ankita Awasthi. 2023. "Significance of Reinforcement in Mg-Based MMCs for Various Applications: A Review." *Materials Today: Proceedings*. https://doi.org/10.1016/j.matpr.2023.02.161

Sankhla, Arvind, and Kaushik M Patel. 2022. "Metal Matrix Composites Fabricated by Stir Casting Process–A Review." *Advances in Materials and Processing Technologies* 8 (2): 1270–91. https://doi.org/10.1080/2374068X.2020.1855404

Sarmah, Pallab, and Kapil Gupta. 2024. "Recent Advancements in Fabrication of Metal Matrix Composites: A Systematic Review." *Materials (Basel, Switzerland)* 17 (18). https://doi.org/10.3390/ma17184635

Sharma, Sachin Kumar, Kuldeep Kumar Saxena, Anil Kumar Dixit, Ramkishore Singh, and Kahtan A Mohammed. 2024a. "Role of Additive Manufacturing and Various Reinforcements in MMCs Related to Biomedical Applications." *Advances in Materials and Processing Technologies* 10 (2): 231–48. https://doi.org/10.1080/2374068X.2022.2122005

Sharma, Sachin Kumar, Kuldeep Kumar Saxena, Karrar Hazim Salem, Kahtan A Mohammed, Rajesh Singh, and Chander Prakash. 2024b. "Effects of Various Fabrication Techniques on the Mechanical Characteristics of Metal Matrix Composites: A Review." *Advances in Materials and Processing Technologies* 10 (2): 277–94. https://doi.org/10.1080/2374068X.2022.2144276

Sharma, Sahil, Farhan Ahmad Shamim, Akshay Dvivedi, Pradeep Kumar, and Tarlochan Singh. 2022. "Hybrid Machining of Metal Matrix Composites." In *Fabrication and Machining of Advanced Materials and Composites*, 235–54. CRC Press.

Sibisi, Thabiso Hopewell, Mxolisi Brendon Shongwe, Lerato C Tshabalala, and Ipfi Mathoho. 2023. "LAM Additive Manufacturing: A Fundamental Review on Mechanical Properties, Common Defects, Dominant Processing Variables, and Its Applications." *The International Journal of Advanced Manufacturing Technology* 128 (7): 2847–61. https://doi.org/10.1007/s00170-023-12139-w

Simões, Sónia. 2024. "High-Performance Advanced Composites in Multifunctional Material Design: State of the Art, Challenges, and Future Directions." *Materials* 17 (23). https://doi.org/10.3390/ma17235997

Sinchai, Ananta, Kunthorn Boonyang, and Thanakorn Simmala. 2024. "Development of a Low-Cost Automated Injection Molding Device for Sustainable Plastic Recycling and Circular Economy Applications." *Inventions* 9 (6). https://doi.org/10.3390/inventions9060124

Sindhu, Nidhi, R K Goyal, and V M Sreekumar. 2024. "Formation, Dispersion and Distribution of TiB2 Particles in In-Situ Al-TiB2 Functionally Graded Nanocomposites." *Metallurgical and Materials Transactions A* 55 (5): 1654–72. https://doi.org/10.1007/s11661-024-07359-y

Sun, Haoran, Bohan Song, Xiaomin Sun, Xinqi Cui, Zexian Liu, Meng Cong, Mingyuan Sun, et al. 2025a. "Recent Representative Progress of Surface Coating Technology." *The Chemical Record* n/a (n/a): e202500054. https://doi.org/10.1002/tcr.202500054

Sun, Xiaoyi, Yubei Zhang, Yanhong Li, Yu Ma, Qiang Zhang, Honglai Lu, Fengshi Li, and Yonghai Yue. 2025b. "Research Progress in Ceramic–Metal Composites: Designing Interface Structures for High Mechanical Performance." *Small Methods* n/a (n/a): 2402100. https://doi.org/10.1002/smtd.202402100

Tengjiao. 2024. "Precision Machining of Hard-to-Cut Materials: Current Status and Future Directions." *International Journal of Advanced Computer Science and Applications* 15 (10). https://doi.org/10.14569/IJACSA.2024.0151088

Wang, Jianqi, Yang Wang, Zhou Chen, Changyang Wang, Jilong Yang, Jian Yang, Jian Gu, and Quan Li. 2025. "Ni@SiC Composites from Ni-Modified PCS-Derived (SiC)p with High-Efficiency Microwave Absorption Properties." *Journal of Materials Science: Materials in Electronics* 36 (5): 356. https://doi.org/10.1007/s10854-025-14448-0

Wang, Yin, Yong Li, Haiyao Wang, Yu Wei, Chen He, Xu Guangming, Jiadong Li, and Hongqun Tang. 2024. "Composition Design and Mechanical Properties of B4C/Al-Zn-Mg-Cu Functionally Graded Materials Prepared by Laser Additive Manufacturing." *Journal of Materials Research and Technology* 32:3620–29. https://doi.org/10.1016/j.jmrt.2024.08.188

Wang, Yunlei, and Tullio Monetta. 2023. "Systematic Study of Preparation Technology, Microstructure Characteristics and Mechanical Behaviors for SiC Particle-Reinforced Metal Matrix Composites." *Journal of Materials Research and Technology* 25:7470–97. https://doi.org/10.1016/j.jmrt.2023.07.145

Wazeer, Adil, Apurba Das, Arijit Sinha, Su Ziyi Kazuaki Inaba, and Amit Karmakar. 2023. "Additive Manufacturing in Biomedical Field: A Critical Review on Fabrication Method, Materials Used, Applications, Challenges, and Future Prospects." *Progress in Additive Manufacturing* 8 (5): 857–89. https://doi.org/10.1007/s40964-022-00362-y

Wu, Jizhan, Peitang Wei, Caichao Zhu, Peng Zhang, and Huaiju Liu. 2024. "Development and Application of High Strength Gears." *The International Journal of Advanced Manufacturing Technology* 132 (7): 3123–48. https://doi.org/10.1007/s00170-024-13479-x

Yavuz, Mustafa Tolga, and Ibrahim Ozkol. 2024. "Thermal Structural Design Aspects of Military Aircraft." *Journal of Aeronautics and Space Technologies* 17 (1): 124–58. https://jast.hho.msu.edu.tr/index.php/JAST/article/view/579

Zou, Taiyong, Zhenqian Lang, Haopeng Chen, Zheng Ye, Tao Wu, Wanli Wang, Jian Yang, and Jihua Huang. 2024. "Rapidly Achieving High-Performance Joining of Single Crystal Superalloy and Powder Metallurgy Superalloy via Partial Transient Liquid Phase Diffusion Bonding." *Journal of Manufacturing Processes* 109:79–88. https://doi.org/10.1016/j.jmapro.2023.12.009

5 Emerging Practices for Physical and Mechanical Characterization of Composite Materials

Jai Inder Preet Singh, Raghu Echempati, Gurpreet Singh Saini, and Ashok Kumar Sharma

5.1 INTRODUCTION

5.1.1 Overview of Composite Materials

Composite materials are purposely developed heterogeneous materials composed of two or more different components—most often, a matrix and the reinforcing component—that provide different properties, giving a greater combination of properties. The matrix is most often a polymer, metal, or ceramic; the matrix holds the structure together and distributes loads evenly. The reinforcement can be fibers, particles, or whiskers and provides stiffness and mechanical strength while possibly providing some functional properties. This ability to tailor properties—such as anisotropic strength, enhanced corrosion resistance, improved thermal/electrical properties, a multifunctional nature, and improved fatigue performance—makes composites the material of choice for aerospace, automotive, marine, energy, and infrastructure applications (Abbass et al., 2021).

For thousands of years, humans innovatively employed composite principles, including ancient brick (mud-straw), laminated wood, and the powerful Mongol bow (~1200 AD) as early examples of material engineering. Composites were revolutionized in the 20th century with the introduction of synthetic matrices to replace natural materials. and are significant milestones in the history of fiber-reinforced polymers (FRPs). During the 1960s, new composites such as carbon fiber and Kevlar were developed that facilitated high-performance structural composites. The Boeing 787, currently in operation, is approximately 80% carbon-fiber-reinforced polymer by weight, demonstrating the total impact of composites on industry today.

5.1.1.1 Types of Matrices and Reinforcement

a. **Types of matrices**:
  - Polymers use either two types of polymers—polymer matrix composites (PMCs) (thermosets, epoxy, polyester, etc.) and thermoplastics (PEEK, PPS, etc.). PMCs are the most widely known and used composite

DOI: 10.1201/9781003620570-5

family primarily because of their processing possibilities and interfacial versatility between matrix and fibrous part design.
- Most metal matrix composites (MMCs) use aluminum, titanium, or magnesium as the matrix, reinforced with fibers or particles. They are typically used at high temperatures and usually for structural purposes.
- Ceramics are either SiC or Al_2O_3 matrix, reinforced with fibers. Used in extreme temperatures or in severe environments (turbine engines, etc.). Table 5.1 represents different applications for the various composite systems.

b. Types of reinforcement:
- Continuous fibers (carbon, glass, and aramid) provide reinforcement in a single direction.
- Discontinuous fibers/particles/whiskers provide reinforcement in all directions (isotropic reinforcement) and in a cost-effective manner.
- Nano-reinforcements (carbon nanotubes (CNTs) and graphene nanoplatelets (GNPs)) possibly provide an ultra-enhanced functional performance improvement.
- Biocomposites can use natural fibers (jute, flax, etc.) that can provide more environmentally responsible options for various applications.

5.1.1.2 Microstructure and Interfaces

The arrangement of a material (fiber orientation, volume fraction (Vf), interface bonding, and voids) dictates the functional use of the product or component and establishes creative and performance limitations to manufacture. Some reports indicate that even a small amount of voids, less than 1–3% in carbon fiber-reinforced plastics (CFRP), could lower strength by nearly ~20%, whereas aerospace applications require less than 1% void content (Abdul Khalil et al., 2014).

5.1.1.3 Functional and Structural Attributes

Composites can provide unique combinations of internal structural and functional properties, which are dependent on their design:

- Mechanical: Carbon fibers give tensile strength of ~3–6 GPa and stiffness of ~230–600 GPa, with density ~1.6 g/cm^3.

TABLE 5.1
Classification of Composite Systems

Composite Type	Matrix Class	Reinforcement Form	Key Applications
PMC	Thermoset/PT	Continuous/ discontinuous fibers	Aerospace, sports, and automotive
MMC	Al, Ti, and Mg	SiC, Al_2O_3, and fibers	Engine components and heat sinks
CMC	SiC and Al_2O_3	SiC fibers	Turbines and refractory linings
Nanocomposite	Polymer	CNT, graphene, and hybrid	EMI shielding, sensors, and batteries
Biocomposite	Biopolymer	Natural fibers	Packaging and consumer goods

- Thermal: Ceramic MM composites can withstand temperatures >600°C; PMCs typically achieve thermal conductivity of around 4 W/mK with aligned nanoparticles.
- Electrical/EMI: CNT, graphene, or nickel/SiC hybrids allow for tuned conductivity and EMI shielding.
- Anisotropy: Orientation adds directionality. For example, vertically aligned CNT arrays show anisotropic thermal conductivity of ~0.65 W/mK at 0.3 wt% loading.
- Multifunctionality: The integration of sensing, energy storage, thermal management, and structural function in advancing—especially in smart composites.

5.1.2 Importance of Characterization in Composite Development

The characterization of composite material systems forms the basis of developing, optimizing, and using them in advanced engineering applications. When composites are made of a matrix and reinforcements, in contrast to a homogeneous material, their compositional design includes distinct phases: the matrix and the reinforcement component. These two elements have very complex intonations that dictate the developed properties that will be seen in a final composite product. Composite materials are engineered by design to meet specific application purposes. Characterization of composite materials is important when mapping for understanding how microscopic features influence macroscopic responses. Aspects of mechanical characterization, such as tensile, flexural, compressive, and impact, provide the first layer of characterization that informs us about strength, stiffness, and toughness. The conventional approaches for mechanical characterization represent only the starting point; for example, to determine information related to the interface between matrix and reinforcement, nanoindentation is needed. To probe viscoelastic behavior, Dynamic Mechanical Analysis (DMA) is suited for this type of analysis, and fracture toughness tests assess real-world crack resistance under load, reflecting the advanced tools previously mentioned (Abdul Khalil et al., 2012). An example highlighting the need for initial composite characterization can be seen in CFRP laminates, where void contents on the order of 1–3% have been shown (contributing in some cases to a 20% drop in tensile strength) for voids in the composite description. Concepts like inclusion/porosity characterization can be performed using ultrasonic C-scan, micro-computed tomography (μCT), and X-ray radiography (Abdul Khalil et al., 2011). The internal angles and orientation influence the mechanical anisotropic properties of a composite based on the Vf of the fibers and fiber orientation. Some of these parameters and properties can be determined using optical microscopy, scanning electron microscopy (SEM), or image processing methods. These microstructural characteristics are vital in assessing micromechanical models, for example, the rule of mixtures, Halpin–Tsai relationships, and finite element simulations. Characterization is also central in assuring manufacturing processes that rely on passive quality assurance and reproducibility. For high-rate production methods, such as automated fiber placement (AFP), resin transfer molding (RTM), or pultrusion, defects, for example, fiber waviness, dry spots, and incomplete cure, can negatively affect performance. Non-destructive evaluation (NDE) through acoustic emission (AE), shearography, digital image correlation (DIC), or similar methods allows for near-real-time evaluation of part health and composite integrity, all without sacrificing a portion of the sample (Abdul Khalil et al., 2011). Thermal analysis techniques such as DSC and TGA allow rapid assessments of resin curing kinetics, degradation profile, and glass transition temperature, which ultimately provide insight into the compounding and application of the materials, processing optimization, or in-service temperature limitations. In safety-critical industries, for example, aerospace, automotive, and civil engineering, rigorous characterization is critical to compliance with certification and to preserving the structural health of parts and assemblies. The initiation of delamination, a primary mode of failure in laminated composites, can arise from low-energy impact, cyclic loading, or a failure that may go undetected on the surface of a composite assembly. Delamination can also occur gradually, which leaves the fabricator with an unreliable product that can impact safety if used in a damaged part. Thankfully, advanced NDE methods such as infrared thermography and ultrasonic phased array scan techniques can be implemented to detect latent subsurface damage occurrences and prevent catastrophic failure (i.e., avoid delamination), all without destructive sampling (Abilash & Sivapragash, 2013). In a similar fashion, characterization of functional properties is highly relevant as the composites industry evolves to create multifunctional composites that utilize woven and fiber reinforcements for embedded sensing, EMI shielding, or protection from thermal loadings outside of the context of mechanical performance. Examples of structured characterization to assess functional integration may include measurements of electrical conductivity, EMI shielding tests, or thermal conductivity tests designed to evaluate composites with polymer nanocomposites (i.e., CNT, graphene, or metallic nanoparticles) as reinforcement, specific to polymer-based composites (Ahankari et al., 2011). Characterization of these functional properties can be used to confirm the percolation threshold, dispersion uniformity, and/or assess the stability of the functional properties after cycling loads or conditioning. In contemporary research and development, characterization can also be leveraged to reduce the design cycle, combined with predictive modelling, or data-driven discovery of materials. Examples of material properties used in multi-scale modeling for design, damage prediction, and virtual prototyping or Twin Digital applications include fiber-matrix interfacial strength, fracture energy, and the matrix modulus (Ahmed et al. 2014). Now, there are machine learning (ML) algorithms being trained using large data sets derived from microscopy images, spectroscopy, and mechanical test outputs to classify defects, predict a sample's performance, or adjust processes moving forward, all in real time (Ahmed & Sanada, 2019). Depending on the method of research, characterization could also refer to environmental and aging characterization of composites, such as salt-spray tests, UV exposure, moisture absorption studies, or fatigue tests, all of which are used to inform understanding of the durability and lifecycle of composites. This is especially relevant for infrastructure, wind turbine blades, or marine structures that will be subjected to environmental exposures for years. Characterization can even assess recyclability or biodegradability for recycled composites or bio-based alternatives, understanding difficult performance degradation changes happening, for example, due to recycled composites (Akhtar et al., 2016).

Life cycle assessments (LCA) and eco-design require raw data to create accurate databases for materials, and this data is collected generally using rigorous protocols for testing and advanced material characterization. To conclude, characterization is not a single laboratory job, but rather an ongoing and iterative part of the composite design process and is relied on for quality assurance, failure mitigation, or facilitating ideation by creating something novel and then maintaining the operation of the part through predictive maintenance (Akil et al. (2011). It is the bridge between processing, structure, property, and performance and provides the scientific basis on which composite technologies can scale and evolve reliably for upcoming applications in aerospace, defense, mobility, healthcare, sustainable infrastructure, etc.

5.1.3 Evolution of Characterization Techniques

The characterization process of composite materials has evolved through several characterizing methods, driven by learning and technology, which have intensified demand from new applications. First came the primitive mechanical testing techniques that are based on standardized tests, such as tensile, compression, flexural, and impact (e.g., ASTM D3039 and D790). These tests provided basic performance structural measures like tensile modulus, tensile strength, elongation at break, flexural stiffness, modulus of resilience, energy absorbing capacity, and mechanical properties pertaining to strength. But these tests provided bulk measures at best, averaged values in connection with mechanical performance, and a lack of consideration of microstructure, heterogeneity, and anisotropy. The dependency of direction shown by fiber-reinforced composites (FRCs) (i.e., values differ across and along fiber orientation) minimized the value of interpretation. Different modes of failure, such as delamination, matrix cracking, or matrix fiber debonding, were invisible without micro-characterization (Aldahdooh et al., 2014), as were material degradation, moisture diffusion, and initiation of microscale damage visible in observational measurements. Thus, there was an urgent need for multi-scaled, spatially resolved measurement techniques.

In addition to bulk methods, and starting in the mid-20th century, researchers started to adapt and utilize microscopy and spectroscopy techniques, in part to remove some of the bulk method limitations. One of the first technologies that significantly transformed how the researchers saw composite materials was SEM. SEM left researchers with an urgent ability to see and distinguish fracture surface features at a detail level in the tens of nanometers—a level of detail not previously achievable, particularly in terms of coloration. SEM examinations not only provided researchers with useful perspectives on the types of failure but also related to mechanical performance features, which included fiber pull-out features, void in the matrix features, or delamination at the matrix-fiber interface. Transmission electron microscopy (TEM) also enabled intramolecular-scale visualization of interfaces in carbon-fiber/epoxy systems, revealing in some cases an interphase region and resin crystallinity in a local descriptive sense. In the later decades, atomic force microscopy (AFM) was introduced to map the local modulus and hardness distribution in micro-laminate cross-sections. The introduction of all three microscopy techniques (SEM, TEM, and AFM) helped improve the understanding of linkages between microstructure and property for use in development of micromechanical models where, for the first time, the models could reflect real morphology instead of feeling required to employ a series of judgments based on previously accepted understanding of curved geometries, based on a mechanical scissors. These models could incorporate the rule of mixtures, Halpin–Tsai equations, or Cox's shear-lag analysis using image-derived inputs to provide suitable quantitative metrics to measure or validate these simple models, especially those that were distinct from the original study in their design approach or interventions (Aldridge et al., 2014).

Driven by increasing industrial demand, NDE methods emerged in the 1970–1980s as an essential quality assurance and in-service inspection technique. The introduction of early ultrasonic pulse-echo techniques (A-scan, C-scan) enabled the detection of delaminations, voids, or inclusions in thick composite laminates, dramatically improving manufacturing quality control (Andre et al., 2016). In addition to ultrasound, radiographic methods (X-ray), which could inspect dense materials, also became available to industry. However, due to safety and equipment investment, radiographic inspection was limited. Recently, thermal imaging techniques have become popular; with a surface being heated and measuring thermal emission, errors such as delaminations create local temperature anomalies on large structural components, making thermal imaging useful for inspection of large turbine blades, pressure vessels, and composite pipes (Aroso et al., 2015). The notable introduction of shearography provided another visual inspection technique based on laser interferometry, along with observing internal strain fields and surface observations, and reduced strain resulting from small mechanical or thermal perturbations Arumuga Prabu et al. (2014). All of these NDE methods can now be introduced into the production lines, such as inline ultrasonic scans, thermographic inspections along conveyors for large structures, or various ways to incorporate inline NDE methods to improve the rate of defect detection without destructive testing of finished structures (Armuga Prabu et al., 2014).

While their success was beneficial, no evaluation technique that could provide 2D images could replicate the 3D internal architecture of composite materials. For that reason, computed tomography (CT) entered the picture, for example, by the 1990s, as a clear and obvious resource to embrace in composite research. Micro-CT scans (voxel sizes under 5 μm) could resolve volumetric distributions of voids in composite materials, reveal fiber bundle waviness, identify whether interlaminar crack propagation can occur or occurs across a laminated composite, and measure the location and volume of resin-rich zones in laminated or woven composites (Aruniit et al., 2012). The introduction of CT provided a nondestructive, repeatable imaging process that could capture the same specimen image pre- and post-loading work application, which represents a notable methodological advancement for studying damage mechanisms, measurement of damage progression, and residual porosity (Asaithambi et al. 2014). In combination with CT scans, DIC represents a high-resolution optical technique to measure deformation across a surface and provide full-field strain maps for a specimen. By applying a speckle pattern to the surface and measuring subsequent shifts in pixels for each image, DIC permits clear visualization of faulted concentration of strain around symmetrical holes, edge delaminations, or specified in-plane shear regions (Asasutjarit et al., 2009). The extension of DIC into volumetric space is the development of digital volume correlation (DVC) as a way of volumetrically mapping strain with respect to CT datasets. All of these imaging modalities enabled the direct readout and comparison of measured changes in structure to those predicted by finite element models (FEMs), which closed the loop between experimental and computational mechanics (Ashori, 2008).

5.1.4 Objectives and Scope of the Chapter

The purpose of this chapter is to investigate and review the advancements of physical and mechanical characterization methodologies applied to composite materials. There is a gap in understanding between the mechanical type testing (traditional) and the various methods/methodologies that are used to better characterize the multiscale nature of the composite system. This chapter will address the following specific objectives: (i) to provide a historical overview of the development of characterization methods, (ii) to present awareness and understanding of current best practices (highlighted with examples), including the traditional "tensile/flexural" load tests to state-of-the-art techniques such as nanoindentation, DIC, and X-ray computed tomography (XCT), and the use of AE, (iii) to discuss the relevance of NDE for defect identification and monitoring the structural integrity of the product, and (iv) to exemplify the incorporation of real-time sensing, ML algorithms, and digital twin concepts as the future of composite materials characterization. The scope will cover both thermoset and thermoplastic matrix composites, continuous and discontinuous fiber-reinforced systems, natural FRCs, and nano-enabled composites. The emphasis is on the complex relationships between the microstructure of the material and its macro behavior, to identify modes of damage and use hybrid methodologies/characterization methods. It will also encompass the role of characterization in quality assurance, product qualification, and LCA. In doing so, the chapter can be used as an encyclopedic reference for researchers, engineers, and material scientists engaged in the design, testing, and overall applications of high-performance composite materials in the aerospace, automotive, marine, civil engineering, and energy sectors (Asiri et al., 2013).

5.2 FUNDAMENTALS OF COMPOSITE MATERIALS

5.2.1 Classification of Composite Materials

Composite materials consist of two or more distinct materials, typically a matrix phase and a reinforcement phase. Composite materials provide improved properties when compared to either of their components. Composite materials are designed to perform a specific functional or structural requirement or combination of requirements. Concurrently, composite materials typically combine light weight with mechanical strength, mechanical stiffness, corrosion resistance, and thermal or electrical resistance. The combination of the matrix and reinforcement phase can be tailored for applications in aerospace, automotive, marine, sports, infrastructure, and biomedical applications (Atiqah et al., 2014). The matrix binds the reinforcement and transfers loads between its components. Meanwhile, the reinforcement provides the strength and stiffness. Understanding the classification of both parts and how they are used will enable the selection and optimization of composite systems to provide the needed performance.

Composite materials can be classified by the matrix type, the type of reinforcement, and the structural arrangement of the composite system. In relation to the matrix phase, composites can be classified as PMCs, MMCs, and ceramic matrix composites (CMCs) (Aydemir et al., 2015). PMCs are the most well-known and used of the composite system types; they are easy to process, have low cost, and have good corrosion resistance. PMCs can be found in applications from aerospace components to sporting equipment. Common matrix materials include thermoset matrix materials like epoxy, polyester, and vinyl esters, and thermoplastic matrix materials like polyetheretherketone (PEEK), polyamide (PA), polypropylene (PP), and more. MMCs use aluminum, titanium, or magnesium matrix materials to achieve superior mechanical performance at elevated temperatures and are suitable for automotive and aerospace engine components. CMCs are composite materials, such as silicon carbide (SiC) or alumina, that have ceramic fibers or whiskers added into them that have high wear resistance and high temp properties that can be used in high temp (nuclear) reactors and turbines (Azevedo et al. 2020).

Composites can also be grouped based on the geometry of reinforcement (reinforced shape) into fibrous, particulate, and laminar composites. However, fibrous composites, specifically FRC, are often more prevalent in structural applications because of their specific strength and stiffness properties. Since they belong to the fibrous classification of composites, FRCs can be categorized as continuous fibers (carbon, glass, and aramid) and discontinuous fibers (randomly oriented or aligned). Continuous fiber composites, namely CFRP, are commonplace in high-performance applications (aerospace wings, automotive body panels, etc.). Discontinuous (short fiber) composites will exhibit moderate stiffness but commonly have moderate processing and moderate mechanical properties and are more routinely found in injection-molded products. In particulate composites, a matrix has small particles (concrete, metal-ceramic dispersions, etc.), most of which can take geometries that have isotropic forms, which are typically used in construction applications and wear applications. In addition to composite categories already mentioned, laminar composites that have layers of different materials and orientations, such as sandwich structures and laminated panels, can provide resistance to damage, vibration dampening, impact behavior, etc.

The matrix material (e.g., polymer matrix) of a composite is very important because it bonds the reinforcement together, redistributes loads, and protects the lengths of fibers from degradation, damage, and environmental conditions. As previously mentioned, polymer matrices can be sorted into thermosets and thermoplastics. Thermosetting resins like epoxy have exceptional chemical resistance, excellent dimensional stability, and adhesion properties to fibers; however, they suffer from brittleness and cannot be recycled due to cross-linking. Epoxy resins are most commonly used in aerospace structures due to their high glass transition temperature and low shrinkage. Examples of ductile, impact-resistant, and recyclable thermoplastics are PEEK or PPS, though these materials usually require high processing temperature/pressure. MMCs typically use an aluminum, titanium, or magnesium matrix with ceramic fiber (e.g., SiC and B4C) or particle reinforcement. MMCs can provide a higher stiffness, greater thermal conductivity, and improved wear resistance property than polymers, but they come with higher weight and require more complex processing methods, such as powder metallurgy or squeeze casting. CMCs have a high compressive strength at a very high temperature, but are brittle. These materials are commonly used in extreme conditions, such as in space shuttle components or cutting tools. For a successfully reinforced matrix, the material must be associated from a chemical standpoint and have sufficient interface adhesion to induce stress transfers while reducing propagation of microcracks while loading.

The reinforcement phase of composite materials enhances mechanical performance by providing stiffness, strength, and functional properties. Fiber reinforcements are the most effective type of reinforcement due to their high aspect ratio, which allows stress to be efficiently transferred along the fiber length. Reinforcements include glass fibers, carbon fibers, aramid fibers (e.g., Kevlar), and natural fibers. Glass fibers, which consist of E-glass or S-glass, are lower in cost, and have good tensile strength and corrosion resistance, though they have lower stiffness than carbon fibers. Carbon fibers are used for many aerospace applications and in sporting goods due to superior stiffness, strength/weight ratio, and fatigue resistance. Carbon fiber and Kevlar have high price points and conductive properties that do not support all applications, though the materials have found their way into aerospace applications through material pairing. As mentioned, aramid fibers (e.g., Kevlar or Twaron) have exceptional impact and abrasion resistance, making them ideal for ballistic equipment or personal protective equipment. Natural fibers, such as jute, hemp, flax, and sisal fibers, have gained attention as a sustainable reinforcement option to use with biodegradable composites. Biodegradable composite solutions encounter many barriers to structural use due to hydrophobicity, and variability of their properties (Azevedo et al. 2023). Particulate reinforcements, for example, silica, alumina, or silicon carbide, are dispersed within the matrix to improve wear resistance, stiffness, and thermal improvement. Nanomaterials like CNTs, GNPs, or nanoclays are used to modify matrices, or as reinforcements within hybrid composites, providing enhanced mechanical, electrical, or barrier properties with a lower filler content.

Hybridization between matrices and reinforcements further increases the design space. Hybrid composites are made with two or more different fibers, such as glass–carbon or carbon–Kevlar, which offer a balance between cost and performance. Similar in type, nano-hybrid composites use both micro- and nano reinforcement combinations for multi-functional performance needs, such as structural reinforcement, integrated sensing, or thermal management (Azevedo et al. 2023). The selection of materials for both the matrix and reinforcements needs to consider processing compatibility, interface strength, load transfer efficiency, environmental susceptibility, and stability. In addition, the fiber length, orientation, Vf, and distribution must be controlled in application, as they affect the mechanical response of composites. In advanced applications, the coupled functionality is used together within the same component; hence, designers are able to use simulated processes, materials informatics, and experimental validation through characterization to optimize the overall performance of the materials through their interactions.

5.2.2 Microstructure and Interfaces in Composite Materials

The performance and reliability of composite materials largely depend on the microstructure, which defines the geometrical arrangement, spatial arrangement, shape, orientation, and interaction of the reinforcement, which is set in the matrix, as well as the properties of the matrix–reinforcement interface, as shown in Figure 5.1. The action of a composite material depends on the interaction of the carbon properties of multiple phases across multiple length scales. These scales can range from the nanostructure of the fiber–matrix interface to the gross geometry of the woven fabrics or laminated plies. Models that predict failure mechanisms, provide uniformity related to the manufacturing process, or tailor properties for specific load-bearing or multifunctional applications require an understanding of microstructure (Azwa et al., 2013).

At the microstructural level, FRCs are either made with aligned continuous fibers that are directionally arranged (e.g., unidirectional, cross-ply, angle-ply, and quasi-isotropic laminated structure) based on the design requirements or use short or chopped fibers, which are often randomly oriented in a matrix to provide isotropic behavior, although with reduced mechanical performance. For instance, unidirectional fiber composites afford superior strength and stiffness in the fiber orientation, but poor transverse and shear properties when contrasted with cross-ply laminates, which exhibit improved isotropic behavior and damage tolerance. For either aligned or non-aligned fibers, the amount of fiber infill and the fiber Vf is defined as the volume of fully infused fiber divided by the total volume of the composite. Fiber Vf is an important microstructural parameter, as it heavily influences stiffness, strength, and even thermal conductivity. Typically, higher Vf yields greater load-carrying performance, but high Vf does not yield superior performance if it causes fiber crowding, poor wetting, or voids (Baheti et al., 2016).

Related to microstructure is fiber orientation distribution, which can be tied to anisotropy in mechanical performance. In injection-molded short fiber composites, the orientation of fibers is very important because fiber orientation induced by flow can create mechanical asymmetry through the thickness of the part. Methods commonly used to characterize orientation tensors and fiber length distribution of the end

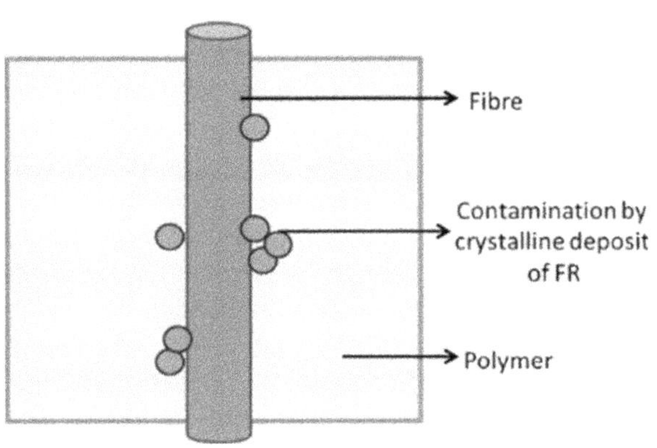

FIGURE 5.1 FR contamination on fiber surface (Azwa et al., 2013) [Open Access].

product are optical microscopy, XCT, and image analysis. The interface (or interphase) between the polymer matrix and the reinforcing material is equally influential for the performance of composites. The interface is very important for stress transfer, and it separately determines the mechanisms by which the polymer matrix can deliver externally applied loads to the reinforcing materials. If the interface is strong and bonded sufficiently well, then loads can be transferred appropriately without pull-out of the fiber and with no cracking of the matrix, as loads are transitioned to the glass transition temperature. Conversely, when there is a weak interface, then the transition of loads can lead to early delamination, debonding from the matrix, and cracking of the matrix when subjected to fatigue or impacts (Bajpai et al., 2013a). The interface can be an indication of mechanical or chemical bonding of the surface of the reinforcement and usually has different quality due to fiber surface treatments such as sizing or use of a coupling agent. For example, carbon fibers are usually oxidized or sized with epoxy-compatible sizing to aid wettability and interfacial adhesion for the polymer matrix. Natural fiber composites are typically treated with alkalis (NaOH) to reduce the high hydrophilic nature of the fibers and enhance bonding to hydrophobic matrices, silanes, or maleic anhydride grafts (Bajpai et al., 2012).

In the case of advanced composites (particularly nanocomposites), the interphase can be only a few nanometers, and can have distinctly different chemical and mechanical properties than both the matrix and reinforcement. In such systems, interphase engineering becomes essential, particularly in regard to high-aspect-ratio nanofillers such as CNTs, graphene, or nanoclays. Poor dispersion or agglomeration of nanofillers can degrade mechanical integrity as well as create a stress concentration area. Also, insufficient matrix interaction can create an ineffective mutual interaction of nanofillers to develop the desired enhancement in modulus, strength, or conductivity (Bajpai et al., 2012). As a result, interfacial bonding is optimized by using strategies such as functionalization of nanofillers, in-situ polymerized nanofillers, or the use of compatibilizers.

The microstructural heterogeneity associated with composites very often creates unique challenges for damage initiation and propagation. The most common modes of failure typically begin with matrix cracking, fiber breakage, delamination, debonding, and fiber pull-out, all of which typically begin at an interface or at some stress concentration area. For example, in a tensile loading condition, microcracks begin to initiate in the resin-rich areas and propagate along the fiber–matrix interface. In laminated composites, delamination is a prominent failure mechanism specific to weak interlaminar bonding that can also be exacerbated by out-of-plane stresses, impact loads, cyclic thermal loading, etc. This is of particular importance when dealing with design requirements related to damage tolerance in aerospace and automotive applications (Bajpai et al., 2013b, Figure 5.2).

Advanced manufacturing techniques such as AFP, RTM, and additive manufacturing are creating opportunities to exert better control over relevant microstructural features such as fiber alignment, resin flow dynamics, and void content. However, they may also introduce process-induced artifacts such as fiber waviness, dry spots, or residual stresses, all of which can adversely impact performance. In this context, microstructure-sensitive design, which is a design methodology combining process simulation, structural analysis, and probabilistic modeling techniques, is becoming accepted as an alternative means of optimizing composite performance.

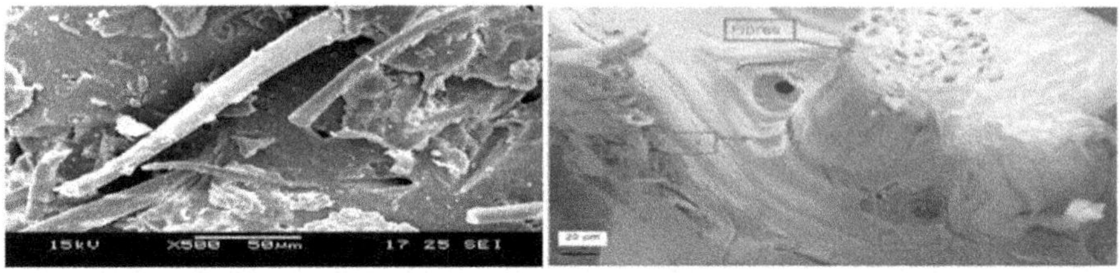

FIGURE 5.2 Fiber pullout and fiber matrix adhesion (Bajai et al., 2013) [Open Access].

Multiscale modeling techniques that utilize finite element analysis (FEA) and employ micromechanics or micromechanical modeling (e.g., Mori–Tanaka and self-consistent methods) can all make use of detailed microstructural data to predict the effective material behavior. In order for these models to converge predictably under an input loading application, reliable input data for fiber/matrix properties, an orientation distribution, an interfacial strength, and flaw content must be provided; the data is often gathered through a combination of microscopy, spectroscopy, and mechanical testing. Experimental validation by using techniques such as SEM, AFM, or X-ray diffraction (XRD) can be used to quantify microstructural uniformity, interfacial bonding, and phase distribution.

A detailed understanding of the microstructure and interfacial properties improves tailored composite design for specific loading environments, failure modes, and durability requirements. The greater the control over fiber arrangement, fiber Vf, and interfacial properties as predetermined variables, the more optimized the resulting structure will be, combining high strength and stiffness at low density with better energy absorption, and improved multi-functionality. Future innovation in composite materials will likely lead toward bio-inspired architectures, hierarchical composites, and self-healing interfaces to further advance the functions and capacities of engineered composite materials.

5.2.3 Properties Influencing Performance

The microstructure of composite materials is important for the performance and reliability metrics, including the placement, shape, orientation, and interactions of the reinforcement with the matrix, along with the character of the matrix–reinforcement interface. The response of composite materials to applied loading can often differ from that of homogenous materials because of the multiple ways that the individual phases will interact across different levels of scale in the manufacturing process, with respect to the size of the nanostructure of the fiber–matrix interface in relation to the overall size and dimensions of the woven fabric or laminated ply. The microstructure is also key to predicting potential failure mechanisms. Microstructure is relevant for a number of reasons, including—but certainly not limited to—ensuring product safety and overall product quality assurance, achieving predictability and repeatability in manufacturing, and optimizing mechanical and/or physical properties as desired for particular loading or multi-functional application.

With respect to microstructure, FRCs can contain continuous aligned fibers placed in unidirectional, cross-ply, angle-ply, or quasi-isotropic orientations that can be designed in any required configuration. A unidirectional fiber composite will typically have its greatest strength and stiffness in the direction of the fibers, whereas the transverse and shear properties are relatively poor. The cross-ply laminate will have more isotropic properties, along with enhanced damage tolerance compared to a unidirectional laminate. Short or chopped fibers will typically have a random orientation in the matrix. In these applications, it could be concluded that, in terms of mechanical performance, the nominal property of isotropy will generally result in less performance. The Vf of fibers is an important microstructural characteristic because it is the ratio of the volume of fibers (in this case) compared to the total composite volume as a whole. Vf is also a critical property because it has been shown that it will influence not only stiffness and strength but, more importantly, thermal conductivity. Higher Vf typically means increased load-bearing capacity, but it is worth noting that fiber bundling may be important when excessive fiber content limits significant characteristics such as the ability to wet, or the risk of creating excessive void space (Bajracharya et al., 2017).

The fiber orientation distribution is another important aspect of microstructure. It is well understood and widely accepted that fiber orientation distribution affects the anisotropic properties in composites. Orientation is critical, especially for injection-molded short fiber composites, where fibers can align in the flow direction, creating mechanical asymmetry as a result of orientation along the part thickness. Generally, either optical microscopy, XCT, image analysis, or a combination of techniques is used to quantify the resulting orientation tensors and fiber length distribution in the final part.

The interface, or interphase, region between the reinforcement and matrix is equally important to the performance of the composite. The interface is vital under load for transfer of stress, since the load is transferred from the matrix to the reinforcing elements with some efficiency dictated by the interface. A strong bonded interface will yield effective sharing of loads, inhibition of fiber pull-out, and delays in crack initiation, while a weak interface will yield early delamination, debonding, and cracking of the matrix, especially under fatigue or impact loads. The interface can be chemically or mechanically bonded and may often be mediated with fiber surface treatments, sizing agents, or coupling agents. For example, carbon fiber surfaces are typically oxidized or coated with epoxy-compatible sizing, with the intent of improving wettability and, consequently, interfacial adhesion. In natural fiber composites, fiber surface treatments have included alkali (NaOH), silane, and grafted maleic anhydrides in efforts to minimize hydrophilicity and improve the bonding with the hydrophobic matrix (Bayerl et al., 2014).

In advanced composites, especially nanocomposites, the interphase region can be only a few nanometers and may be chemically and mechanically different than both the matrix and reinforcement phases.

In such composite systems, interphase engineering is vital, especially in scenarios with a rational degree of nanofiller aspect ratio, such as CNT, graphene, or nanoclays. A bad dispersion and agglomeration of the nanofiller will detract from the mechanical integrity of the composite because it will also cause stress concentration areas. If the interphase is weak and the interaction is not optimally effective in the matrix, then it is likely that effective modulus, strength, or conductivity would be more difficult to achieve. Thus, it is a common procedure for researchers and engineers to use methods such as the functionalization of the nanofiller, in situ polymerization, or compatibilizers (Bhattacharyya et al., 2015).

The microstructural heterogeneity associated with composites often presents problems in the initiation and propagation of damage. The common composite failure modes of matrix cracking, fiber breakage, delamination, debonding, and fiber pull-out tend to be initiated at interfaces or areas of stress concentration. For example, under tensile loading, microcracks that may form inside the composite will generally first nucleate in the resin-rich zones and propagate along the fiber–matrix interface. In laminated composites, delamination of the composite is a primary failure mechanism, attributed to weak interlaminar bonding, and is influenced by out-of-plane stresses, impact, or thermal cycling. Sustained focus on these phenomena will remain significant, especially as design requirements for the aerospace or automotive industry challenge corrosion and damage tolerance (Boronat et al., 2015).

More recent advancements in manufacturing techniques like AFP, RTM, and additive manufacturing do provide enhanced delivery of predictable microstructural features like fiber orientation, resin flow, or void content. However, in many cases, process-induced artifacts such as fiber waviness, dry spots, and residual stresses contributed to reduced performance. In this sense, microstructure-sensitive design that incorporates predictive process simulation, structural analysis, and probabilistic modeling is gaining popularity when seeking to optimize composite performance.

Multiscale modeling approaches such as FEA used in conjunction with micromechanical models (e.g., Mori–Tanaka and self-consistent methods) are establishing effective material potential and behaviors, based on our detailed knowledge of the microstructure. For these types of modeling to be useful, they rely on the proper input data, such as the provision of composite fiber and matrix properties, relevant orientational distribution (OD), relevant interfacial strength, and flaw content, which are then assessed using relevant microscopy, mechanistic process, and/or spectroscopy. Experimental approaches [using SEM, AFM, and XRD] can give quantifiable information on microstructural consistency, interfacial bonding, and phases present during mechanical analysis (Bourmaud et al., 2015). Table 5.2 represents the key factors influencing the properties and composite performance.

In the end, researching and then developing a solid knowledge of the types of microstructure and the highly controllable properties of the interface allows engineers or designers to design composites of specific tailored loading environments, failure modes, and durability parameters. Cycle control over fiber arrangement, fiber Vf weights, and interfacial patterned properties enables the optimization of composite with respect to maximizing structures that can have high strength or stiffness attributes with low density, enhanced energy absorption, and possible multifunctional capabilities. In developments moving forward, the union of biophysically inspired designed architectures, multilayered and hierarchical composites, and self-healing interfaces stands to show promise in extending the functional and structural capabilities of engineered composite materials (Bourmaud et al., 2016).

Composite materials represent a class of engineering materials that have the potential to change the landscape of manufacturing. Individuals and designers are no longer restricted to finding the ultimate material that performs multiple functions at minimal weight; the unique combination of high strength-to-weight ratio, capability to be designed in a tailored fashion, and capability to function in multiple manners produces fortified products in a variety of value-added applications ranging from aerospace, automotive, civil infrastructure, biomedical devices,

TABLE 5.2
Summary of Key Properties Influencing Composite Performance

Property Category	Key Factors	Influence on Performance	Typical Enhancements or Considerations
Mechanical	Fiber type, volume fraction, orientation, matrix type, and interface quality	Affects tensile strength, stiffness, impact resistance, fatigue, and failure modes	Optimized fiber alignment, surface treatments, and void control
Fracture and Fatigue	Interfacial bonding, stacking sequence, and flaw distribution	Influences damage tolerance, crack propagation, delamination, and fatigue life	Interleaving, z-pinning, and hybrid fiber systems
Thermal	Fiber and matrix conductivity, coefficient of thermal expansion (CTE), and glass transition temperature	Determines heat dissipation, dimensional stability, and operating temperature limits	Use of carbon fibers, nano-fillers, and high Tg resins
Electrical	Reinforcement type (e.g., carbon and CNTs), filler dispersion, and percolation threshold	Enables conductivity, EMI shielding, lightning protection, and self-sensing	Addition of CNTs, GNPs, carbon fibers, and surface functionalization
Environmental Durability	Moisture absorption, UV exposure, and chemical resistance	Impacts longevity, strength retention, and interfacial stability in harsh conditions	Coatings, coupling agents, fiber treatments, and the use of hybrid fibers
Creep and Relaxation	Matrix type (especially thermoplastics), loading time, and temperature	Affects dimensional stability under prolonged stress or heat	Reinforcement with high aspect ratio fibers and thermoset matrices
Fire Resistance	Resin type and the presence of additives or retardants	Determines flammability, smoke emission, and thermal degradation	Use of phenolic resins, flame-retardant additives, and nanoclays
Sustainability	Recyclability of matrix, renewable fibers, and environmental regulations	Impacts eco-friendliness, LCA performance, and end-of-life disposal	Bio-based resins, natural fibers, recycled fibers, and thermoplastic matrix selection

consumer goods, etc. This chapter has introduced you to the foundational aspects of composite materials by identifying the composite material categories based on the matrix type (polymer, metal, and ceramic matrix) and geometry of the reinforcement (fibers, particulates, and laminates) and the structure of the material (notably as a layered structure). Each category or combination of categories has distinct mechanical, thermal, and chemical properties that provide an engineered material selection so the behavior of the material matches the functional performance requirement (Bulota & Budtova 2015).

The role of a matrix and reinforcements will be critical to understanding in assembling a performance composite where the composite is specified as a whole. While thermoset and thermoplastic polymers are currently the most common structural composites, innovations in natural fibers and nano-reinforcements are developing alternative pathways for sustainable and high-performance composites. The microstructural behavior or arrangement of reinforcement materials (e.g., fiber orientation and distribution and fibre–matrix interface attachment) will determine how loading is transferred, where damage will initiate, etc., and measure the quality of recovery or performance of the composite. The interface between the matrix and reinforcement will measure the quality of the interfacial region for mechanical strength, fatigue performance, and durability in a service life and under environmental conditions.

As reviewed in this chapter, there are a variety of mechanical, thermal, electrical, and environmental properties of composite materials to scrutinize to ascertain performance reliability. There are properties like tensile strength, stiffness, impact resistance, thermal conductivity, and fire resistance that can be enhanced or tailored to external energy inputs by adjusting contributions of constituent phases, controlling processing parameters, or by adding a functional filler. Challenges to the implementation of composites include ensuring durability over their service life on the end-user side and achieving recyclability and cost-effective production for the manufacturers. Also, research is being actively performed in the area of multi-functional composites, incorporating nanostructured reinforcements, and in the development of biodegradable, bio-based, or recyclable matrices (Cao et al., 2020).

In conclusion, knowing what composite materials are structured from is not only important for material scientists and engineers but also for their industries that manage and involve product innovation or product qualifiers. As a materials scientist and composites engineer, the developments and changes in the understanding and applications of composite materials have changed beyond just the consideration of composite materials, as the average material scientist or manufacturer has established a formal structure to examine and utilize composites. What should remain central to knowing about composite materials in the smart, sustainable, high-performance engineering for the next generation of products is a microstructure-informed, property-neutral approach to design and characterization.

5.3 FUNDAMENTALS OF COMPOSITE MATERIALS

5.3.1 Physical Properties

In composite materials, it is necessary to appreciate some of the physical properties, such as density, porosity, and water absorption, as they significantly affect mechanical performance, durability, and consistency of manufacture.

5.3.1.1 Density

Density is a base material property and affects performance as a building component when weight is an item of consideration, as well as structural efficiency. The density of a composite of any type is a function of the density and Vf of its constituents: the matrix (ρ_m), reinforcement (ρ_r), and to a lesser degree, porosity (ρ_v). Density can be determined using:

- Pycnometry: Air- or gas-displacement techniques, to measure volume at a high degree of precision. This is a critical property to measure density for low-porosity high-performance composites.
- Archimedes' Principle: Simple buoyancy principle; very simple but only considers buoyancy in terms of the size of the piece and does not properly consider porosity (even if it is just air stuck in voids).
- Computed tomography (CT): Non-destructive density mapping. There are artifacts presented in terms of both materials and component geometry, and details of the construction within may influence the density value.

5.3.1.2 Porosity

Porosity is the voids or holes that have formed within the laminate; porosity may have a significant limiting effect on strength, fatigue life, and cumulative effect on out-of-plane thermal conductivity. There are methods to measure volume porosity. For example, using the first example, just by taking the density measurement and the difference between the theoretical and measured density, you can determine porosity as shown below:

Volume porosity is derived via density comparison:

$$P_v = \rho_{\text{theoretical } \rho\text{theoretical}} - \rho_{\text{measured}}$$

- Mercury intrusion porosimetry: A destructive test method using toxic mercury to quantify micro-porosity.
- Gas adsorption (BET method): The method is typically used to measure surface area, looking for nanopores, and they examine the results using a composite modified with nanoparticles.
- X-ray computed tomography (XCT): This provides non-destructive images of the x, y, and z distribution of volume porosity and provides percent porosity with resolution up to ~10 μm.

Carbon fiber laminates with only 1–3% porosity will have an estimated 10–20% loss of mechanical strength. Thus, control of porosity should be primary for carbon fiber composites for aerospace applications and/or high-performance sports equipment. Table 5.3 represents the physical properties and the method to evaluate.

5.3.1.3 Water Absorption

Water absorption testing (e.g., ASTM D570) measures moisture absorption limits by recording weight gain at given intervals until saturation is achieved. The absorption of water into either composite, polymer, and/or fiber can lead to:

- Plasticization—polymer matrix is mechanically disabled due to water uptake.

TABLE 5.3
Physical Properties Overview

Property	Measurement Method	Significance	Typical Observations
Density	Pycnometry, Archimedes, and XCT	Calculates fiber volume and detects internal flaws	Deviations indicating voids or incomplete cure
Porosity	Archimedes, MIP, and CT	Impacts strength, fatigue, thermal conductivity, and defect detection	1–3% porosity reduces strength by ~10–20%
Water Absorption	ASTM D570 immersion, TGA, and MEA	Evaluates long-term durability in moisture-rich environments	1–3% weight gain in marine composites

- Interfacial debonding—ability to transfer mechanical load is compromised due to debonding.
- Residual stress—swelling may lead to warpage of the component.
- Microcracking to lamination—it continues to provide pathways and conditions that allow for delamination.

In marine cases, it is expected that an epoxy/glass composite might absorb 1–3% of its weight in water, which they expected to result in an associated 5–15% loss in interlaminar shear and/or flexural strength (Ebrahimnezhad-Khaljiri & Eslami-Farsani, 2020). In general, composite conditioning would be in marine environments where the only consideration would be seawater or a humid location. As in, for composites used in pipelines, civil infrastructure would expect sampling of the same types of absorption occurrence.

5.3.2 Mechanical Properties

5.3.2.1 Tensile Strength

Tensile testing is a basic mechanical characterization method used to determine the strength and stiffness of composite materials under uniaxial loading conditions. Tensile testing, when required, is performed according to ASTM D3039, using a flat laminate test specimen. The laminate is held securely in gripping devices, while an increasing axial force is applied to the specimen until failure occurs. The output obtained from tensile testing is considered key engineering responses: ultimate tensile strength, tensile modulus, and strain-to-failure. The properties obtained from tensile testing are dependent on fiber orientation and stacking sequence because composites are inherently anisotropic materials. Proper tabbing and alignment of the specimen are important to achieve proper tensile testing because improper tabbing and alignment can lead to premature gripping and extensometer failure. Tensile data is typically useful for investigations about aeronautical, automotive, and civil engineering structures to assess load-bearing structural design. Table 5.4 shows the tensile properties of the various reinforced composite materials.

TABLE 5.4
Summary of Tensile Properties of Common Composite Materials

Composite Type	Fiber Orientation	Tensile Strength (MPa)	Tensile Modulus (GPa)	Strain at Break (%)	References
Carbon/Epoxy (UD laminate)	0°	1500–1800	130–150	1.2–1.8	Azevedo et al. (2020)
Glass/Epoxy (UD laminate)	0°	900–1200	35–45	2.5–4.0	Azwa et al. (2013)
Aramid/Epoxy (e.g., Kevlar/ epoxy)	0°	1000–1400	60–80	2.0–3.5	Bajpai et al. (2013a)
Natural Fiber/ Epoxy (e.g., jute)	Random	50–150	2–6	1.5–4.5	Bajracharya et al. (2017)
GFRP (Woven bidirectional)	0°/90°	400–600	20–25	1.5–2.5	Bhattacharyya et al. (2015)
CFRP (Quasi-isotropic laminate)	Multiple angles	800–1200	70–100	1.0–1.5	Bourmaud et al. (2015)

5.3.2.2 Flexural Strength

Flexural tests are used to evaluate the resistance of a composite material to bending and provide the flexural strength and flexural modulus of the material. The predominant and preferred tests are three- and four-point bending tests as established by ASTM D790 and ISO 178. In three-point bending, a load is applied at the midpoint of a simply supported specimen. In four-point bending, the load is applied at two separate points on the specimen to obtain more uniform loading to avoid localized stresses within the specimen. Flexural testing is often very relevant for any structural component subjected to bending, such as beams, panels, and automotive parts. The results of flexural tests can vary with directional fiber orientations, stacking sequences, ductility of the matrix, and thickness of the laminate. The common modes of failure can vary between a cracked matrix, a broken fiber, or an interlaminar shear failure (which may also become an issue in indentation tests). Composite failure is generally more brittle than metals, and the material has a low strain-to-failure value, as composite laminated materials can have as much as 4 to 5 times less strain-to-failure from various thermoset-based laminated materials. While ductility improves with laminated matrix materials used in hybrid or thermoplastic composites, the ductility associated with laminated matrix materials lends a useful ability toward energy absorption capacity in bending (Katoueizadeh et al., 2020).

5.3.2.3 Impact Strength

Impact testing is used to develop a composite material's resistance to sudden or dynamic load applications while collecting information on energy absorption capacity and the damage tolerance of a notched specimen in both the Izod and Charpy tests (ASTM D256, ISO 179). The specimen is fractured using a swinging pendulum distributed along the height of the pendulum to simply demonstrate damage tolerance, while a drop-weight test (ASTM D7136) represents a much greater, typical, real-life situation, for example, when dropping a tool or an object or from an impact with some type of debris. In general, as compared with metals in impact or shock-absorbing situations, composite materials will trend more toward internal types of failure (i.e., internal delaminating and cracked matrix failure). The amount of energy absorbed prior to failure is established to signify its toughness as well as the overall safety of the structure. The impact behavior of a composite material will likely depend on the type of fiber (carbon, glass, or aramid), the ductility of the matrix, the laminate architecture, and laminate thickness. In impact situations, aramid fibers (i.e., Kevlar) provide excellent impact resistance; on the other hand, carbon fibers often provide low energy absorption values despite possessing a high-strength fiber, and they are brittle and degrade the energy-absorbing capacities of a composite. Impact testing is primarily used in the aerospace, automotive, marine, and sports equipment markets (Khan & Halder, 2020). Table 5.5 represents the flexural, impact, and compression strength of the various composites.

5.3.2.4 Compression Strength

Compression testing is used to measure the actions and performance of a composite material subjected to an axial compressive load. Compression testing is especially important for columns, stiffeners, and sandwich panels. The standards for compression testing require conforming to ASTM D695 for polymer materials or ASTM D3410 or ASTM D6641 for composite laminated materials. Composites do not behave the same way as metals in compression; composites have a tendency to micro buckle, kink, and delaminate when subjected to compression, where the load is perpendicular to the

TABLE 5.5
Summary of Flexural, Impact, and Compression Properties of Common Composite Materials

Composite Type	Flexural Strength (MPa)	Flexural Modulus (GPa)	Impact Strength (kJ/m^2)	Compressive Strength (MPa)	References
Carbon/Epoxy (UD laminate)	1000–1200	110–130	35–50	1000–1200	Bulota & Budtova (2015)
Glass/Epoxy (UD laminate)	600–800	25–35	80–120	450–600	Cao et al. (2020)
Aramid/Epoxy	500–650	50–65	200–300	250–400	Khan & Halder (2020)
Natural Fiber/ Epoxy (Jute)	60–150	3–6	15–40	40–100	Khorasani & Neisiany (2020)
GFRP (Woven bidirectional)	300–450	15–25	80–110	300–500	Loganathan et al. (2020), Ma et al. (2021)
CFRP (Quasi-isotropic)	700–900	60–80	30–45	700–1000	Romero-Sabat et al. (2021)

fiber direction. Generally, but not always, compressive strength and modulus will be lower than tensile strength and modulus, as compressive failures are less stable and allow interlaminar shear failures to occur in gemstones. The successful completion of compression testing of composite laminate requires careful preparation of specimen geometry, anti-buckling fixtures, and specimen alignment to ensure specimens are not damaged or destroyed prematurely. As with other testing modes of failure in FRCs, failure will generally need to begin at a defect or defect from a minimized performance interface to then propagate through fiber waviness, matrix yielding, interfacial decohesion, or another failure mode. Knowing how raw or composite materials behave in compression mode is critical for the capacity design of structural integrity for use in aircraft fuselage sections and in automotive safety components.

5.3.3 Thermal and Thermo-mechanical Analysis (TGA, DSC, and DMA)

Thermal and thermo-mechanical performance of composite materials is important in applications that will be exposed to varying temperatures, such as aerospace, automotive, electronics, and structures. If we understand how composites respond to heat and temperature changes, and long-term thermal cycling, we can make better decisions on materials, predict failures, and increase reliability in the end application. Common methods for thermal and thermo-mechanical characterization include thermogravimetric analysis (TGA), differential scanning calorimetry (DSC), and DMA tests. Each of the tests will identify material transitions (e.g., Tg), degradation temperatures, and viscoelastic responses, and will provide essential background for understanding structural integrity under service conditions (Khorasani & Neisiany, 2020).

5.3.3.1 Thermogravimetric Analysis (TGA)

TGA is an analytical technique designed to measure the change in mass of a material or specimen when it is heated, either in an inert (nitrogen/argon) environment or in an oxidative (air) environment. TGA is commonly used to characterize thermal stability, decomposition temperatures, and filler content in composites. In PMCs, TGA can also detect the onset of matrix degradation, and after the complete burn-off of resin, it can give an estimate of the fiber or filler content for mass fractions.

For example, in the case of carbon/epoxy composites (T650-35% carbon), the degradation of the matrix may start at ~350°C. With glass/epoxy composites, the matrix may show similar degradation from 300 to 350°C, depending in part on the particular resin formulation. Likewise, degradation occurs at an even earlier temperature (~200–250°C) in natural fiber composites, due to the presence of hemicellulose and lignin. TGA is similarly useful for determining the weight fraction of various phases in hybrid composites or for providing verification of consistent manufacturing.

5.3.3.2 Differential Scanning Calorimetry (DSC)

DSC measures the heat added to, or extracted from, a material during thermal transitions such as glass transition temperature (Tg), melting temperature (Tm), and crystallization. For composite materials in particular, Tg is an important temperature because it represents the moment when the matrix begins to be malleable, below which the mechanical properties drop significantly (Kosarli et al., 2020, Figure 5.3).

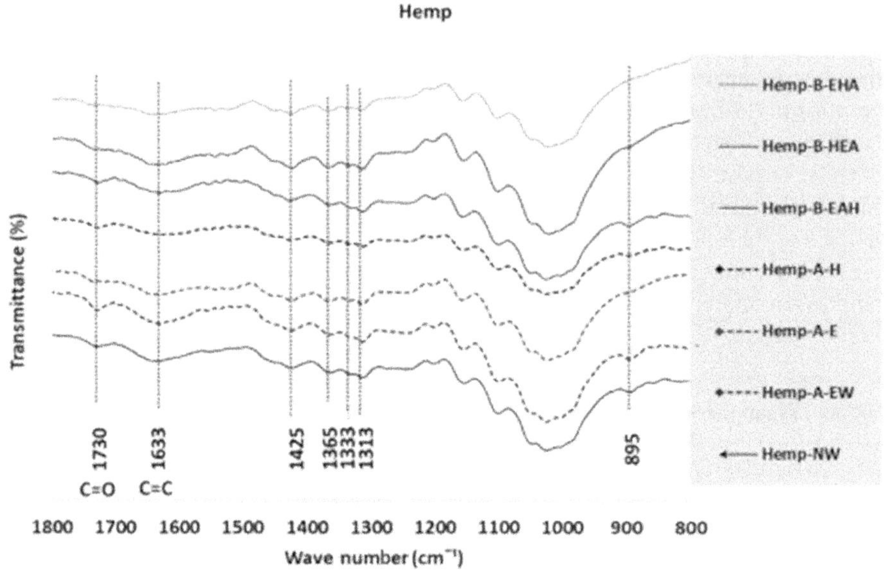

FIGURE 5.3 Spectra of hemp fibers before washing (NW) and after washing (Safri et al., 2020) [Open Access].

5.3.3.3 Dynamic Mechanical Analysis (DMA)

For thermosetting composites (epoxy), Tg will be dependent on curing conditions, but will typically fall between 120 and 200°C. In thermoplastic composites, DSC can offer you information on crystallinity as well, which also contributes to the strength, stiffness, and thermal resistance of the material. For example, a composite made of polyethylene has a Tm of near 130°C, whereas a composite made with PEEK has a melting temperature above 330°C. Further, you can use the DSC's useful information to create a post-curing schedule in order to fully crosslink the matrix.

DMA measures a material's viscoelastic response to oscillatory stress as a function of temperature, time, or frequency. DMA can provide complex data like storage modulus (E'), loss modulus (E"), and damping factor (tan δ), which are essential to understand the stiffness of the material and energy dissipation and/or internal friction. When a material is in a rubber state (i.e., viscoelastic), the oscillatory behavior tends to lean toward pure damping. The glass transition temperature is typically defined as the peak in the tan δ curve. Additionally, DMA is generally more sensitive than DSC to changes in molecular mobility and is much more definitive in measuring Tg. DMA is directly informing us about matrix softening onset, delamination behavior, and iron matrix interface behavior. Table 5.6 shows the key parameters used for TGA, DCS, and DMA characterization techniques.

As an example, in a carbon/epoxy laminate, DMA may demonstrate the storage modulus of ~10–20 GPa at room temperature, with limited decrease until approaching Tg, when it will drop rapidly. DMA is also effective with respect to frequency-dependent behavior, which is critical in the case of components under vibration or cyclic loading (Kovačič & Slugovc, 2020).

TABLE 5.6
Summary of Key Parameters Measured in TGA, DSC, and DMA for the Composite Materials

Technique	Purpose	Key Parameters Measured	Typical Applications	References
TGA	Measure weight loss with temperature	Degradation temperature and filler content	Thermal stability and matrix/fiber content	Radovic et al. (2020)
DSC	Measure heat flow during thermal transitions	Glass transition (Tg), melting point (Tm), and crystallinity	Cure analysis, crystallization behavior, and Tg identification	Kosarli et al. (2020)
DMA	Assess viscoelastic behavior under oscillating stress	Storage modulus (E′), Loss modulus (E″), and tan δ	Damping behavior, Tg detection, and modulus versus temperature	Kosarli et al. (2020)

5.3.4 Limitations of Conventional Methods

Despite being valuable for characterizing the behavior of composites, typical characterization methods, that is, tensile, flexural, impact, TGA, DSC, and DMA, have limitations. In the case of the methods presented here, the concern is that they are commonly destructive characterization methods, so that specimens are often not suitable for one or more tests, which represents concerns for composite material use and test time. Further, while these characterization methods provide bulk-averaged properties, they do not provide information about local defects, that is, micro-voids, fiber misalignment, and interfacial debonding.

As the community begins to assess in-situ durability or engage in longevity assessments, whether for traditional materials or existing structural health monitoring (SHM) techniques, there are again common limitations of traditional materials, where traditional health monitoring often has too low resolution or sensitivity. An additional speculative concern with the tests recommended or prescribed in standards, especially in determining a sampling interval, or testing under static/quasi-static testing, is that tests mainly assume isotropic or homogeneous mathematical functions, when behavior for advanced composites typically has an anisotropic or heterogeneous definition. Additionally, the materials or behavior may differ due to differing sample preparation, fiber orientation, or environmental conditioning (Loganathan et al., 2020).

New challenges arising from advanced composites (where the novel functions amount to nano-enhanced composites or multifunctional materials) increasingly discount the robustness of conventional, adequate resolution or sensitivity tests or protocols for characterizing the behavior of composites. Integral to more generally adopting/assessing newer sciences and changes in solutions within existing applicable design criteria or policy proof of concept frameworks will most likely be integrating advanced methods, for example, micro-CT, SEM, AE, or DR, into mechanical performance and property assessments.

5.4 EMERGING PHYSICAL CHARACTERIZATION PRACTICES

The evaluation of composite materials has advanced significantly from historical bulk testing to more modern and informative methods, allowing for multiscale, nondestructive, and high-resolution evaluation. These newer processes—digital image analysis and morphological characterization, XCT, nanoindentation and micro-mechanical testing, and enhanced NDE—allow for a better evaluation of microstructure, defect populations, interphase mechanisms, and functional performance and are central to developing the next generation of composites (Ma Y et al., 2021).

5.4.1 Digital Image Analysis and Morphological Characterization

Digital image analysis is an effective means of characterizing microstructural features in composite materials. The power of combining optical or electron microscopy with software-based image analysis, or digital image processing (DIP), allows researchers to study the distribution of voids, fiber orientations, surface roughness, and distribution of particles on images that are often created at very high resolutions. Morphological parameters such as aspect ratio, roundness or sphericity, or inter-fiber spacing can also be obtained through image processing or segmentation and thresholding.

DIC is a relatively new technology that has allowed researchers a standard tool to study strain distribution on the surface of composites while being mechanically loaded. DIC provides researchers with full-field deformation maps, revealing local variations of deformation that cannot be detected using a strain gauge or extensometer to monitor strain. Furthermore, DIC can be extended into three dimensions using DVC and, coupled with XCT data, can be used to monitor internal strain and damage development in a composite in real time.

These methods have been particularly useful for assessing and evaluating the quality of manufactured composite products, assessing any fiber misalignment, and assessing early signs of delamination or matrix cracking. Furthermore, high-resolution imaging can be used to monitor the consistency of woven architectures, regions rich in resin, and any waviness in the fiber, which all relate to microstructural features that significantly impact mechanical performance (Radovic et al., 2020).

5.4.1.1 X-Ray Computed Tomography (XCT)

XCT is a non-invasive imaging technique that produces three-dimensional reconstructions of composite components internally, with a resolution anywhere from several microns (micro-CT) to nanometers (nano-CT). XCT is the recognized standard of accuracy for voids, cracks, fiber/alignment orientation, and interlaminar defects without cutting the material apart.

A major benefit of XCT is its ability to characterize the porosity in more detail based on volume, distribution, and morphology to help gauge process-induced defects in FRCs, such as defects associated with the resin infusion or autoclave process. XCT data can be used further in higher development applications by incorporating the volumetric results into FEM of a material in order to simulate stress distribution and failure modes based on realistic geometry instead of ideal conditions or assumptions.

XCT can also make in situ testing possible, where a sample is loaded in a controlled mechanical manner inside the CT scanner. Documentation of crack propagation, fiber pull-out, and the debonding at the interface of constituents can be visualized under a certain level of stress in this context—that is, a powerful function for a failure analysis or validating computational models (Romero-Sabat et al., 2021).

5.4.1.2 Nanoindentation and Micro-mechanical Testing

Nanoindentation is a method for measuring local mechanical properties—hardness and elastic modulus—from sub-micrometer scales. This is especially important for composites, where the fiber, matrix, and interphase may all respond very differently. Using a sharp indenter tip, a normal force is applied, and the depth of penetration is measured. From the obtained load-displacement curve, the mechanical properties can be reported.

For composites, nanoindentation can be used to quantify mechanical properties in the matrix region, interphase, and single fibers, too. The method has proven particularly useful to evaluate the effects of aging, moisture ingress, or UV degradation on matrix properties. Further, it can distinguish chemically modified surfaces and distinguish or qualitatively quantify nanoparticle reinforcement in interface zones.

Micro-mechanical tests, such as micro-pillar compression or micro-beam bending testing, provide an additional resource to gain insight into localized failure mechanisms. When conducted in a scanning electron microscope (SEM), these types of tests will allow for direct imaging and capture of elemental crack nucleation, fiber/matrix debonding, or interfacial shear strength, which are critical for microstructural modeling of composites, particularly regarding failure.

5.4.1.3 Non-Destructive Evaluation (NDE) Techniques

Advanced NDE techniques provide valuable tools for inspecting, monitoring, and maintaining composite structures without affecting structural integrity. Traditional NDE techniques, including ultrasonic C-scan and radiography, are augmented with more sensitive and higher-resolution methods, such as thermography, AE, laser shearography, and eddy current testing.

Infrared thermography allows for quantifying subsurface defects such as delamination or voids by observing the surface temperature change during thermal excitation. It is rapid and can be applied to large surface areas. This is ideally suited for aerospace or wind turbine blade inspections. AE analysis allows for real-time monitoring of the damage evolution. AE signals can be classified by the type of damage—fiber breakage, matrix cracking, or debonding—allowing for comprehensive information on the health of the structure.

Recently, NDE practices have incorporated ML and deep learning algorithms. This provides automated detection and classification of defects from large datasets. This technology is being adopted as part of Industry 4.0 manufacturing and enables real-time quality assurance and predictive maintenance of composite components (Table 5.7).

TABLE 5.7
Emerging Physical Characterization Techniques

Technique	Key Capabilities	Applications	References
Digital Image Analysis and DIC	Quantify voids, fiber orientation, and full-field strain mapping	Surface damage and strain localization	Katoueizadeh et al. (2020)
X-ray Computed Tomography (XCT)	3D imaging of internal structure and defects	Porosity, in-situ damage, and FEM model input	Khan & Halder (2020)
Nanoindentation and Micro-testing	Local hardness/modulus and interphase strength	Interphase analysis and aging studies	Kosarli et al. (2020)
Advanced NDE (AE and Thermography)	Real-time and non-invasive monitoring	Structural health monitoring and defect detection	Kovačič & Slugovc (2020)

5.5 EMERGING MECHANICAL CHARACTERIZATION PRACTICES

As composite materials have progressed, we can now use, in addition to traditional static tests, dynamic, multiscale, and real-time measures to investigate complex behaviors. For example, we can use AE Monitoring, Instrumented Indentation Testing, Fracture Toughness and Crack Propagation Analysis, and Multi-scale/Multi-axial Testing as reliable techniques that will contribute to our understanding of the failure mechanisms, structural integrity, and durability of composites in ways not previously available.

5.5.1 Acoustic Emission and Real-Time Monitoring

AE monitoring detects elastic waves produced by damage due to events of matrix cracking, fiber fracture, and interface debonding. AE can use piezoelectric sensors bonded to composite structures to initiate micro-damage detection far before the catastrophic failure, making it one of the leading technologies available in SHM and predictive maintenance applications in aerospace, wind turbine, and automotive safety features through methods such as descriptive statistics, parameter tracking, and modeling.

AE monitoring capabilities have advanced beyond the original findings of cumulative energy and raw data recounting; modern AE practice now typically reviews an event count, amplitude, energy, and frequency content. The advent of advanced signal processing, notably the wavelet transform, has enabled a unique combination of detection, classification, and analysis for individual events and multiple events through the introduction of damage mode pattern recognition classifiers. An example of classifying an event would be low-frequency bursts from matrix cracking, medium-frequency events for delamination, and high-frequency signals for fiber breakage. Real-time systems provide alerts and damage maps while implementing mechanical or environmental stress simulations that provide diagnostics. Coupling AE with other types of sensors, such as strain gauges and thermography, will provide a more complete picture of damage propagation for composite materials viewed within their defined in-use loading and limits. For applied research purposes, it is also possible to embed continuous AE sensors into composite laminates during manufacturing, which provides lifetime damage monitoring capabilities (Safri et al., 2020).

5.5.2 Instrumented Indentation Testing

The historical evolution of instrumented (or continuous stiffness) indentation testing has progressed from passive hardness testing to dynamic mapping of mechanical property assessments with depth. In the case of composites, this test aims for micro-scale evaluation of the modulus, hardness, and plastic responses of test materials with respect to the matrix, fiber, and interphase. Instrumented indentation systems are very sophisticated in this regard, as they can incorporate grid-based indentation arrays, which create contour maps of viscoelastic properties in microstructural domains integrated in the test. This method provides the following advantages: quantifiable values for both modulus and hardness to compare across regions; curve fitting algorithms (e.g., Oliver-Pharr) to derive contact stiffness; and the ability to interrogate creep performance through force-hold differentials. FRCs may also provide important values for interfacial shear strength through indentation in the fiber–matrix interphase combined with finite-element modeling. If the method is performed in situ within an SEM, users are able to visually observe the in situ behavior of microcrack initiation, elastic recovery, and plastic deformation, thereby complementing their mechanical data with insight into various behaviors.

5.5.3 Fracture Toughness and Crack Propagation Analysis

Ideally, performing fracture toughness tests and studying crack propagation is instrumental in determining the expected damage tolerance and residual life in composites. The established methodologies, including double cantilever beam (DCB) or end-notched flexure (ENF), measure the mode I and II interlaminar fracture energy. However, newer approaches now include advanced imaging and feedback/monitoring of load-displacement paths to determine crack initiation and growth rates under mixed-mode loading in a controlled environment.

DIC yields full-field strain mapping at the crack tip and permits visualizing subcritical damage and damage process zone evolution at its inception. Correlating DIC to optical or scanning electronic microscopy allows researchers to investigate microscopic processes such as fiber bridging and kinking of fiber at an advancing crack front. New fatigue crack growth protocols include systematic changes in loading frequency, loading amplitude, and environment, all of which try to simulate the realities of service conditions. Composite fatigue crack growth models, such as Paris' law, can be improved with data from empirical observations to create life prediction curves.

An emerging area of interest is investigating the bonding of in-situ XCT with mechanical testing apparatus, such as servohydraulic rigs, to provide 3D insights into tortuous crack paths and delamination propagation from the time steps between load increments. This development allows for simultaneous loading and optical monitoring of the internal damage that occurs during the mechanical test and the external forced-loading conditions experienced. Table 5.8 shows the list of emerging techniques used for the mechanical characterization of composite materials.

TABLE 5.8
List of Emerging Mechanical Characterization Techniques Used for Composite Materials

Technique	Key Insights	Applications
Acoustic Emission and Real-Time Monitoring	Early detection of damage events; source classification using waveform analysis	SHM, fatigue life prediction, and real-time alerts
Instrumented Indentation Testing	Modulus/hardness mapping at micro-scale; interphase characterization	Material optimization, aging studies, and interface analysis
Fracture Toughness and Crack Propagation	Mixed-mode toughness; fatigue crack growth tracking; in-situ 3D crack observation	Damage tolerance and delamination resistance studies
Multi-Scale and Multi-Axial Testing	Stress-state realism; scale bridging; complex loading conditions	Structural integrity for aerospace, rotors, and pressure vessels

5.5.4 Multi-scale and Multi-axial Testing Techniques

Structural applications place composites in complex, multi-directional loading states. Multi-axial testing has evolved and advanced well beyond the uni-axial tension/compression stage. Current test setups incorporate quasi-isotropic tubes undergoing tension–torsion combined loadings, biaxial membrane-biaxial flexure rigs, and cruciform specimens with four corner grips to enable the users to apply multi-axial stress states. These relatively recent testing protocols mimic the in-service conditions subjected to multi-directional loading more realistically than using standard coupons. Multi-scale testing continues to expand the link between component and microstructure. For example, microscale tensile testing of microfabricated coupon specimens is used to obtain the stress–strain behavior at sub-laminate levels. Linking from macro to micro involves micro-computed tomography (micro-CT) in combination with DIC to capture phenomena like fiber bridging, local buckling, and resin pooling under axial off-axis loading. Combining biaxial testing with high-frequency cycling has also become more available and is useful where the testing application can simulate fatigue on an aero-engine turbine blade subject to complex loading. This often saw mechanical rigs used in combination with in-situ thermography, AE monitoring, or DIC to capture the progression of damage and energy dissipation.

5.6 HYBRID AND ADVANCED TESTING APPROACHES

The demands of composites today—from safety in aerospace to durability for renewable energy—have encouraged the development of hybrid methods and testing that combine physical and mechanical understanding and artificial intelligence and incorporate smart sensors to advance real-time monitoring of material behavior. These methodologies increase resolution, anticipate the material behavior, and have realistic environments under inspection, while also decreasing the time and cost of inspection. Developments include Physical and Mechanical Tests Integration, AI/ML-Assisted Characterization, In-Situ Testing Under Environmental or Loaded Conditions, and Smart/Embedded Monitoring Systems (Radovic et al., 2020).

5.6.1 Integration of Physical and Mechanical Tests

Integrating physical imaging (e.g., XCT and DVC) with a mechanical load has opened the possibility of a multi-modal perspective of composite action. In such test setups, a specimen will be subjected to tensile, compression, flexural, or fatigue loading as it is being scanned or optically analyzed. For instance, in-situ XCT with cyclic loading can visualize predetermined crack nucleation sites and accumulated strain maps in three dimensions, providing better insights for fracture prediction models. Similarly, optically tracked DIC over a specimen's surface can be correlated to internal porosity or fiber misorientation derived from XCT, allowing for full-field strain localization to be mapped to microscale defects. Hybrid experimental methods could also link nanoindentation hardness of the interphase as an average value to local waviness or fiber clustering to assess how microscale variations influence macroscale damage mechanisms. Through this type of interdisciplinary experimental approach, the boundary between "see" and "load" disappears and provides comprehensive validation data for multiscale and multiphysics models.

5.6.2 AI and Machine Learning-Assisted Characterization

The sheer volume of data generated in today's hybrid approaches requires automatic analytics. ML, in any definition and especially deep learning, has been applied to the evaluation of images, signals, and/or spectroscopy for the detection and classification of composite defects. Convolutional neural networks (CNNs) applied to XCT reconstructions were able to identify voids, delaminations, fiber breaks, and matrix fractures with >95% accuracy in pilot tests. ML has also been applied to analyze AE feature sets (e.g., amplitude and frequency) for online classification of damage events while they occur during fatigue loading, which is critical for real-time failure modes and condition predictions. Regression methods, including random forests or support vector machines, can provide predictions of mechanical properties based on descriptors (e.g., fiber orientation tensor, void fractions, or networking connectivity) as descriptors of properties derived from hybrid tests. AI permits faster identification of problems, less chance of subjectivity, and the potential for autonomous quality assurance. Table 5.9 lists emerging testing approaches for the composite materials.

5.6.3 In-Situ Testing under Environmental or Load Conditions

The behavior of composite materials is often significantly influenced by environmental conditions. The best chance for understanding the effects of combined mechanical loads and certain environmental conditions is through in-situ tests, which induce

TABLE 5.9
List of Emerging Testing Approaches for the Composite Materials

Approach	Key Technologies	Capabilities
Integration of Physical + Mechanical Tests	In-situ XCT, DIC, and nanoindentation with load rig	Damage mapping; multiscale model validation
AI/ML-Assisted Characterization	CNNs, regression trees, and acoustic signal analytics	Automated defect detection; property prediction
In-Situ Environmental/ Load Testing	Fatigue/thermal/moisture cycling; multi-physics rigs	Real-world aging and damage interaction analysis
Smart Sensors and Embedded Monitoring	FBGs, PWAS, conductive networks, and MEMS wireless sensors	Continuous, real-time structural health monitoring

simultaneous mechanical loads and subject the material to realistic environmental conditions. Below are some examples:

- Combined mechanical and thermal cycling: Mechanical testing using tensile or fatigue rigs inside a temperature chamber can simulate the flight cycles. Incorporating in-situ tools to log strain, AE, and thermography data can allow the observation of damage zones evolving as a consequence of a mismatch in thermal expansion between the composite and its interface to the rigid support material.
- Mechanical testing after moisture conditioning: Fatigue tests on composites saturated with water can show the effects of plasticization and degradation at inter-fiber interfaces subjected to a constant load for durations that mimic marine exposure or high humidity.
- Rigs that expose composites to UV and frost: Exposing composites to freeze–thaw cycles while performing tensile tests can show the degradation that might occur outdoors, with respect to mechanical loss and damage accumulation.
- Multi-physics rigs: These create test setups that integrate ultrasonic inspection, thermal imaging, and mechanical loading into a single test sequence. The ability to characterize the mechanical response with its subsurface condition in real-time improves the analytical capabilities of the composite material testing field. The complexity of these testing regimens can contribute to a deeper understanding of the synergies that emerge from the damage mechanisms and service environments in composites, leading to better durability prediction and composite material design opportunities.

5.6.4 Smart Sensors and Embedded Monitoring Systems

The embedding of sensors in manufacturing composite layers is an original thought from post-fabrication testing to real SHM. Methods of embedding include:

- Fiber Bragg Grating (FBG) sensors: Thin optical fibers can be woven or layered between plies to monitor strain, temperature, and damage initiation by utilizing wavelength shift. FBGs have been embedded during

manufacturing and shown to survive thermal cycling while enabling real-time mapping of area strain.
- Piezoelectric wafer active sensors (PWAS): Lightweight piezo patches can be affixed to the surfaces of the laminate components to monitor acoustic waves, hailstone impacts, and delaminations on a sub-millisecond time scale.
- Integrated conductive networks: Composites using integrated conductive networks (e.g., layers infused with CNTs) can be a "self-sensing" material that responds to induction resistance change relative to crack opening or surface delamination.
- Wireless MEMS sensor systems: Small, MEMS-based inertial, temperature, humidity, or piezo sensors can be embedded in remote or inaccessible structures or comply with military applications and wirelessly communicate to enable real-time reporting with minimal additional mass.

These embedded systems further enable composites to monitor strain, damage, or environmental exposure during service life while remaining internally dormant until activated, thereby mitigating operational risk and inspection downtime.

5.7 CASE STUDIES AND APPLICATIONS

Composite materials have historically given rise to superior strength-to-weight ratios, as well as controlled and separate properties that add to differing levels of durability. Due to their improved properties, they have been used across a number of sectors of modern engineering and design. These applications have varied widely, from aerospace to automotive, construction, and, more recently, sustainable applications. Due to other practical advances related to efficiency, structural durability, and environmental responsibility, the versatility and advantages of composites compared to conventional materials can be derived across industries. Global pressures to reduce weight, improve durability, and meet sustainability commitments will fluctuate; however, composites will continue to present new possibilities that meet structural and environmental requirements. This section presents useful examples and relevant case studies where advanced composite materials are influencing whole areas of innovation from multiple sectors, improving fuel economy and reducing emissions, accelerating construction, and increasing the use and reduction of renewable resources in design or structure.

5.7.1 Aerospace Composites

Aerospace is the most advanced sector that has developed composites primarily to achieve lower-weight aircraft with better fuel efficiency, while at the same time providing structural performance with fatigue resistance. The Boeing 787 Dreamliner is the first commercial airliner with 50% of the combined weight of the aircraft coming from the CFRP airframe. Reducing aircraft weight by 20% compared to aluminum with the CFRP fuselage and wings will reduce fuel burn by approximately 15–20%. The Airbus A350 XWB now has CFRP for 53% of its structure and has CFRP wing spars and a fuselage to provide better aerodynamic efficiency. The CFRP fuselage will also maintain comfort levels due to the more significant structural improvements from corrosion resistance. In addition, CMCs have been applied to jet engines, such as C/SiC materials used in brake discs and other turbine parts. CMCs can achieve thermal stability up to approximately 1,600°C, and save approximately 300 kg of material in the Boeing 737 NG brake system. The developments and research in CFRP and CMCs have been leading to quieter and longer-range aircraft and emphasize the need for composites in the aviation technologies of today.

5.7.2 Automotive and Transportation

The automotive industry has increasingly incorporated composite technologies to meet performance requirements while also balancing cost, scalability, and sustainability. High-performance brands such as Lamborghini, BMW, and Mercedes-Benz have adopted carbon fiber sheet molding compounds (CF-SMCs) in chassis, monocoques, and body panels to address weight and stiffness. For example, Lamborghini's Aventador chassis was the first out-of-autoclave CFRP monocoque using Boeing-style building block techniques adopted from aerospace. Biocomposites (e.g., made with wheat straw, jute, and flax) are used in passenger vehicles (e.g., the Toyota Prius PHV and Ford Flex) for interior trim components, which can enable up to 20% weight savings and reduce energy use by approximately 80% when compared to standard petroleum-based composites. Recent reviews have documented that CFRP and GFRP can achieve 40–70% and 25–35% weight savings, respectively, when used in body-in-white structures, although high production costs and fast cycle times remain significant challenges. Although OEMs are using composites in the manufacture of roofs, pillars, and bumpers, more widespread adoption into production vehicles would likely depend on automated manufacturing advances and improved recyclability.

5.7.3 Construction and Infrastructure

The emergence of corrosion-resistant composite materials—notably glass-fiber-reinforced polymer (GFRP) pultruded profiles—has brought several advances that can now be used in a range of structures in construction, including bridges, decks, pipes, and marine situations. For example, in Denmark, the FRP footbridge was assembled in less than 24 hours in 1997, and in the United Kingdom, the Standen Hey bridge illustrated FRP's unique ability to be assembled with great durability and strength. In addition, since FRP retains a lower dead load compared to steel or other materials, FRP can offer a much simpler support structure. FRP typically has a much lower lifetime cost than steel, just for the fact that FRP usually requires no maintenance. There are several new composites developed for use in construction, including engineered cementitious composites (ECC), which have been used in dam maintenance and seismic retrofitting in Japan; ECCs rely on strain-hardening behavior and micro-crack bridging to offer durability and serviceability. Furthermore, there is growth in bio-based natural fiber composites, such as hemp- or linseed-fiber-reinforced epoxy sheets, which are entering the construction markets for insulation materials, wall panels, and flooring; these bio-based materials are proving to be similar to GFRP in terms of acoustic and thermal properties, while exhibiting significantly less embodied energy. Sustainability assessments of these materials show a range of lifecycle benefits and a range of challenges associated with durability, resistance to weather, and flame retardancy.

5.7.4 Bio-based and Sustainable Composites

As we move toward achieving climate goals and resource constraints and limits, biocomposites can be a new and sustainable alternative. Mycelium-based products can take shape as blocks or bricks or insulation blocks that are hydrophobic and stiff and are developed with agricultural waste (with mushrooms and fungal networks) and do not require high-energy processes to function. Cardboard-timber sandwich panels for emergency housing and structures printed at the University of Maine as part of their wood-fiber bio-home not only demonstrate that biocomposites have the ability to lower costs and reduce carbon emissions, but they can also be designed for recyclability and repurposed for centuries.

In transportation, natural fibers such as flax, hemp, jute, sisal, and abaca have been utilized as part of vehicle interiors by Ford, Mercedes, Toyota, Volkswagen, and GM to achieve weight reduction of 10–20% with an average oil saving of ~ 5 million lbs/year, while dropping energy use across the life cycle of the vehicle by ~80%. We still observe challenges with moisture sensitivity in combination with the mechanical properties of this process. Cutting-edge work included hybrid composites—such as jute fibre composites that have been coated in graphene—indicating a ~324% increase in elastic modulus and ~110% increase in tensile strength, indicating potential for future structural applications. Sustainability assessments highlight considerations of sourcing agricultural by-products, and cycle-in-place materials to enable the best possible outcomes for the environment.

5.8 CHALLENGES AND FUTURE TRENDS

The emergence of composite materials has propelled incredible change in many industries; however, there are still numerous hurdles preventing composites from being accepted on a large scale. These challenges include a lack of standardization of the test standards, high costs, and the scalability of composites, all of which complicate composite characterization. Just as industry is now undergoing a digital transformation, with new ways to innovate, including digital twins, real-time monitoring, and automation for materials design, testing, and maintenance, there are also new solutions to key challenges to enhance standardization, reduce costs, and improve Industry 4.0 technologies, which make composite's future characterization smarter, faster, and more predictive.

5.8.1 Standardization and Benchmarking Issues

One of the essential problems of composite characterization is that there are no unified standards or benchmarks pertaining to the performance of either an established standard technique or new techniques. Typical testing serves to provide uniformity within an established standard technique (e.g., ASTM D3039 for tensile, D790 for flexural, or D5528 for fracture toughness). Nonetheless, when contrasting novel modalities (e.g., in-situ XCT, DIC, or AE), there aren't universal testing protocols. This subgrouping adds complexity for materials qualification in highly regulated sectors (e.g., aerospace and nuclear), and matters where traceability of data and repeatability across laboratories are of critical importance.

Benchmarking between institutions, while possible, stumbles upon different dialects defined by the practical performance of equipment (fidelity, software algorithm (e.g., segmentation or feature extraction) standard, calibration values, etc.). A micro-indentation study from one lab could report local modulus values that differ by 10–20% from another—this can be driven by tip calibration, rate of loading, and/or surface preparation. As composites increasingly contain nanofillers, multifunctional components, or bio-based phase ingredients, the specification of metrics and tolerance limits for characterization will require comprehensive round-robin studies, open-access databases, and standard reference materials. Creating shared datasets, with best-practice recommendations, will be key in developing confidence in new characterization results.

5.8.2 Scalability and Cost-Effectiveness

Many emerging characterization methods are still hard or costly to achieve at scale. While lab-based XCT, SEM, or nanoindentation equipment is very beneficial, they are too slow, too costly, and generally not designed to quickly assess anything other than small specimens. For example, a micro-CT scan of a typical 50 mm^3 composite coupon can take hours and cost hundreds of dollars to run. From a quality assurance perspective, performing this for thousands of parts is simply infeasible.

Conventional NDE methods (e.g., ultrasonic A-scan and thermography) are better suited for production inspections, but there is still insufficient resolution and repeatability for complex geometries, and growing costs to apply NDE methods in production using emerging technologies (e.g., automated DIC, sensors embedded in materials, and AE), which require infrastructure adjustments in production lines and data analytics applications. Similarly, the overwhelming availability of quality data in real time in the manufacturing process lends itself to possibilities of feedback loops to refine processing parameters, which has been attempted infrequently in composite manufacturing shops.

Cost-effective scale-up will require:

a. Low-cost emerging sensor technology (e.g., printed electronics on prepreg and passive RFID strain gauges on composite parts).
b. Modular measurement systems (e.g., optical scanners mounted on gantries).
c. Multi-method processes using online monitoring (e.g., low-resolution IR) and offline (high-resolution) analyses, and only on flagged items.

Current economic models indicate a generational opportunity to invest in the infrastructure required for a transformative step-change in NDE. Justifying infrastructure intensity investment would be justified with the sharing of incurred costs during three major phases: design (part) development, manufacture (part) development, and service.

5.8.3 Integration with Digital Twins and Industry 4.0

Digital twin technology—that is, dynamic, virtual replicas of real-world systems—represents a significant opportunity for composite manufacturing and health monitoring. Digital twins can be used to connect process parameters, material behavior, sensor inputs, and predictive simulations, predicting residual stress, the interfacial health of layered components, and safety margins over time.

For instance, the composite section consisting of an AFP-wound wing spar might digitally log details of fiber orientation, cure cycle, and the fabrication in-situ strain history. Strain measurements collected in real-time from embedded FBG sensors or AE would continuously feed into the twin to refresh stress maps and time-dependent aging estimates. This would enable condition-based maintenance and focused inspections, reducing downtime.

To realize our vision, we will require a harmonized data architecture, standard APIs, and real-time analytics, which are sorely lacking in many composite supply chains beyond discrete intelligence-like monitoring. New capabilities in scalable edge computing, sensor fusion, and AI-enabled predictive models are all needed to start solving the feedback loop, or "closing the loop," between the physical structure and its twin. Finally, the fidelity of the twin itself will need to be verified against real-time data collected in the field. Demonstrated correlation with real-time field data would drive higher confidence associated with predicted simulation models.

5.8.4 Prospects for Automated and Real-Time Characterization

The future will be based on automated inspection and testing for manufacturing and service life health assessment. Inspection will no longer be random, premium periodic inspections; it will become a continuous health assessment.

- Machine vision systems: Through high-speed optical imaging and artificial intelligence in manufacturing, machine vision can conduct real-time surface inspection, fiber orientation assessment, and resin pooling detection.
- Robotic scanners: Many NDE methods—ultrasonic, thermography, and shearography—are either developing or have developed autonomous mobile robotic versions. These robotic scanners can scan any instrumentable part in minutes, even a complex, curved, and contoured part.
- Embedded sensor networks: Fiber optic sensors or printed piezoelectric (or piezo polymer) arrays can be either integrated into the part or otherwise embedded into the mold during manufacturing. This method allows for spatially distributed monitoring of a structure with little or no additional weight.
- Cloud analytics and anomaly detection: Online cloud data storage and analytics of large datasets of individual sensors over time allow for fleet-level analytics of health and maintenance, allowing predictive maintenance scheduling.

Possible challenges include the volume of data and data acquisition and management, the lifetime of sensors, power supplies for embedded networks, and minimizing false alarms from faulty AI prediction. Table 5.10 shows the future trends in the

TABLE 5.10
Enlist the Future Trends in the Field of Composite Materials

Challenge/Trend	Current State	Path Forward
Standardization and Benchmarking	Fragmented metrics, limited labs, and manual comparisons	Open data standards, inter-lab studies, and reference samples
Scalability and Cost	High costs for high-resolution tools and limited inline capability	Low-cost sensors, hybrid analysis pipelines, and shared infrastructure
Digital Twins and Industry 4.0	Isolated twins and limited real-time feedback loops	Connected process data, sensor fusion, edge computing, and validated predictive models
Automated Real-time Characterization	Manual inspections and limited embedded monitoring	Vision + NDE robots, sensor-rich composite parts, AI-driven health analysis, and real-time QC

field of composite materials. However, the trends toward real-time SHM systems that are lightweight, cost-effective, and designed to be manufactured sustainably, under the umbrella of smart manufacturing and asset digitalization, are aligned with the modern goal of development.

5.9 CONCLUSION

Engaging with new and evolving practices related to physical and mechanical characterization of composite materials has uncovered a rapidly evolving domain, which is technocentric and highly relevant to material science development.

5.9.1 Key Insights

This chapter has provided a wide-ranging collection of conventional (i.e., "standard") and newly emerging characterization practices so that the physical and mechanical performance of composite materials can be assessed. Conventional (mechanical) characterization test methods (i.e., tensile, flexural, impact, and thermal) are relevant and clearly established (used) for material characterization, but conventional test methodologies are merging with more studied advanced characterization test methods such as digital image analysis, XCT, nano-indentation, and NDE methods. With the inclusion of advanced mechanical characterization methods, conventional mechanical testing methods are extending to include real-time AE test methods, instrumented indentation, fracture toughness testing, and multi-scale or multi-axial testing methods. The inclusion of conventional mechanical testing methods and advanced (mechanical) characterization methods supports better defect detection, improved understanding of failure mechanisms, and improved predictive modeling. Additionally, hybrid methods, AI system-based methods, sensor networks, and digital twin perspectives inspired by Industry 4.0 will enable automated, real-time, and intelligent characterization of composites.

5.9.2 Importance to Research and Industry

These new and emerging practices have different significance to academic research and industry. It opens possibilities to advance understanding of the structure–property relationship and enhance optimization, material design, and modeling/computational modeling to predict and ascertain the complex behaviors in composites across actual length and time scales. New and emerging practices, with high-resolution and in-situ aspects, have the potential to reveal new data regarding damage evolution and degradation of materials, capture multi-dimensional composite response, and support fundamental research. In industrially relevant sectors such as aerospace, automotive, building and construction, or anything else linked with sustainability, it will help achieve better and optimized performance, safety, lifecycle monitoring and qualification, predictive health management, and quality assurance that leads to better costs, more reliable and safer products and processes, faster development, and advances in new technology.

5.9.3 Recommendations for Future Work

There are several areas that should be actively explored, but the following are some key areas:

- Standardization and benchmarking: The first priority is to create standard test methods; reference materials are needed that are applicable across numerous labs and industries.
- Scalability and embedment of characterization: Beyond creating new technologies, the goal is to move them from the lab to real-time, cost-effective industrial applications that can be seamlessly embedded into manufacturing and construction to deliver effective services.
- AI and data mining: Data mining and ML will be paramount to automate defect detection, failure prediction, and the move to smarter real-time decision making.
- Sustainability: Advocate for the development of bio-based, recyclable, self-sensing composites or in characterization approaches.
- Cross-Disciplinary Approach: Material science and mechanical engineering need to collaborate with data scientists and industry partners to assist in building tools that are scientifically valid and practically applicable.

To summarize, we are currently at a point in time where the convergence of new and emerging experimental methods, intelligent systems, and digital technologies is disrupting the materials characterization process for composites. The emerging practices potentially offer an increase in the ability to suitably and comprehensively understand composites and maximize their potential and more broadly represent signals toward the future of materials engineering and a smarter, lighter, and sustainable way to respond to the significant challenges we face globally.

REFERENCES

Abbass, A., Paiva, M. C., Oliveira, D. V., Lourenço, P. B., & Fangueiro, R. (2021). Insight into the effects of solvent treatment of natural fibers prior to structural composite casting: Chemical, physical and mechanical evaluation. *Fibers*, 9(9), 1–18. https://doi.org/10.3390/fib9090054

Abdul Khalil, H. P. S., Alwani, M. S., Islam, M. N., Suhaily, S. S., Dungani, R., H'ng, Y. M., & Jawaid, M. (2014). The use of bamboo fibres as reinforcements in composites. In *Biofiber Reinforcements in Composite Materials*. https://doi.org/10.1533/9781782421276.4.488

Abdul Khalil, H. P. S., Bhat, A. H., & Ireana Yusra, A. F. (2012). Green composites from sustainable cellulose nanofibrils: A review. *Carbohydrate Polymers*, 87(2), 963–979. https://doi.org/10.1016/j.carbpol.2011.08.078

Abdul Khalil, H. P. S., Marliana, M. M., Issam, A. M., & Bakare, I. O. (2011). Exploring isolated lignin material from oil palm biomass waste in green composites. *Materials & Design*, 32(5), 2604–2610. https://doi.org/10.1016/j.matdes.2011.01.035

Abilash, N., & Sivapragash, M. (2013). Optimizing the delamination failure in bamboo fiber reinforced polyester composite. *Journal of King Saud University - Engineering Sciences* 2013, 1–11. https://doi.org/10.1016/j.jksues.2013.09.004

Ahankari, S. S., Mohanty, A. K., & Misra, M. (2011). Mechanical behaviour of agro-residue reinforced poly(3-hydroxybutyrate-co-3-hydroxyvalerate), (PHBV) green composites: A comparison with traditional polypropylene composites. *Composites Science and Technology*, 71(5), 653–657. https://doi.org/10.1016/j.compscitech.2011.01.007

Ahmed, A., & Sanada, K. (2019). Micromechanical modeling and experimental verification of self-healing microcapsules-based composites. *Mechanics of Materials*, 131, 84–92. https://doi.org/10.1016/j.mechmat.2019.01.020

Ahmed, A. S., Islam, M. S., Hassan, A., Mohamad Haafiz, M. K., Islam, K. N., & Arjmandi, R. (2014). Impact of succinic anhydride on the properties of jute fiber/polypropylene biocomposites. *Fibers and Polymers*, 15(2), 307–314. https://doi.org/10.1007/s12221-014-0307-8

Akhtar, M. N., Sulong, A. B., Radzi, M. K. F., Ismail, N. F., Raza, M. R., Muhamad, N., & Khan, M. A. (2016). Influence of alkaline treatment and fiber loading on the physical and mechanical properties of kenaf/polypropylene composites for variety of applications. *Progress in Natural Science: Materials International*, 26(6), 657–664. https://doi.org/10.1016/j.pnsc.2016.12.004

Akil, H. M., Omar, M. F., Mazuki, A. A. M., Safiee, S., Ishak, Z. A. M., & Bakar, A. A. (2011). Kenaf fiber reinforced composites: A review. *Materials and Design*, 32(8–9), 4107–4121. https://doi.org/10.1016/j.matdes.2011.04.008

Aldahdooh, M. A. A., Muhamad Bunnori, N., & Megat Johari, M. A. (2014). Influence of palm oil fuel ash on ultimate flexural and uniaxial tensile strength of green ultra-high performance fiber reinforced cementitious composites. *Materials & Design*, 54, 694–701. https://doi.org/10.1016/j.matdes.2013.08.094

Aldridge, M., Waas, A., & Kieffer, J. (2014). Spatially resolved, in situ elastic modulus of thermoset polymer amidst carbon fibers in a polymer matrix composite. *Composites Science and Technology*, 98, 22–27. https://doi.org/10.1016/j.compscitech.2014.03.002

Andre, N. G., & Ishak, Z. M.(2016). Predicting the Tensile Modulus of Randomly Oriented Nonwoven Kenaf/Epoxy Composites. *Procedia Chemistry*, 19, 419–425. https://doi.org/10.1016/j.proche.2016.03.033

Aroso, I. M., Fernandes, E. M., Pires, R. A., Mano, J. F., & Reis, R. L. (2015). Cork extractives exhibit thermo-oxidative protection properties in polypropylene-cork composites and as direct additives for polypropylene. *Polymer Degradation and Stability*, 116, 45–52. https://doi.org/10.1016/j.polymdegradstab.2015.03.006

Arumuga Prabu, V., Uthayakumar, M., Manikandan, V., Rajini, N., & Jeyaraj, P. (2014). Influence of redmud on the mechanical, damping and chemical resistance properties of banana/polyester hybrid composites. *Materials & Design*, 64, 270–279. https://doi.org/10.1016/j.matdes.2014.07.020

Aruniit, A., Kers, J., Krumme, A., Poltimäe, T., & Tall, K. (2012). Preliminary study of the influence of post curing parameters to the particle reinforced composite' s mechanical and physical properties. *Materials Science*, 18(3), 256–261.

Asaithambi, B., Ganesan, G., & Ananda Kumar, S. (2014). Bio-composites: Development and mechanical characterization of banana/sisal fibre reinforced poly lactic acid (PLA) hybrid composites. *Fibers and Polymers*, 15(4), 847–854. https://doi.org/10.1007/s12221-014-0847-y

Asasutjarit, C., Charoenvai, S., Hirunlabh, J., & Khedari, J. (2009). Materials and mechanical properties of pretreated coir-based green composites. *Composites Part B: Engineering*, 40(7), 633–637. https://doi.org/10.1016/j.compositesb.2009.04.009

Ashori, A. (2008). Wood-plastic composites as promising green-composites for automotive industries! *Bioresource Technology*, 99(11), 4661–4667. https://doi.org/10.1016/j.biortech.2007.09.043

Asiri, A. M., Khan, S. B., Rahman, M. M., Al-Sehemi, A. G., Al-Sayari, S. A., & Al-Assiri, M. S. (2013). Green material: ecological importance of imperative and sensitive chemi-sensor based on Ag/Ag2O3/ZnO composite nanorods. *Nanoscale Research Letters*, 8(1), 380. https://doi.org/10.1186/1556-276X-8-380

Atiqah, A., Maleque, M. A., Jawaid, M., & Iqbal, M. (2014). Development of kenaf-glass reinforced unsaturated polyester hybrid composite for structural applications. *Composites Part B: Engineering*, 56, 68–73. https://doi.org/10.1016/j.compositesb.2013.08.019

Aydemir, D., Kiziltas, A., Erbas Kiziltas, E., Gardner, D. J., & Gunduz, G. (2015). Heat treated wood-nylon 6 composites. *Composites Part B: Engineering*, 68, 414–423. https://doi.org/10.1016/j.compositesb.2014.08.040

Azevedo, A., Fernandez, F., Fábio, S., Ferreira, E. P. C., Daniel, J., Melo, D., & Cysne, A. P. (2020). Addition of poly (ethylene-co-methacrylic acid) (EMAA) as self-healing agent to carbon-epoxy composites. *Composites Part A*, 137(February), 106016. https://doi.org/10.1016/j.compositesa.2020.106016

Azevedo, A., Trappe, V., Paula, A., & Barbosa, C. (2023). Fatigue behavior of self-healing glass fiber/epoxy composites with addition of poly (ethylene-co-methacrylic acid) (EMAA). *Polymer Testing*, 117(June 2022). https://doi.org/10.1016/j.polymertesting.2022.107863

Azwa, Z. N., Yousif, B. F., Manalo, A. C., & Karunasena, W. (2013). A review on the degradability of polymeric composites based on natural fibres. *Materials & Design*, 47, 424–442. https://doi.org/10.1016/j.matdes.2012.11.025

Baheti, V., Maqsood, H. S., Wiener, J., & Militky, J. (2016). Reinforcement of ozone pretreated and enzyme hydrolyzed longer jute micro crystals in poly lactic acid composite films. *Composites Part B: Engineering*, 95, 9–17. https://doi.org/10.1016/j.compositesb.2016.03.093

Bajpai, P. K., Meena, D., Vatsa, S., & Singh, I. (2013a). Tensile behavior of nettle fiber composites exposed to various environments. *Journal of Natural Fibers*, 10(3), 244–256. https://doi.org/10.1080/15440478.2013.791912

Bajpai, P. K., Singh, I., & Madaan, J. (2012). Comparative studies of mechanical and morphological properties of polylactic acid and polypropylene based natural fiber composites. *Journal of Reinforced Plastics and Composites*, 31(24), 1712–1724. https://doi.org/10.1177/0731684412447992

Bajpai, P. K., Singh, I., & Madaan, J. (2013b). Tribological behavior of natural fiber reinforced PLA composites. *Wear*, 297(1–2), 829–840. https://doi.org/10.1016/j.wear.2012.10.019

Bajracharya, R. M., Bajwa, D. S., & Bajwa, S. G. (2017). Mechanical properties of polylactic acid composites reinforced with cotton gin waste and flax fibers. *Procedia Engineering*, 200, 370–376. https://doi.org/10.1016/j.proeng.2017.07.052

Bayerl, T., Geith, M., Somashekar, A. A., & Bhattacharyya, D. (2014). Influence of fibre architecture on the biodegradability of FLAX/PLA composites. *International Biodeterioration and Biodegradation*, 96, 18–25. https://doi.org/10.1016/j.ibiod.2014.08.005

Bhattacharyya, D., Subasinghe, A., & Kim, N. K. (2015). Natural fibers: Their composites and flammability characterizations. In *Multifunctionality of Polymer Composites: Challenges and New Solutions*. Elsevier Inc. https://doi.org/10.1016/B978-0-323-26434-1.00004-0

Boronat, T., Fombuena, V., Garcia-Sanoguera, D., Sanchez-Nacher, L., & Balart, R. (2015). Development of a biocomposite based on green polyethylene biopolymer and eggshell. *Materials and Design*, 68, 177–185. https://doi.org/10.1016/j.matdes.2014.12.027

Bourmaud, A., Corre, Y. M., & Baley, C. (2015). Fully biodegradable composites: Use of poly-(butylene-succinate) as a matrix and to plasticize l-poly-(lactide)-flax blends. *Industrial Crops and Products*, 64(1), 251–257. https://doi.org/10.1016/j.indcrop.2014.09.033

Bourmaud, A., Le Duigou, A., Gourier, C., & Baley, C. (2016). Influence of processing temperature on mechanical performance of unidirectional polyamide 11-flax fibre composites. *Industrial Crops and Products*, 84, 151–165. https://doi.org/10.1016/j.indcrop.2016.02.007

Bulota, M., & Budtova, T. (2015). PLA/algae composites: Morphology and mechanical properties. *Composites Part A: Applied Science and Manufacturing*, 73, 109–115. https://doi.org/10.1016/j.compositesa.2015.03.001

Cao, S., Zhu, W., & Liu, T. (2020). Bio-inspired self-healing polymer foams with bilayered capsule systems. *Composites Science and Technology*, 195(February), 108189. https://doi.org/10.1016/j.compscitech.2020.108189

Ebrahimnezhad-Khaljiri, H., & Eslami-Farsani, R. (2020). The tensile properties and interlaminar shear strength of microcapsules-glass fibers/epoxy self-healable composites. *Engineering Fracture Mechanics*, 230(February), 106937. https://doi.org/10.1016/j.engfracmech.2020.106937

Katoueizadeh, E., Zebarjad, S. M., & Janghorban, K. (2020). A practical analytic model for predicting the performance of an encapsulated polymer composite. *Applied Mathematical Modelling*, 78, 418–432. https://doi.org/10.1016/j.apm.2019.09.042

Khan, N. I., & Halder, S. (2020). Self-healing fiber-reinforced polymer composites for their potential structural applications. *Self-Healing Polymer-Based Systems*, Elsevier, 455–472. https://doi.org/10.1016/B978-0-12-818450-9.00015-5

Khorasani, S. N., & Neisiany, R. E. (2020). Characterization of self-healing polymeric materials. *Self-Healing Polymer-Based Systems*, Elsevier. 123–140. https://doi.org/10.1016/B978-0-12-818450-9.00005-2

Kosarli, M., Bekas, D., Tsirka, K., & Paipetis, A. S. (2020). Capsule-based self-healing polymers and composites. In *Self-Healing Polymer-Based Systems*. INC. https://doi.org/10.1016/B978-0-12-818450-9.00010-6

Kovačič, S., & Slugovc, C. (2020). Ring-opening Metathesis Polymerisation derived poly (dicyclopentadiene) based materials. *Materials Chemistry Frontiers*, 4(8), 2235–2255. https://doi.org/10.1039/d0qm00296h

Loganathan, T. M., Sultan, M. T. H., Ahsan, Q., Jawaid, M., Naveen, J., Md Shah, A. U., & Hua, L. S. (2020). Characterization of alkali treated new cellulosic fibre from Cyrtostachys renda. *Journal of Materials Research and Technology*, 9(3), 3537–3546. https://doi.org/10.1016/j.jmrt.2020.01.091

Ma, Y., Liu, J., Zhang, Y., Ge, Y., Wu, R., Song, X., Zhang, P., & Wu, J. (2021). Mechanical behavior and self-healing mechanism of polyurea-based double-walled microcapsule/epoxy composite films. *Progress in Organic Coatings*, 157(December 2020), 106283. https://doi.org/10.1016/j.porgcoat.2021.106283

Radovic, I., Stajcic, A., Radisavljevic, A., Veljkovic, F., Cebela, M., Mitic, V. V., & Radojevic, V. (2020). Solvent effects on structural changes in self-healing epoxy composites. *Materials Chemistry and Physics*, 256(August), 123761. https://doi.org/10.1016/j.matchemphys.2020.123761

Romero-Sabat, G., Gago-Benedí, E., Roa Rovira, J. J., González-Gálvez, D., Mateo, A., Medel, S., & Tolentino Chivite, A. (2021). Development of a highly efficient extrinsic and autonomous self-healing polymeric system at low and ultra-low temperatures for high-performance applications. *Composites Part A: Applied Science and Manufacturing*, 145(February). https://doi.org/10.1016/j.compositesa.2021.106335

Safri, S. N. A., Sultan, M. T. H., & Shah, A. U. M. (2020). Characterization of benzoyl treated sugar palm/glass fibre hybrid composites. *Journal of Materials Research and Technology*, 9(5), 11563–11573. https://doi.org/10.1016/j.jmrt.2020.08.057

6 Overview of Optimization Processes by Statistical and AI Techniques

Neeraj Sharma, Shaman Gupta, Indira Bhati, and Rishu Arora

6.1 INTRODUCTION

Nowadays, many engineers and researchers are engaged in optimizing manufacturing processes, which encompass various operations such as machining, welding, forming, casting, and industrial manufacturing systems. The need for optimization arises from the necessity to utilize limited resources efficiently. Optimization can involve maximizing or minimizing specific responses. In some cases, achieving an intermediate value rather than an extreme is preferred, following the "nominal is best" criterion.

Maximizing responses is crucial for parameters such as production rate, cutting rate (CR) (Sharma, Khanna, and Gupta 2015; Dutta et al. 2019), quality, dimensional accuracy, efficiency, and strength (Sarıkaya et al. 2021; Khanna, Sharma, and Gupta 2019). On the other hand, minimization applies to factors like surface roughness, different types of wear, including flank wear (Abbas et al. 2020, 2023) and abrasive wear (Nur et al. 2021; Sharma et al. 2020; Valizade and Farhat 2024), as well as dimensional deviations (Sharma et al. 2022), coefficient of friction (Etri et al. 2023), production costs, residual stresses, and defects. The "nominal is best" approach applies to variables such as coating thickness, where maintaining a specific value is crucial.

Manufacturing processes convert raw materials into finished products. This may involve the extraction of minerals, such as producing aluminum or iron from their ores, or the transformation of raw materials into final goods, such as shaping iron into beams. In all manufacturing processes, multiple parameters influence response characteristics (Ali et al. 2019; Kumar and Sharma 2020; Sharma et al. 2019). Therefore, optimizing these parameters is essential for improving process efficiency.

Optimization problems may involve single-response or multi-response criteria. Single-response optimization focuses on improving a single variable, whereas multi-response optimization or multi-criteria decision-making (MCDM) techniques handle multiple responses simultaneously (Sharma et al. 2020; Ahuja et al. 2019).

This research aims to review various planning and optimization techniques utilized in manufacturing. Initially, the study discusses experimental design methods.

 DOI: 10.1201/9781003620570-6

Subsequently, statistical optimization techniques are examined, followed by an exploration of advanced optimization methods. Finally, key challenges in manufacturing optimization are presented prior to the conclusion.

6.2 PLANNING OF EXPERIMENTS/DESIGN OF EXPERIMENTS (DOE)

A systematic investigation method of processes or systems operates under the structure of DoE (Montgomery, 2017). The method includes performing multiple controlled tests to reveal the relationships between manipulation variables and measured results. Figure 6.1 depicts the screenshot of DoE captured from Minitab. The key concepts of DoE are:

The experiment uses variables known as factors, which researchers modify as part of their experimental procedures.

Planned outputs called responses serve as the measured variables to detect influences of experimental factors. The factors contain different distinct values or operational settings that represent their levels in the experimental system.

The experimental design outlines every step of conducting the experiments by specifying which factors to use, along with their possible levels and their test order.

The statistical analysis method called analysis of variance (ANOVA) helps researchers study experimental results to demonstrate important factors within the experiment.

The benefits of DoE are that with the help of DoE, researchers can determine which factors demonstrate maximum influence on responses. It also serves as a tool

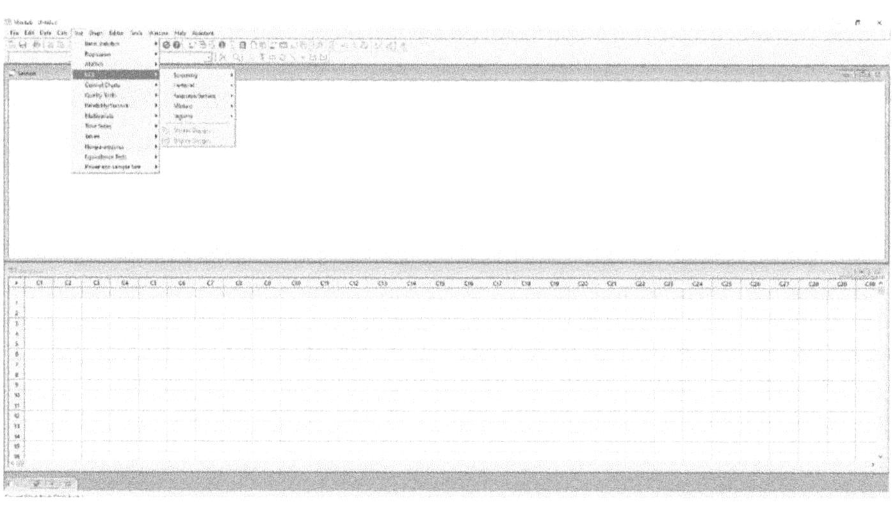

FIGURE 6.1 Design of experiments from Minitab.

to determine the best factor settings that maximize process performance goals. The reduction of response variability becomes possible through DoE, which leads to more consistent and reliable results. There are different types of DoE, that is, Full Factorial Designs, Fractional Factorial Designs, and Response Surface Methodology (RSM) (Roy 2001). A detailed explanation of Taguchi and RSM methodologies is provided in the following section.

6.2.1 Taguchi Method

The Taguchi Method, developed by Dr. Genichi Taguchi, is a robust design methodology that aims to minimize the variation in a product or process's performance. It focuses on making the design insensitive to uncontrollable factors (noise) by designing quality into the product or process from the start (Roy 2001). Figure 6.2 shows the process of choosing the Taguchi method from Minitab.

- **Key Principles**
    - The practice of designing products along with processes ensures they remain unaffected by manufacturing or environmental changes and use conditions.
    - The Loss Function measures the financial expense that arises from straying away from quality benchmark values because every minor deviation from target specifications represents a cost.
    - Orthogonal Arrays (OAs) provide researchers with an efficient testing system that investigates multiple factors through minimal experimental runs.
    - Design assessment through signal-to-noise ratio (SNR) involves finding the optimal condition that enhances the ratio between desired output signals and unwanted variation noise.

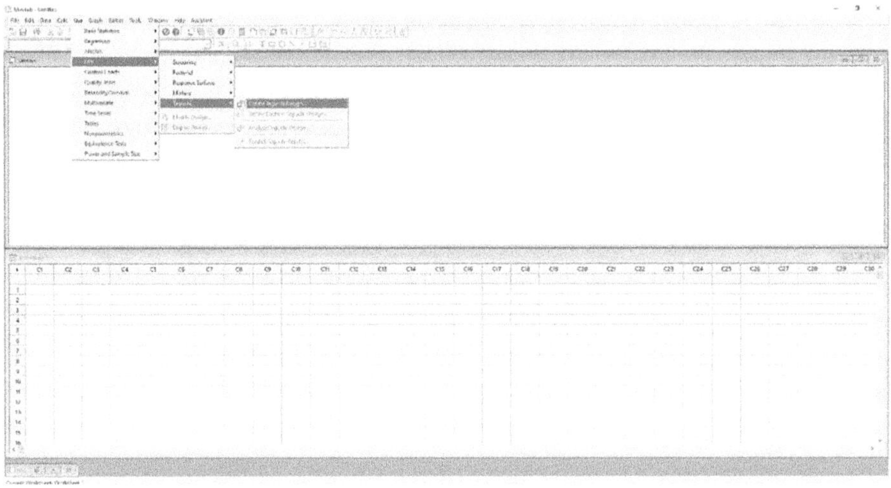

FIGURE 6.2 Implementation of the Taguchi technique using Minitab.

- The method for achieving robust design through finding optimal control factor settings is known as parameter design.
- The implementation of tolerance design should be the last step because it involves tightening critical parameter tolerances only after completing parameter design.

- **Orthogonal Array Selection**
  The Taguchi Method requires OAs to build effective experimental designs. The selection of an OA (Figure 6.3) hinges on both the number and design levels of the studied factors. Here's how it works:
  1. Specific Control Factors: Identify all elements that influence the product or process, followed by the determination of the attribute number for each element.
  2. Compute the degrees of freedom for each factor by subtracting one from the number of levels, then determine the total number of degrees of freedom needed for the experiment.
  3. Pick an OA that suits the required degrees of freedom from Taguchi's standard arrays. The array needs to contain columns that display all factors together with their interactions, and its rows need to maintain the needed confidence level.
  4. The experimental factors need to be placed into the selected OA columns.
  5. Conduct the experiments as per the array's layout, varying the levels of factors as specified in each row.
- **Key Considerations in Array Selection**:
  Number of Factors and Their Levels: The array must accommodate all factors and their levels.
  Interactions: If interactions between factors are important, choose an array that allows for their estimation.

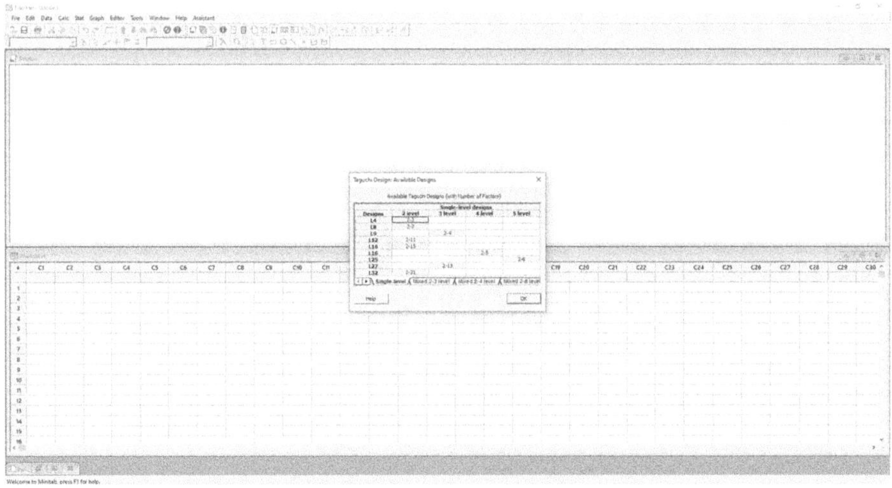

FIGURE 6.3 Selection of an orthogonal array using Minitab.

Resolution: The resolution of the array determines the ability to estimate main effects and interactions separately.

Number of Runs: The number of rows in the array determines the number of experimental runs.

Numerous researchers (Singh et al. 2018, 2021, 2024) have conducted studies on experimental planning using the Taguchi method in various manufacturing processes (Jangra, Grover, and Aggarwal 2011; Mia et al. 2018). The choice of an appropriate OA is primarily dependent on the number of factors, their levels, and the interactions considered in the experiment.

6.2.2 Response Surface Methodology (RSM)

RSM is a subset of statistical analysis that aims at evaluating a relationship between one or many output variables and more than one input variable. In this chapter, the following outlines a concise set of fundamental concepts and theory on RSM:

- The Goal of RSM

  The purpose of RSM is to determine which level of the input variables will give the best results in terms of the response variable. This could be done through an approach that focuses on the largest, the smallest, or a particular value of the response.
- Response Surface

  RSM believes that the input factors are related to the response through a polynomial of a certain degree.

  This can be depicted by a response surface, which is a graphical diagram of the relationship between the response and the variables that are used as inputs. Figure 6.4 shows the interface of Design Expert, from which the design matrix was selected.

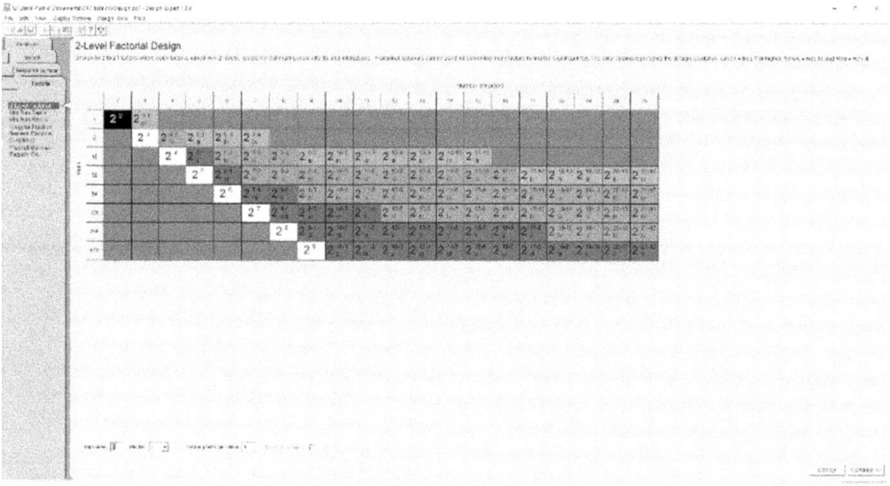

FIGURE 6.4 Interface of design expert for the selection of design.

- Mathematical Models
  RSM replaces the original function with other equally effective mathematical equations known as polynomial equations, revealing the relationship between the input variables and the response. The most common models are as follows:
  i. First-order model: This is also known as a linear model since it models the relationship between the output variable, which is the response, and the input variables. It is utilized to define the initial value of the response surface analysis.
  ii. The second order: In this model, terms such as x1 2, x2 2, and x1x2 are included in the equation, thus allowing curving of the response surface. It is used when the relation between the input variables and the response is more flexible and complicated.
- Experimental Design
  Therefore, the RSM uses well-planned experiments that investigate the response surface. Such experiments imply manipulation of the input variables and quantification of the output ones. Some of the conventional designs that are applicable in RSM are as follows:
  It also entails estimation of first-order and second-order effects; this makes it famous and widely used, known as CCD (central composite design). Figure 6.5 represents the screenshot of Design Expert, where the values of input parameters are incorporated. Another design used for creating second-order models is the Box-Behnken Design, which is good when confronted with many factors.
- Statistical Analysis
  The collected experimental data will be analyzed with statistical means to fit the chosen model and get the model coefficients. The terms regression analysis and ANOVA are used for this process.

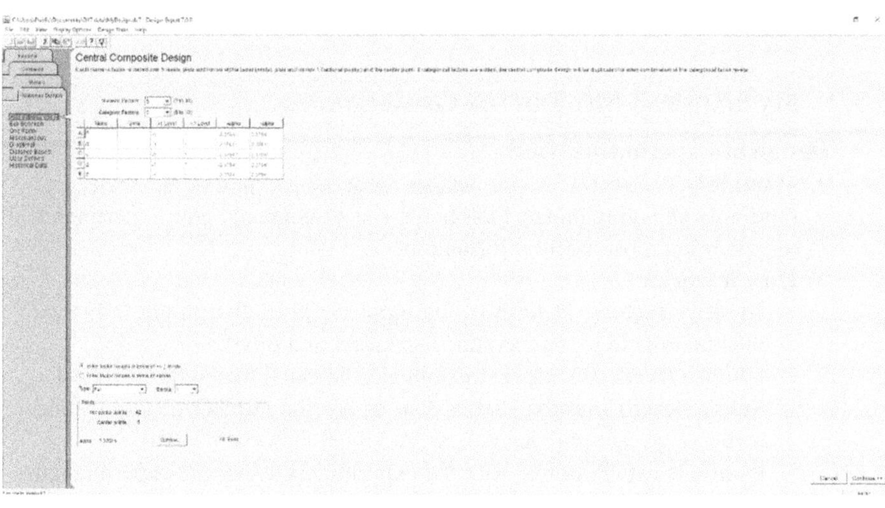

FIGURE 6.5 Central composite design for different parameters.

- Optimization
  RSM serves as a tool after fitting the model to find the optimal settings of the input variables. This can involve:
  i. Response surface as a series of contour lines: Each line of the contour plot corresponds to a constant value of the response, and they plot the response surface.
  ii. Response surface: These 3D plots illustrate the visual presentation of the response surface.
  iii. Response can be optimized by these algorithms; that is, find the maximum, minimum, or target value of response.

6.3 OPTIMIZATION THROUGH STATISTICAL METHODS

Optimization is about finding the *best* solution from a set of possibilities. "Best" depends on what you're trying to achieve:

- **Maximize**: Profit, yield, efficiency, etc.
- **Minimize**: Cost, waste, risk, etc.
- **Target**: A specific value for a quality characteristic.

6.3.1 Why Use Statistical Methods for Optimization?

Real-world problems are often complex and involve uncertainty. Statistical methods help us to:

- **Understand Relationships**: Identify how different factors influence the outcome.
- **Quantify Uncertainty**: Account for random variations and errors.
- **Make Informed Decisions**: Choose the best course of action based on data and analysis.

6.3.2 Key Statistical Methods for Optimization

- **Design of Experiments (DoE)**
  - **What it is**: A structured approach to conducting experiments. Instead of randomly changing things, DoE helps you strategically plan experiments to gather the most useful information.
  - **How it works**:
    - **Identify factors**: Determine the input variables that might influence the outcome (e.g., temperature, pressure, and time).
    - **Choose levels**: Decide on the range of values for each factor to test.
    - **Select design matrix**: Pick a plan as per the number of factors and levels (e.g., factorial design and fractional factorial design).
    - **Conduct experiments**: Run the experiments as per the experimental array and record the outcomes.

  - **Analyze results**: Use statistical methods to determine which factors have the biggest influence on the responses and how they affect them.
- **Goal**: To understand the cause-and-effect relationships between the factors and outcomes.

6.3.3 Response Surface M(RSM)

- **What it is**: Builds on DoE by creating a mathematical model (response surface) that describes the relationship between factors and outcomes.
- **How it works**:
  - **Fit a model**: Uses statistical techniques (like regression) to create an equation that predicts the outcome based on the factors.
  - **Visualize the response surface**: Creates graphs (3D surface plots and contour plots) to show how the outcome changes as you vary the factors.
  - **Optimize**: Uses mathematical techniques to find the factor settings that lead to the best outcome (e.g., maximum yield and minimum cost).
- **Goal**: To understand the shape of the relationship between input and output and find the optimal settings.

6.3.4 Statistical Process Control (SPC)

- **What it is**: Uses statistical methods to monitor and control processes. SPC helps identify and eliminate sources of variation to improve quality and consistency.
- **How it works**:
  - **Control charts**: Graphs that represent how a process is performing over time. Control limits are used to identify when the process is out of control.
  - **Process capability analysis**: Measures how well a process can meet specifications.
  - **Root cause analysis**: Investigates the causes of problems and implements corrective actions.
- **Goal**: To maintain process stability and produce consistent outputs.

6.3.5 Simulation

- **What it is**: Uses computer models to simulate complex systems. Simulation allows you to test different scenarios and optimize system performance.
- **How it works**:
  - **Build a model**: Create a mathematical or computer model of the system.
  - **Run simulations**: Vary the input parameters and observe the outputs.
  - **Analyze results**: Use statistical methods to study the simulation results and identify optimal solutions.
- **Goal**: To understand system behavior and optimize performance in a virtual environment.

6.4 TAGUCHI METHOD FOR OPTIMIZATION

The Taguchi concentration technique is a process of tolerant product or process design (Roy 2001), which is, on the contrary, insensitive to the noise. As for the utilization of this term, optimization in this context implies seeking a set of parameters that would result in this noise having the least influence on the desired goal.

Establish Objective: To determine what needs to be achieved (e.g., the quality of a particular product or yield of a specific process) and get acquainted with control and noise factors, which are the variables that can and cannot be influenced.

Design Experiments (Orthogonal Arrays): It is also possible to use Taguchi's OAs to develop a good experimental design. These arrays enable the assessment of a number of factors in the fewest possible runs, as noted in this section.

Chaotic Experimentation: Execute the experiments as directed by the OA by substituting control factors while bearing in mind the noise factors; if possible, they should also be included in the design matrix.

Analyze Results (Signal-to-Noise (S/N) Ratio): Determine the S/N Ratio of all the runs that were conducted in the experiment. The S/N Ratio represents the ratio of signal to noise, where the signal is the useful signal and the noise is the unwanted signal. There are different measures of S/N Ratio based on what is to be achieved: greater, lesser, or around the mean or average (as it is commonly referred).

Optimal settings should correspond to the control factor level that yields the maximum SNR. These settings are regarded as being quite strong because they reduce the influence of noise on the results.

Confirmation Run: Repeat an experiment, carrying out all necessary procedures at the best determined conditions.

6.5 RSM FOR OPTIMIZATION

The goal of RSM (Box and Draper 2007) is to describe the relationship between the factors and the response and improve this response.

- Define What to Optimize: Same as Taguchi, the analyst is required to clearly state what it is that he or she wants to optimize in the process.
- Experiments with Factors: Select an adequate experimental design to develop the set of experiments (e.g., Central Composite Design and Box-Behnken).
- Carry Out Experiments: With the help of a researcher, the experiments should be conducted, and data on the responses should be taken.
- Fit a Response Surface Model: Use regression analysis to estimate N, where N is the coefficients for a mathematical model that can be a polynomial of a specific degree. In this model, all the factors are included, and the responses of their interaction are depicted.

- Investigate the Response Surface: Explore the model response surface shape with contour and 3D surface plots, and perform ANOVA analysis to determine the optimal region.
- Maximize or minimize the response: Use optimization techniques (such as steepest ascent, algorithms, and optimization) in order to pick which factor settings cause the response to clearly maximize or minimize.
- Validation: Verify the optimal settings through additional experiments.

6.6 MOORA FOR OPTIMIZATION

Multi-Objective Optimization using Ratio Analysis (MOORA) (Brauers and Zavadskas 2006) is employed when you have more than one objective to balance out.

- Objectives and Alternatives. List all the objectives and your various alternatives (options).
- Develop a Decision Matrix: Make a matrix in which columns are criteria (objectives), rows are alternatives, and the entries are weights and values of the criteria for the alternatives. The performance of each alternative on each objective is contained in the cells.
- The data within the matrix should be normalized to a common scale. This allows for the comparison between different objectives of different units. Each normalized value is multiplied by the weights to identify the weighted normalized values.
- Overall score: Aggregate the weighted ratios for each alternative to obtain an overall next ratio.
- Score Alternatives: Score all the alternatives as they appear on the slide, then rank them. The best alternative is the one with the highest (or lowest, depending on whether one is maximizing or minimizing) score.

6.7 TOPSIS FOR OPTIMIZATION

Technique for Order of Preference by Similarity to Ideal Solution (TOPSIS) (M. K. Gupta et al. 2018) also handles multi-objective optimization.

- Same as MOORA objectives and alternatives to be defined.
- MOORA is the same as creating a Decision Matrix.
- Same as MOORA.
- **Identify Ideal and Worst Solutions**: Find the best and worst performance that can be possible on each objective.
- **Separate Measures**: Separation measures calculate the distance (usually Euclidean distance) of an alternative from ideal and worst solutions.
- **Calculate Relative Closeness to Ideal Solution**: Calculate a closeness coefficient for each alternative, which represents its relative distance to the ideal solution compared to the worst solution.
- **Rank Alternatives**: Rank the alternatives based on their closeness coefficients. The alternative closest to the ideal solution is considered the best.

6.8 GREY RELATIONAL ANALYSIS (GRA) FOR OPTIMIZATION

GRA (Işık, Çakıroğlu, and Günay 2023; Arora, Kapoor, and Sharma 2021; Sharma et al. 2021) is useful when dealing with incomplete or uncertain information.

- **Define Objectives and Alternatives**: Same as MOORA and TOPSIS.
- **Create a Decision Matrix**: Same as MOORA and TOPSIS.
- **Data Pre-processing**: Normalize the data and perform grey relational analysis if needed.
- **Calculate Grey Relational Coefficients**: Calculate the grey relational coefficients, which represent the relationship between each alternative and the ideal solution for each objective.
- **Calculate Grey Relational Grade**: Aggregate the grey relational coefficients for each alternative to get an overall grey relational grade.
- **Rank Alternatives**: Rank the alternatives based on their grey relational grades.

Key Differences and When to Use Which Method

- **Taguchi**: Focuses on robustness and minimizing variation. Best for designing products or processes that are insensitive to noise.
- **RSM**: Models and optimizes the relationship between factors and responses. Best for understanding and fine-tuning complex systems.
- **MOORA, TOPSIS, and GRA**: These are MCDM methods. Best when you have multiple, potentially conflicting objectives and need to rank alternatives. TOPSIS is more widely used and often preferred. GRA is suitable where the data is incomplete or vague. MOORA is easier to calculate.

Several statistical optimization methods, such as Taguchi, MOORA, TOPSIS, and VIseKriterijumska Optimizacija I Kompromisno Resenje (VIKOR) (Abbas et al. 2023), play a crucial role in manufacturing optimization. These techniques can be broadly classified into single-response optimization and multi-objective optimization.

Optimization has a significant impact on manufacturing industries. For instance, in semiconductor manufacturing, where machines are highly expensive and infrequently used, experimentation is limited. As a result, a comprehensive analysis is conducted, and the resulting data is compiled into a database for engineers. This database enables the prediction of process parameters and response relationships, facilitating the determination of optimal settings. A study by Khanna et al. (2015) involved optimizing the wire electrical discharge machining (WEDM) process for NiTi alloys using desirability functions. The key input parameters included pulse-on time, pulse-off time, servo voltage, and wire feed, while response characteristics such as CR, material removal rate (MRR), and dimensional shift were analyzed.

6.9 ARTIFICIAL INTELLIGENCE TECHNIQUES FOR OPTIMIZATION

Optimization, the process of finding the best solution among a set of possibilities, is a fundamental challenge across various fields. Artificial intelligence (AI) has emerged as a powerful toolkit for tackling complex optimization problems, offering a range of techniques that can learn, adapt, and efficiently navigate vast search spaces (P. Gupta and Singh 2020). This detailed exploration delves into the various AI techniques employed for optimization, highlighting their strengths, weaknesses, and applications.

6.9.1 Machine Learning (ML) for Optimization

ML algorithms (Hart et al. 2021) consider the pattern in the data to make predictions. In optimization, ML can approximate complex functions, guide search processes, or learn optimal control strategies.

- **Supervised Learning**: This paradigm involves training a model on the data, which is labeled data, to predict outputs for unseen inputs.
  - **Regression**: Used when the output is continuous. In optimization, regression models predict the objective function's value for different input parameter combinations, enabling efficient exploration of the search space.
    - **Linear Regression**: Suitable for the data if there is a linear relationship between input and output. Less effective for complex, nonlinear relationships.
    - **Polynomial Regression**: Extends linear regression to capture curved relationships by introducing polynomial terms. It can overfit if the polynomial degree is too high.
    - **Support Vector Regression (SVR)**: SVR aims to find a hyperplane that not only is a best fit but also maximizes the margin. Computationally intensive for large datasets.
    - **Neural Networks**: Powerful function approximators capable of learning highly complex, nonlinear relationships. Require substantial data and computational resources for training.
  - **Classification**: Used when the output is categorical. In optimization, classification can categorize solutions as "good" or "bad" or select the best option from a set of discrete choices.
    - **Logistic Regression**: When working with binary classification data, logistic regression calculates instance probability distribution for specific classes. This model maintains that all features develop a direct linear link with the calculated log-odds numbers for the outcome.
    - **Support Vector Machines (SVM)**: Support Vector Machines (SVM) achieve their goal by finding a maximum-margin hyperplane optimizer to partition classes within high-dimensional areas.
    - **Decision Trees**: Decision Trees apply hierarchical structures that use feature values for making decisions. It is easy to understand while bearing the risk of fitting data incorrectly.
    - **Random Forests**: Random Forests use multiple decision trees assembled in an ensemble, which creates a robust system that does not easily overfit data. Computationally intensive for large datasets.
    - **Neural Networks**: It can handle complex classification tasks by learning intricate patterns in the data. Require significant data and computational resources.

- **Reinforcement Learning (RL)**: An agent learns to interact with an environment by taking actions and receiving rewards or penalties. The goal is to learn a policy that maximizes the cumulative reward. RL is particularly useful for sequential decision-making problems, such as optimizing a process over time.
  - **Q-learning**: Q-learning operates as a model-free RL method that calculates Q-values across all state-action pairs for forecasting the reward potential through each state-action combination. RL algorithms using Q-learning take a prolonged time to reach convergence when the number of state-action pairs remains high.
  - **SARSA**: Another model-free RL method known as SARSA functions like Q-learning yet uses the actual agent movement to modify Q-values.
  - **Deep Q-Network (DQN)**: A neural network structure in DQN provides Q-value approximation to enable RL applications in environments with multiple state variables. The algorithm needs precise adjustment to function optimally because of instability.
  - **Policy Gradient Methods**: The algorithms learn a policy function between states and actions while seeking to maximize expected rewards. Q-learning requires more efficiency for continuous action spaces compared to policy gradient methods in handling such systems.
- **Unsupervised Learning**: Unsupervised learning, as the name suggests, learns patterns from unlabeled data.
  - **Clustering**: Grouping similar data points together. In optimization, clustering can identify clusters of similar solutions, which can then be explored further.
    - **K-means Clustering**: The K-means Clustering technique breaks data into k clusters through the reduction of within-cluster diverse elements. Sensitive to the initial cluster centers and the value of k.
    - **Hierarchical Clustering**: A hierarchy of clusters has been built, either agglomerative or divisive. Computationally expensive for large datasets.
    - **DBSCAN**: DBSCAN performs density-based clustering to identify cluster formations of any shape. The method tolerates outliers while needing manual adjustment of its parameters.
  - **Dimensionality Reduction**: It means to omit some variables, keeping the important information preserved. This can simplify optimization by reducing the size of the search space.
    - **Principal Component Analysis (PCA)**: PCA extracts principal data components through linear variable combinations that describe maximum data diversity. Assumes linear relationships between variables.
    - **t-SNE**: The t-SNE technique shows high-dimensional data in lower dimensions while retaining nearby data resemblance. The computation burden is intense, and result production might vary for each run due to algorithmic variations.

6.9.1 Deep Learning for Optimization

Deep learning, a subfield of ML, uses ANN (Saffaran, Azadi Moghaddam, and Kolahan 2020) with multiple layers so as to learn complex patterns of the data. Deep learning is specifically effective for optimization problems with high-dimensional input spaces or complex relationships between variables.

- **Neural Networks**: Can be used for regression, classification, or RL tasks in optimization. Different architectures exist, which are as follows:
  - **Feedforward Neural Networks**: Information flows in one direction from input to output. Suitable for a wide range of tasks, but can be computationally expensive for very deep networks.
  - **Recurrent Neural Networks (RNNs)**: Handle sequential data by incorporating feedback loops. Useful for time-series optimization problems. It can be difficult to train due to vanishing or exploding gradients.
  - **Convolutional Neural Networks (CNNs)**: Process image data by using convolutional filters. Effective for image-based optimization problems.

6.9.2 Evolutionary Algorithms (EAs)

Evolutionary Algorithms (EAs) are inspired by natural selection and use principles like mutation, crossover, and selection to find the optimum solution.

- **Genetic Algorithms (GAs)**: Represent solutions as "chromosomes" and use genetic operators (mutation and crossover) to evolve them. Effective for complex, nonlinear problems, but can be slow to converge.
- **Genetic Programming (GP)**: Evolves computer programs to solve a problem, automatically finding optimal algorithms or strategies. It is computationally intensive.
- **Evolution Strategies (ES)**: Use real-valued parameters and Gaussian mutations to evolve solutions.
- **Particle Swarm Optimization (PSO)**: PSO is based on a population where particles "swarm" toward the best solution found by the swarm. Easy to implement, but can get stuck in local optima.

6.9.3 Natural Language Processing (NLP) for Optimization

NLP techniques can optimize text-based content or processes.

- **Text Analysis**: Can be used to analyze social media data, customer feedback, or other text sources to identify areas for improvement, optimize marketing copy, or automate report generation.
- **Sentiment Analysis**: Can be used to measure public opinion about a product or service, which can inform optimization strategies.
- **Topic Modeling**: Can be used to identify key themes in a set of documents, which can be useful for organizing information or optimizing search results.

6.9.4 Computer Vision for Optimization

Computer vision techniques can optimize image-related tasks.

- **Image Recognition**: Can be used to recognize objects in images, which can be useful for optimizing manufacturing processes or quality control.
- **Object Detection**: Can be used to locate objects in images, which can be useful for optimizing logistics or robotics.
- **Image Segmentation**: Can be used to divide an image into regions, which can be useful for optimizing medical imaging or remote sensing.

6.9.5 Key Considerations When Using AI for Optimization

- **Data Quantity and Quality**: AI models require high-quality data so as to train efficiently. Data preprocessing and cleaning are crucial steps.
- **Computational Resources**: Training the AI models, specifically in the deep learning models, not only requires specialized hardware such as GPUs but is also computationally expensive.
- **Model Selection**: Choosing the right AI technique for a specific optimization problem is crucial. The choice depends on the availability of data, the nature of the problem, and the level of accuracy required.
- **Explainability and Interpretability**: Understanding why an AI model made a particular decision can be challenging, especially for complex models such as deep neural networks. Interpretability is decisive for efficient modeling and building trust.
- **Bias**: The training data of AI models can pass discriminatory biases into models, which results in wrong or unfair outcomes. Careful data collection and model evaluation are necessary to mitigate bias.
- **Validation and Testing**: It's crucial to validate and test AI models thoroughly before deploying them to ensure that they are performing as expected and generalize well to new data. Cross-validation, hold-out sets, and other techniques are used for this purpose.
- **Hyperparameter Tuning**: Numerous AI models require hyperparameter adjustment because this step produces optimal results. ANOVA provides three basic parameter tuning techniques: grid search, random search, and Bayesian optimization.

Several AI-based optimization techniques are applied in manufacturing, including GA, Non-dominated Sorting Genetic Algorithm (NSGA), PSO, Grey Wolf Optimization (GWO), Teaching-Learning-Based Optimization (TLBO), and Cuckoo Search Algorithm. These techniques, often implemented using MATLAB or Python, allow for faster and more effective problem-solving. AI-based hybrid models, combining multiple statistical and AI approaches, are increasingly explored for manufacturing optimization.

6.10 CONCLUSIONS

This review explores different optimization techniques used in manufacturing:

- Experimental design can be effectively planned using Taguchi or RSM, depending on parameter levels and ranges.
- Single-response optimization can be performed using Taguchi or RSM, while multi-response optimization requires methods such as GRA, MOORA, and TOPSIS.
- AI approaches present significant potential for future research, particularly in hybrid optimization methods.

Further studies should investigate unexplored statistical and AI combinations for enhanced manufacturing process optimization.

REFERENCES

Abbas, Adel T., Magdy M. El Rayes, Monis Luqman, Noha Naeim, Hussien Hegab, and Ahmed Elkaseer. 2020. "On the Assessment of Surface Quality and Productivity Aspects in Precision Hard Turning of AISI 4340 Steel Alloy: Relative Performance of Wiper vs. Conventional Inserts." *Materials* 13 (9): 2036.

Abbas, Adel T., Neeraj Sharma, Mahmoud S. Soliman, Magdy M. El Rayes, Rakesh Chandmal Sharma, and Ahmed Elkaseer. 2023. "Effect of Wiper Edge Geometry on Machining Performance While Turning AISI 1045 Steel in Dry Conditions Using the VIKOR-ML Approach." *Machines* 11 (7): 719.

Ahuja, Neeraj, Kamal Kumar, Uma Batra, and Sudhir Kumar Garg. 2019. "Fabrication of Biodegradable Mg Alloy Bone Scaffold through Electrical Discharge M-Drilling Route." In *International Scientific-Technical Conference Manufacturing*, 145–155. Springer.

Ali, Raneen Abd, Mozammel Mia, Aqib Mashood Khan, Wenliang Chen, Munish Kumar Gupta, and Catalin Iulian Pruncu. 2019. "Multi-Response Optimization of Face Milling Performance Considering Tool Path Strategies in Machining of Al-2024." *Materials* 12 (7): 1013.

Arora, Ravish, Jatinder Kapoor, and Rakesh Chandmal Sharma. 2021. "Development of Microwave Hybrid Heating Welded Joints of Inconel-600 Superalloys Using Grey Relational Analysis." *International Journal of Vehicle Structures and Systems* 13 (1): 53–59. https://doi.org/10.4273/ijvss.13.1.11

Box, George E. P., and Norman R. Draper. 2007. *Response Surfaces, Mixtures, and Ridge Analyses*. John Wiley & Sons.

Brauers, Willem Karel, and Edmundas Kazimieras Zavadskas. 2006. "The MOORA Method and Its Application to Privatization in a Transition Economy." *Control and Cybernetics* 35 (2): 445–469.

Dutta, Sourav, K. Bavya Devi, Santanu Mandal, Arnab Mahato, Sanjay Gupta, Biswanath Kundu, Vamsi Krishna Balla, and Mangal Roy. 2019. "In Vitro Corrosion and Cytocompatibility Studies of Hot Press Sintered Magnesium-Bioactive Glass Composite." *Materialia* 5:100245.

Etri, Hamza E. L., Anil Kumar Singla, Mehmet Tayyip Özdemir, Mehmet Erdi Korkmaz, Recep Demirsöz, Munish Kumar Gupta, J. B. Krolczyk, and Nimel Sworna Ross. 2023. "Wear Performance of Ti-6Al-4 V Titanium Alloy through Nano-Doped Lubricants." *Archives of Civil and Mechanical Engineering* 23 (3): 147. https://doi.org/10.1007/s43452-023-00685-9

Gupta, Pankaj, and Bhagat Singh. 2020. “Local Mean Decomposition and Artificial Neural Network Approach to Mitigate Tool Chatter and Improve Material Removal Rate in Turning Operation.” *Applied Soft Computing* 96:106714.
Gupta, Munish Kumar, P. K. Sood, Gurraj Singh, and Vishal S. Sharma. 2018. “Investigations of Performance Parameters in NFMQL Assisted Turning of Titanium Alloy Using TOPSIS and Particle Swarm Optimisation Method.” *International Journal of Materials and Product Technology* 57 (4): 299–321.
Hart, Gus L. W., Tim Mueller, Cormac Toher, and Stefano Curtarolo. 2021. “Machine Learning for Alloys.” *Nature Reviews Materials* 6 (8): 730–755.
Işık, Ahmet Tolunay, Ramazan Çakıroğlu, and Mustafa Günay. 2023. “Multiresponse Optimization of Performance Indicators through Taguchi-Grey Relational Analysis in EDM of Cemented Carbide.” *CIRP Journal of Manufacturing Science and Technology* 41:490–500.
Jangra, Kamal, Sandeep Grover, and Aman Aggarwal. 2011. “Simultaneous Optimization of Material Removal Rate and Surface Roughness for WEDM of WC-Co Composite Using Grey Relational Analysis along with Taguchi Method.” *International Journal of Industrial Engineering Computations* 2 (3): 479–490.
Khanna, Rajesh, Anish Kumar, Mohinder Pal Garg, Ajit Singh, and Neeraj Sharma. 2015. “Multiple Performance Characteristics Optimization for Al 7075 on Electric Discharge Drilling by Taguchi Grey Relational Theory.” *Journal of Industrial Engineering International* 11:459–472.
Khanna, Rajesh, Neeraj Sharma, and Rahul Dev Gupta. 2019. “Investigations of the Effect of Parameters on the CR and SR during Al6063/SiC/Ti at WEDM.” *Applied Mechanics and Materials* 895:181–186.
Kumar, Anish, and Renu Sharma. 2020. “Multi-Response Optimization of Magnetic Field Assisted EDM through Desirability Function Using Response Surface Methodology.” *Journal of the Mechanical Behavior of Materials* 29 (1): 19–35.
Mia, Mozammel, Md Asaduzzaman Rifat Farhan Tanvir, Munish Kumar Gupta, Md Jubayer Hossain, and Anjan Goswami. 2018. “Multi-Objective Optimization of Chip-Tool Interaction Parameters Using Grey-Taguchi Method in MQL-Assisted Turning.” *Measurement* 129:156–166.
Montgomery, D. C. 2017. *Design and analysis of experiments*. John Wiley & Sons: New York.
Nur, Rusdi, Noordin Mohd Yusof, Izman Sudin, Fethma M. Nor, and Denni Kurniawan. 2021. “Determination of Energy Consumption during Turning of Hardened Stainless Steel Using Resultant Cutting Force.” *Metals* 11 (4): 565.
Roy, R. S.. 2001. *Design of Experiments (DOE) Using the Taguchi Approach:: 16 steps to Product and Process Improvement*. John Wiley & Sons: New York.
Saffaran, Ali, Masoud Azadi Moghaddam, and Farhad Kolahan. 2020. “Optimization of Backpropagation Neural Network-Based Models in EDM Process Using Particle Swarm Optimization and Simulated Annealing Algorithms.” *Journal of the Brazilian Society of Mechanical Sciences and Engineering* 42:1–14.
Sarıkaya, Murat, Munish Kumar Gupta, Italo Tomaz, Danil Yu Pimenov, Mustafa Kuntoğlu, Navneet Khanna, Çağrı Vakkas Yıldırım, and Grzegorz M. Krolczyk. 2021. “A State-of-the-Art Review on Tool Wear and Surface Integrity Characteristics in Machining of Superalloys.” *CIRP Journal of Manufacturing Science and Technology* 35:624–658.
Sharma, Neeraj, Neeraj Ahuja, Rachin Goyal, and Vinod Rohilla. 2020. “Parametric Optimization of EDD Using RSM-Grey-TLBO-Based MCDM Approach for Commercially Pure Titanium.” *Grey Systems: Theory and Application* 10 (2): 231–245.
Sharma, Neeraj, Rajesh Khanna, and Rahul Dev Gupta. 2015. “WEDM Process Variables Investigation for HSLA by Response Surface Methodology and Genetic Algorithm.” *Engineering Science and Technology, an International Journal* 18 (2): 171–177.

Sharma, N., R. Khanna, Y.K. Sharma, and R.D. Gupta. 2019. "Multi-Quality Characteristics Optimisation on Wedm for Ti-6al-4v Using Taguchi-Grey Relational Theory." *International Journal of Machining and Machinability of Materials* 21 (1–2): 66–81.

Sharma, G., K. Kumar, P. S. Satsangi, and N. Sharma. 2021. "Surface Modification of Biodegradable Mg-4Zn Alloy Using PMEDM: An Experimental Investigation, Optimization and Corrosion Analysis." *Irbm* 1:1–14. https://doi.org/10.1016/j.irbm.2021.02.003

Sharma, Neeraj, Vishal S. Sharma, Rakesh Chandmal Sharma, Rishu Arora, and Abhishek Sharma. 2022. "Development of Quality Microholes by Electrical Discharge Drilling on Al/SiC Composite Using of Grey-Desirability Approach." *International Journal of Lightweight Materials and Manufacture* 5 (2): 267–277.

Sharma, Pardeep, Neeraj Sharma, Gurpreet Singh, and Vishal Dabra. 2020. "Abrasive Wear Study of AA7075/ZrB 2 Reinforced Composites." *Refractories and Industrial Ceramics* 60:506–509.

Singh, Gurpreet, Sanjeev Goyal, Georgina Miranda, and Neeraj Sharma. 2018. "Parametric Study of the Dry Sliding Wear Behaviour of AA6082-T6/SiC and AA6082-T6/B 4 C Composites Using RSM." *Journal of Mechanical Science and Technology* 32:579–592.

Singh, Gurpreet, Neeraj Sharma, Sanjeev Goyal, and Rakesh Chandmal Sharma. 2021. "Comparative Measurements of Physical and Mechanical Properties of AA6082 Based Composites Reinforced with B 4 C and SiC Particulates Produced via Stir Casting." *Metals and Materials International* 27:4333–4345.

Singh, Gurpreet, Neeraj Sharma, Pankaj Kumar Singh, Ismail Hossain, and Shaman Gupta. 2024. "Development and Tribological Investigation of B4C Reinforced AA5082 Aluminum Alloy Cast Composites for Automotive Applications." *Engineering Research Express* 6 (1): 15073.

Valizade, Nima, and Zoheir Farhat. 2024. "A Review on Abrasive Wear of Aluminum Composites: Mechanisms and Influencing Factors." *Journal of Composites Science* 8 (4): 149.

7 Tribology of Metal Matrix Composites

A Comprehensive Review

Kanwarjeet Singh, Gaurav Arora, Pankaj Kumar Singh, Vaibhav Chandra, Praveen Kumar, and Kunal Arora

7.1 INTRODUCTION

Composites are materials that are formed by compounding two or more different constituent materials with different physical or chemical properties to obtain better performance compared to the individual materials. They are classified into several categories based on their matrix material as polymer matrix composites (PMCs), ceramic matrix composites (CMCs), and metal matrix composites (MMCs). MMCs are a class of advanced materials that are made up of a tough metal or metallic alloy matrix, for example, aluminum, copper, magnesium, or nickel alloys, which are reinforced with hard particles, whiskers, or fibers like ceramics (Al_2O_3, SiC, and TiC) or solid lubricants (MoS_2 and graphite). These materials are replacing the pure metals and alloys, as they can be tailored and possess high wear resistance, hardness, and tensile strength, with a high strength-to-weight ratio. Additionally, the MMCs have gained suitable applications in the fields of defense, sports, automotive, marine, construction, aerospace, and various engineering sectors. (Rohatgi 1993, Ramnath et al. 2014, Harrigan 1998, Canter 2016). Nowadays, the emergence of novel materials and their characterization has gained significant interest from researchers in the field of material science. The characterization of novel materials involves analyzing their mechanical, physical, thermal, corrosion, and tribological properties to determine a range of suitable applications.

7.2 PRINCIPLES OF TRIBOLOGY

Tribology is the science and technology of surfaces that come into contact with each other while in relative motion, with particular attention to problems of friction, wear, and lubrication. It is an interdisciplinary field that blends concepts from material science, mechanical engineering, physics, and chemistry to understand how surfaces behave when they slide, rub, or roll against each other. At its core, tribology emphasizes three phenomena: friction, wear, and lubrication. Friction can be defined as the resistive force that arises when two contacting surfaces slide against one another.

DOI: 10.1201/9781003620570-7

On the other hand, the deformation or gradual removal of any material due to mechanical action is known as its wear. The third element, lubrication, includes the application of a substance, generally a solid or liquid, to minimize the wear and friction between two contacting surfaces (Kajdas et al. 1990, Miyoshi and Chung 1993).

Tribology also considers the effect of types of contact between surfaces (such as rolling or sliding), material compatibility, surface roughness, and surrounding conditions, namely, humidity and temperature, on the performance of the material. Further, whether the contact surfaces are dry or lubricated also plays a critical role in the tribological characteristics of contacting materials (Booser 1997). The selection of the right materials to enhance performance and reduce the failure rate relies on the tribological principles. The implementation of surface coating techniques is widely utilized to boost the tribological characteristics, particularly for fast speeds or high-load environments. So, understanding the concepts of tribology is essential for designing energy-efficient, durable, and cost-effective systems. With the advancement of technology, the field of tribology keeps advancing and plays a vital role in areas like micro- and nanotechnology (Bhushan 1999).

Frictional behavior in MMC materials is more complicated as it is impacted by the type of reinforcement (particulate or fiber, etc.) and matrix metal. The friction coefficient in MMCs depends upon the volume fraction, orientation, and interface of the reinforcement with the metal matrix. Further, the wear of the metal matrix involves the wear of the matrix and reinforcement, along with the fatigue and delamination. In MMC tribology, lubrication plays a vital role in reducing the value of wear and friction, especially under heavy load and speed conditions (Wilson and Ball 1993, Hutchings 1994). Numerous studies in the literature examined the tribological behavior of MMCs, emphasizing the importance of choosing the best combination of matrix metal and reinforcement for the intended applications.

7.3 TRIBOLOGICAL BEHAVIOR OF METAL MATRIX COMPOSITES (MMCS)

MMCs are advanced materials composed of a metallic alloy matrix, such as magnesium, aluminum, copper, and titanium, reinforced with hard ceramic particulates or fibers such as alumina, silicon carbide, and boron nitride. One of the most important characteristics of MMCs is their tribological behavior, which refers to their performance under rolling, sliding, or abrasive contact conditions involving wear and friction. The tribological behavior of MMCs is intricate, with dependencies on parameters such as the matrix type, reinforcement nature, fabrication method, and operating conditions, such as load and temperature. Literature does contain extensive studies on the tribological performance of MMCs.

The evaluation of tribological characteristics of Al-Cu/SiCw and Al-Si/SiCp has been carried out by Prasad and McConnell (1991). The composites reinforced with SiCw and SiCp were fabricated by powder metallurgy and casting techniques, respectively. A graphite film was burnished on the polished samples of fabricated composites and corresponding matrix alloys. The results of the study exhibited that a substantial reduction in wear of developed composites was observed with the inclusion of reinforcements. The lubricant graphite films formed on Al-Cu alloys revealed a higher load-bearing capacity in comparison to those formed on Al-Si alloys. In another study, Lo et al. (1992) examined the mechanical and tribological performance of ZA-27/Al_2O_3f developed by the squeeze casting route. The wear behavior of unreinforced alloy and fabricated composites was evaluated under lubricated conditions. The outcome of research exhibited an improved wear of fabricated composites in comparison to ZA-27 alloy.

Saka et al. (1992) studied the wear and friction characteristics of Al/Gr, Al/steel, and Al-Li/alumina composites. The author performed wear testing under dry sliding conditions across normal, longitudinal, and transverse orientations. The best wear-resistant properties were exhibited by Al/Gr composites due to the presence of graphite. The wear of Al-Li/Al_2O_3 composites was due to the abrasion and pulling out of fiber phenomena. The tribological study of varying percentages of Al_2O_3 fiber-reinforced aluminum and magnesium-based MMC was conducted by Alahelisten et al. (1993). The author stated that the wear behavior of composites was affected by the matrix metal and tribosystem. An optimum value of sliding wear resistance was obtained with 10% of fiber content. The presence of fibers led to a reduction in adhesive wear, but an increase in delamination wear was noted by the author. The results also exhibited no effect of fiber reinforcement on the abrasive wear rate of aluminum-based composites, whereas abrasive wear resistance was improved with the inclusion of fiber content.

The wear behavior of Zn-Al-Cu/carbon fiber composites developed by the squeeze casting technique was analyzed by Zhu et al. (1996). The present study carried out dry and lubricated wear testing of fabricated composites, considering different loads and fiber volume contents. Under the dry sliding environment, the wear and friction of developed composites improved with the increased fraction of carbon fibers; in contrast, the coefficient of friction of Zn/Cf composites increased with the inclusion of fiber content for lubricated environment testing.

Since the MMCs possess improved mechanical and microstructural characteristics, the understanding of wear and friction mechanisms of MMCs is essential to decide the final performance of fabricated composites. Applications can be selected based on the final characteristic properties of MMCs.

7.4 MECHANISMS OF FRICTION AND WEAR IN METAL MATRIX COMPOSITES (MMCS)

MMCs are engineered materials that have suitable applications in several industrial and engineering sectors. The analysis of the influence of wear and friction plays a critical role in determining the life of a product and improving the product's performance in a sliding environment. Primarily, the mechanism of wear and friction in MMCs is a complex phenomenon, as it is impacted by the way the matrix and reinforcement respond under the contact conditions. Additionally, it is affected by the type of matrix and reinforcement, composition, speed of sliding, load applied, and environment.

The friction characteristics in MMCs are controlled by abrasion, adhesion, ploughing, and tribo-layer mechanisms. Generally, the reinforcements are harder than the matrix materials or counter surface elements. During the sliding conditions, these hard reinforcements sometimes wear out the counter surface (two-body abrasive mechanism) or behave as free particles between the contact surfaces (three-body abrasive mechanism). Both of these mechanisms resulted in an enhancement of friction and led to the improved resistance to wear.

Moreover, the adhesive friction mechanism includes the tiny atomic bonding between the contacting surfaces. The presence of the hard reinforcements in MMCs may decrease the contact area, resulting in a reduction in the adhesive atomic bonding of contacting surfaces. The ploughing mechanisms involve the inclusion of hard reinforcements into mating surfaces, which contributes to deformation or micro cutting of the counter surface or softer matrix, and also results in increased wear and friction. When two surfaces slide with each other, due to the combined effect of the squeezing of debris and oxidation, a tribo layer, or mechanically mixed layer (MML) is formed on the contact surface. This layer reduced the contact area and acts as a protective layer, and finally reduces the coefficient of friction.

Furthermore, the various types of wear mechanisms that occur in MMCs are adhesive, abrasive, oxidative, and delamination wear. Adhesive wear comes into play when there is the formation of small junctions between the counter surface and matrix under the applied load, resulting in tearing or material transfer. The abrasive wear mechanism in MMCs contains the cutting of matrix material due to the sliding of hard particles of counter surface material. On the other hand, at elevated temperatures, due to the oxidation phenomena, there is a formation of an oxide layer on the surface of the composite. The frequent formation and breaking of this oxide layer resulted in oxide wear debris known as oxidative wear. Moreover, delamination wear in MMCs results from the formation and development of cracks by the repeated cyclic loading. This type of wear is affected by the dissimilar characteristics (strength, modulus, and thermal expansion) of reinforcement and alloy matrix. The literature has shown various research findings demonstrating the different wear and friction mechanisms that occur in MMCs during investigations.

The examination of wear and friction characteristic properties of Al/Gr, Al/SS, and Al-Li/Al_2O_3 MMCs was performed by Saka et al. (1992). The testing samples were fabricated with transverse, longitudinal, and normal orientations of fiber reinforcements and tested for wear and friction under a dry sliding environment. The results exhibited the least wear and friction by Al/Gr MMC due to the formation of a graphite layer at the contact surface. The dominant wear mechanism was fiber pull-out for Al/Gr and Al/SS composites, whereas for Al/Al_2O_3 MMC, it was abrasive and fiber pull-out. The microstructural analysis showed an uneven distribution and weak interfacial bonding, resulting in elevated friction and wear.

The influence of counter surface material and microstructure on the wear behavior of Al2024/SiC and Al6061/Al_2O_3 MMCs was investigated by Alpas and Zhang (1994). Dry sliding wear testing with varying load and constant sliding velocity conditions was carried out. The outcome of research revealed that with the increase in volume fraction and particle size of reinforcements, an enhancement in wear resistance of fabricated composites was observed. The composite samples tested against the counter surface elements having steel have shown a better wear resistance in comparison to the counter surface elements having mullite due to the development of iron-rich oxide film. Further, the rate of wear of MMCs is enhanced with the increasing values of load, as it leads to the fracturing of Al_2O_3 and SiC particulates.

Furthermore, Narayan et al. (1995) studied the friction and wear properties of Al2024/Al_2O_3 MMCs fabricated via the stir casting technique, hot extruded, and T6 heat treated. A dry sliding test was carried out at room temperature and constant sliding velocity. The hot extruded composite has shown a poor resistance to wear, whereas T6 heat-treated composites have shown an improved wear resistance as compared to the unreinforced matrix alloy. This was due to the fracturing of particulates of Al_2O_3, which resulted in the development of cracks and material removal during sliding tests.

The wear mechanism transformations of Al/SiC_w composites developed via the squeeze casting route were analyzed by Wang et al. (1995). The sliding wear testing of fabricated composites was carried out without any lubrication. The results exhibited a wear transformation of fabricated composites from mild to severe and finally seizing with the increment of load and speed of sliding. The fracturing of SiC_w was observed for high values of load and severe wear scenarios. However, the wear and friction characteristics of Al-Cu/SiC composites under the varying load and reciprocal velocity were examined by Bai et al. (1996). The dry sliding wear testing against three $W_{18}Cr_4V$, SN, and 4Cr13 types of counter surface was carried out. The delamination, adhesive (intermediate, mild, and severe), and mild melt mechanism of wear were observed for the Al-Cu-SiC/T1 developed composites during testing. Furthermore, mild melt and adhesive (intermediate, mild, and severe) wear mechanisms for Al-Ci-SiC/SN composites were shown in the results. The results also exhibited that the dominant wear mechanism was the abrasive one.

The impact of matrix hardness and varying applied load on the tribological properties of alumina particulate-reinforced Al6061 composites was studied by Straffelini et al. (1997). The matrix hardness was varied by exposing the composite samples to forging and thermal treatments. During testing, a high wear resistance was shown by the extruded composite samples in comparison to the heat-treated ones due to the occurrence of subsurface hardening phenomena. The heat-treated and forged composite samples have shown a delamination wear mechanism in the research findings. Moreover, the wear characteristics of Al6061/ Al_2O_3 and Al359/SiC extruded composites were analyzed by Ye et al. (1998). The wear testing of the manufactured composite samples was carried out under a lubricated environment. The research findings revealed polishing, fatigue, and abrasive wear mechanisms in the MMCs. Further, the Al6061/Al_2O_3 composites have shown a better resistance to wear in contrast to Al359/SiC composites.

The study of the wear mechanism of the tool cutting the various types of Al356/SiC composites was performed by Yanming and Zehua (2000). The outcome has shown abrasive wear on the edge and tool flank of tools while cutting the composites. Additionally, composites with a higher fraction and coarser size of SiC have shown severe wear on the flank and corner of the tool. Moreover, the composites having the fine size of SiC reinforcement were adaptable for cutting at high speeds. Besides this, the wear characteristics of granite-reinforced ZA-27 composites manufactured via the stir casting approach were investigated by Ranganath et al. (2001). The slide wear testing without any lubrication environment against an EN24 counter surface element was conducted. The outcome of the investigation states that the developed composites indicated an improved wear resistance with the added content of granite.

The micrographs of tested composite samples at high values of load demonstrated that subsurface cracks began, multiplied, developed, and finally joined to contribute to delamination wear. Additionally, at low and high values of applied load, composite samples indicated abrasive and delamination wear mechanisms, respectively.

The wear attributes of AZ91/SiC_p fabricated utilizing the powder metallurgy technique were studied by Lim et al. (2003). The dry sliding wear tests were performed by varying the values of load and speed of sliding. At the low values of load, the oxidation wear mechanism was dominant for the testing composite specimens. With the increased value of load, the oxidation wear mechanism transformed into abrasive and delamination wear mechanisms and resulted in more wear of composite samples. Furthermore, the wear mechanism transformed to adhesive for higher values of speed and applied load values. Finally, with the further increased value of high speeds, softening and melting wear mechanisms of composite samples were observed. In addition to that, Basavarajappa et al. (2006) compared the wear properties of SiC and graphite particulate-reinforced Al2219 hybrid composites with the base alloy matrix developed through the stir casting method. The ASTM G99-95 standards were followed for the wear testing of composite specimens. The results indicated an improved wear resistance of hybrid composites in contrast to the base alloy matrix due to the formation of a graphite layer at the contact surface. The wear of the hybrid composite reduced due to the presence of hard SiC particulates and the formation of MML with the increasing values of load and speed. The wear testing of composite samples also showed the abrasive wear mechanism at low load and speed values, whereas the delamination mechanism of wear was observed at high speed and load values, respectively.

The tribological attributes of turning of A356/SiC composites were analyzed and discussed by Kannan and Kishawy (2008). A carbide tool coated with tungsten was utilized for the turning of fabricated composite samples. The wear study was carried out for both wet and dry environments. The abrasive wear of the tool was higher for dry conditions, whereas with the usage of cutting fluid, the value of surface roughness of composites declined. This diminishing of surface roughness was due to the formation of voids and pitholes, as the cutting fluids contributed to clearing out of debonded particles. To add to this, the investigation of wear mechanisms of Al/AlB_2 flakes in situ manufactured composites was performed by Ficici et al. (2011). A dry environment was maintained for wear testing experiments of the developed composite specimens. The results demonstrated an enhancement in resistance to wear of composite samples due to the addition of reinforcement. The rate of wear declined and increased with the increment of speed of sliding and distance of sliding, respectively. As per the images of the worn surface, the adhesive and abrasive mechanism of wear was dominant.

In another research work, the wear characteristics of fine and coarse rutile mineral-incorporated LM13 composites fabricated by the stir casting approach were examined by Arora et al. (2015). The wear testing experiments were performed at varying temperature values. The composite samples having fine-sized reinforcements have shown superior wear resistance properties due to the occurrence of morphological refinement of silicon. Besides this, the wear transition for both types of composites was delayed because of the formation of an oxide layer at the interacting surfaces. The dominance of oxidative wear was prominent in the micrographs of SEM.

Moreover, the effect of varying temperature on the wear attributes of nano-sized Al2219/B4C and Al2219/B_4C/MoS_2 hybrid composites developed via the stir casting method was evaluated by Kumar et al. (2020). The micrograph images of SEM have shown an even distribution of reinforcements in the alloy matrix. The research findings of the study indicated a reduction in wear rate of composites with the increment of temperature values, as oxide films were developed at the contact surfaces. Deep and wide tracks of wear were observed for the worn surfaces of the base alloy matrix; in contrast, it was less deep for composite specimens, as shown in the images. Delamination and plastic deformation as mechanisms of wear were observed for the Al2219 alloy at low values of temperature, whereas at high values of temperature, abrasive wear was the dominant mechanism for fabricated composites.

From the above analysis, it can be concluded that wear and friction attributes of MMCs need a perfect balance of tribological analysis and material engineering. Out of the available parameters, the role of reinforcement is very crucial in improving the tribological behavior of MMCs.

7.5 INFLUENCE OF REINFORCEMENTS ON TRIBOLOGICAL PERFORMANCE OF METAL MATRIX COMPOSITES (MMCS)

Technically, it is the type of reinforcement that decides the overall performance of the developed MMCs. The inclusion of these hard reinforcements in MMCs principally improves the mechanical attributes of MMCs, which is very much required for enhancing the tribological performance under abrasive and sliding environments. Generally, the various carbides, nitrides, and oxides of various elements in the form of particulates, fibers, whiskers, and flakes are utilized as reinforcements to enhance the superiority of MMCs.

The reduction of wear of MMCs is the most significant effect of the incorporation of reinforcements, as these reinforcements behave as the load-bearing elements in it. Additionally, these reinforcements act as a barrier between the contacting surfaces and reduce the wear rate of MMCs. Moreover, the coefficient of friction is also affected by the inclusion of reinforcements, as sometimes it contributes to the development of a lubricating layer or films among the interacting surfaces and finally results in an improved wear resistance. Basically, the degree of this enhanced performance totally depends on the type, size, structure, and volume content of included reinforcements.

The impact of the inclusion of TiB_2 particulate reinforcements on the wear characteristics of SS304/TiB_2 composites manufactured via hot isostatic processing was investigated by Tjong and Lau (1999). The results clearly indicated that the wear resistance of composites was enhanced with the addition of TiB_2 particulates and an increase in the values of applied load and velocity of sliding. SEM micrographs depicted the fracturing and distribution of TiB_2 particles, resulting in the increment of microhardness of sub-surfaces, which finally contributed to the reduction of wear rate of MMCs. In addition to that, the effect of the incorporation of nano-sized SiC and B_4C on the wear performance of AA6061 hybrid composites developed through the stir casting route was examined by Das et al. (2019). The outcome of the research demonstrated a decrease in friction coefficient and an increase in the rate of wear of composites with the increased load values. Abrasive, adhesive, and oxidative wear mechanisms were dominant in MMCs as per the micrographs.

As per the above literature, the friction and wear mechanism of MMCs is fully dependent on the composition of the material and operating environment. It is significant that for the development of MMCs for a particular application, the selection of type, size, volume, and structure of reinforcement is a crucial part of the material design process. Apart from the selection of suitable reinforcement, the treatment of the surface is another popular approach to improve the wear characteristics.

7.6 SURFACE ENGINEERING TECHNIQUE TO ENHANCE TRIBOLOGICAL PERFORMANCE

The treatment of the surface, or surface engineering, is a popular technique utilized to modify surface attributes to enhance its performance under various environments. Basically, all the tribological interactions are carried out at the contact surface; therefore, enhancing the surface properties only is the right approach to increase the durability and performance of products used in various engineering and industrial sectors.

Surface engineering covers all the scientific and technical aspects involved in creating special surface layers on a substrate, whether these layers are made before use, develop during service, lie just beneath the surface, or are applied as coatings and cladding on a substrate material. These hard surface coatings/claddings often have different properties than the base material, tailored to meet specific service needs. The field also involves studying how these surface layers perform and addressing challenges related to their design. By using surface engineering techniques, engineers can extend the life of components, reduce initial costs by using less expensive base materials protected by hard coatings on the surface of soft base materials, and restore worn parts through resurfacing. This is especially important since many engineering components of machines need to function reliably in harsh, demanding, and wear-intensive environments. The surface treatment of MMCs and engineering components can be performed using three different ways, as presented in Figure 7.1. The figure clearly demonstrated that the surface characteristics can be enhanced by

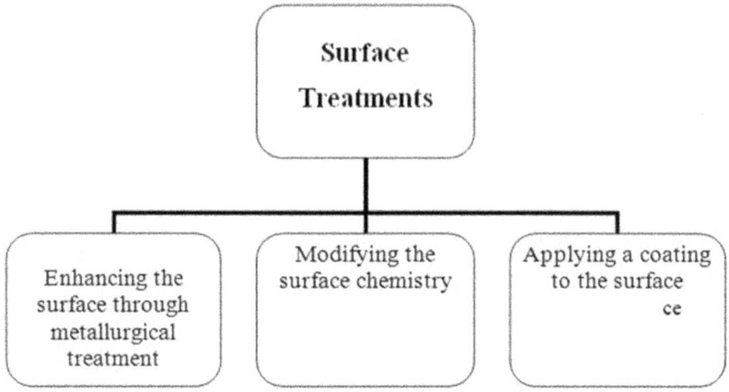

FIGURE 7.1 Surface treatment classification.

altering the metallurgy, surface chemistry, and application of a suitable coating on or of the required surface.

Surface coating and cladding techniques are widely used in industries to improve and strengthen the surface functions without changing the bulk material's composition. These coating methods include physical vapor deposition (PVD), chemical vapor deposition (CVD), and thermal spraying, among others. These coating techniques are also widely used in the tool and die industry, where dies are essential for mass production. The utilization of MMCs and applying such coatings helps extend the service life of the dies. By using these processes, it is possible to create engineered surfaces with superior quality and enhanced properties. Sometimes, surface coatings are combined with other treatments like laser cladding to further boost surface performance (Mo et al. 2025 and Nikam et al. 2025). In addition to these, the microwave cladding by the microwave hybrid heating method is also used for the development of surface coating (Singh and Sharma 2018, 2019). The final surface features—such as texture, chemical makeup, impurity distribution, porosity, hardness, and resistance to heat—depend heavily on the chosen treatment method and how well it achieves the desired surface characteristics.

REFERENCES

Alahelisten, A., F. Bergman, M. Olsson, and S. Hogmark. "On the wear of aluminium and magnesium metal matrix composites." *Wear* 165, no. 2 (1993): 221–226.

Alpas, A. T., and J. Zhang. "Effect of microstructure (particulate size and volume fraction) and counterface material on the sliding wear resistance of particulate-reinforced aluminum matrix composites." *Metallurgical and Materials Transactions A* 25 (1994): 969–983.

Arora, R., S. Kumar, G. Singh, and O. P. Pandey. "Influence of particle size and temperature on the wear properties of rutile-reinforced aluminium metal matrix composite." *Journal of Composite Materials* 49, no. 7 (2015): 843–852.

Bai, M., Q. Xue, and H. Guo. "Reciprocal sliding wear of SiC particle-reinforced Al-Cu aluminium matrix composites against stainless steel, high speed tool steel and ceramics II. Wear mechanisms." *Wear* 194, no. 1–2 (1996): 126–136.

Basavarajappa, S., G. Chandramohan, K. Mukund, M. Ashwin, and M. Prabu. "Dry sliding wear behavior of Al 2219/SiCp-Gr hybrid metal matrix composites." *Journal of Materials Engineering and Performance* 15 (2006): 668–674.

Bhushan, B. *Micro/Nano Tribology. CRC Press* 79 (1999): 7–831.

Booser, E. R. *Tribology Data Handbook: An Excellent Friction, Lubrication, and Wear Resource*. CRC Press, 1997.

Canter, N. "Lightweight self-lubricating metal matrix composites." *Tribology & Lubrication Technology* 72, no. 6 (2016): 18.

Das, S., S. Samanta, and P. Kayaroganam. "Fabrication and tribological study of AA6061 hybrid metal matrix composites reinforced with SiC/B_4C nanoparticles." *Industrial Lubrication and Tribology* 71, no. 1 (2019): 83–93.

Ficici, F., S. Koksal, R. Kayikci, and O. Savas. "Investigation of unlubricated sliding wear behaviours of in-situ AlB_2/Al metal matrix composite." *Advanced Composites Letters* 20, no. 4 (2011): 096369351102000404.

Harrigan Jr, W. C. "Commercial processing of metal matrix composites." *Materials Science and Engineering: A* 244, no. 1 (1998): 75–79.

Hutchings, I. M. "Tribological properties of metal matrix composites." *Materials Science and Technology* 10, no. 6 (1994): 513–517.

Kajdas, C., E. Wilusz, and S. Harvey. *Encyclopedia of Tribology*. Vol. 15. Elsevier, 1990.

Kannan, S., and H. A. Kishawy. "Tribological aspects of machining aluminium metal matrix composites." *Journal of Materials Processing Technology* 198, no. 1–3 (2008): 399–406.

Kumar, N. G. S., R. Suresh, and G. S. Shiva Shankar. "High temperature wear behavior of Al2219/n-B4C/MoS2 hybrid metal matrix composites." *Composites Communications* 19 (2020): 61–73.

Lim, C. Y. H., S. C. Lim, and M. Gupta. "Wear behaviour of SiCp-reinforced magnesium matrix composites." *Wear* 255, no. 1–6 (2003): 629–637.

Lo, S. H. J., S. Dionne, M. Sahoo, and H. M. Hawthorne. "Mechanical and tribological properties of zinc-aluminium metal-matrix composites." *Journal of Materials Science* 27 (1992): 5681–5691.

Miyoshi, K., and Y.-W. Chung. *Surface Diagnostics in Tribology: Fundamental Principles and Applications*. Vol. 1. World Scientific, 1993.

Mo, K., L. Cao, D. Zhao, Y. Long, J. Zhong, J. Zhong, and G. Bi. "Study on the microstructure and wear resistance of laser clad martensitic stainless steel 420 with different content of TiC." *Materials Today Communications* 26 (2025): 112615.

Narayan, M., M. K. Surappa, and B. N. Pramila Bai. "Dry sliding wear of Al alloy 2024-Al_2O_3 particle metal matrix composites." *Wear* 181 (1995): 563–570.

Nikam, N., B. S. Shenoy, K. N. Chethan, L. G. Keni, S. Shetty, and N. S. Bhat. "Advancements in surface coatings for enhancing longevity in hip implants: A review." *Prosthesis* 7, no. 1 (2025): 21.

Prasad, S. V., and B. D. McConnell. "Tribology of aluminum metal-matrix composites: Lubrication by graphite." *Wear* 149, no. 1–2 (1991): 241–253.

Ramnath, B. V., C. Elanchezhian, R. M. Annamalai, S. Aravind, T. S. A. Atreya, V. Vignesh, and C. Subramanian. "Aluminium metal matrix composites: A review." *Reviews on Advanced Materials Science* 38, no. 5 (2014): 55–60.

Ranganath, G., S. C. Sharma, and M. Krishna. "Dry sliding wear of garnet reinforced zinc/aluminium metal matrix composites." *Wear* 251, no. 1–12 (2001): 1408–1413.

Rohatgi, P. K. Metal matrix composites." *Defence Science Journal* 43, no. 4 (1993): 323.

Saka, N., N. K. Szeto, and T. Erturk. "Friction and wear of fiber-reinforced metal: Matrix composites." *Wear* 157, no. 2 (1992): 339–357.

Singh, K., and S. Sharma. "Development of Ni-based and CeO_2-modified coatings by microwave heating." *Materials and Manufacturing Processes* 33, no. 1 (2018): 50–57.

Singh, K., and S. Sharma. "Fabrication and investigation of co-based and CeO 2-modified microwave coatings." *Protection of Metals and Physical Chemistry of Surfaces* 55 (2019): 352–358.

Straffelini, G., F. Bonollo, A. Molinari, and A. Tiziani. "Influence of matrix hardness on the dry sliding behaviour of 20 vol.% Al_2O_3-particulate-reinforced 6061 Al metal matrix composite." *Wear* 211, no. 2 (1997): 192–197.

Tjong, S. C., and K. C. Lau. "Sliding wear of stainless steel matrix composite reinforced with TiB_2 particles." *Materials Letters* 41, no. 4 (1999): 153–158.

Wang, D. Z., H. X. Peng, Jun Liu, and C. K. Yao. "Wear behaviour and microstructural changes of SiCw-Al composite under unlubricated sliding friction." *Wear* 184, no. 2 (1995): 187–192.

Wilson, S., and A. Ball. "Performance of metal matrix composites in various tribological conditions." In *Composite Materials Series*, vol. 8, pp. 311–366. Elsevier, 1993.

Yanming, Q., and Z. Zehua. "Tool wear and its mechanism for cutting SiC particle-reinforced aluminium matrix composites." *Journal of Materials Processing Technology* 100, no. 1–3 (2000): 194–199.

Ye, Z., H. S. Cheng, and N. S. Chang. "Wear characteristics of particle-reinforced aluminum matrix composites in lubricated reciprocating contacts." *Tribology Transactions* 41, no. 3 (1998): 359–367.

Zhu, H. X., S. K. Liu, X. J. Bai, and Z. J. Li. "Sliding wear behavior of planar random zinc-alloy matrix composites." *Wear* 199, no. 1 (1996): 82–88.

8 Lubrication Technology, Additives, and Lubricants

Harjit Singh, Maninder Singh, Amarjeet Kaur, and Bharat Bajaj

8.1 INTRODUCTION

Lubrication technology is at the heart of mechanical engineering and plant operations, the linchpin of industrial development (Mortier and Orszulik 2010). Lubrication, at a basic level, is the process of applying a substance called a lubricant to two or more moving parts to minimize friction and wear. This process helps ensure the strength and durability of mechanical systems, as friction can wear down materials, generate heat, and affect performance negatively. Indeed, studies show that overcoming frictional forces accounts for about 30% of total energy used during production processes, making an efficient lubricating strategy essential (Mang and Dresel 2007, Czichos 2009, Ahmed and Nassar 2013).

Lubricants serve two main roles: first, they generate a protective film between surfaces in contact, and second, they prevent direct metal-to-metal contact. Hydrodynamic, boundary, and mixed lubrication are regimes that create this film (Spikes and Olver 2003, Zhang 2006, Brandão et al. 2012). This hydrodynamic lubrication occurs when a full fluid film separates the surfaces, and boundary lubrication occurs when a thin film is present and does not come in full contact, hence reducing wear. The specific applications and requirements for each regime will depend on the operating conditions, such as load, speed, and temperature (Totten 2017, Khonsari and Booser 2017).

Lubricants are classified into different types based on their physical state and composition, as shown in Figure 8.1 (Rudnick 2020). Liquid lubricants (also known as oils) are the most common and widely used for continuous lubrication applications. However, they can be obtained from mineral, synthetic, or vegetable sources, each offering different benefits in terms of functional properties and ecological footprint. In extreme conditions where liquid lubricants fail, solid lubricants, like graphite and molybdenum disulfide (MoS_2), offer viable solutions (Vazirisereshk et al. 2019; Clauss 2012). However, greases, semi-solid lubricants formed by mixing oils with thickening agents, are particularly well-suited for applications in hard-to-reach areas or locations where minimum leakage is required (Kayode et al. 2023).

 DOI: 10.1201/9781003620570-8

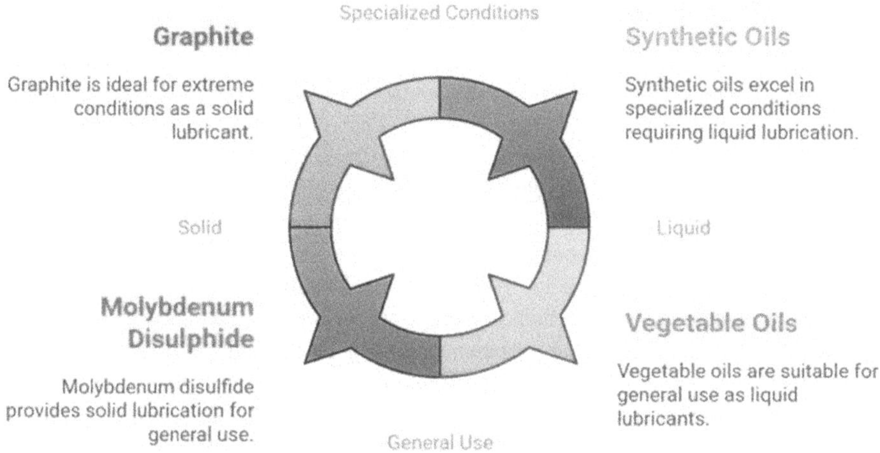

FIGURE 8.1 Lubricant categorization by state and application.

Lubricant application efficiency has been further improved by the development of advanced lubrication systems. However, automated systems facilitate accurate distribution and tracking of lubricants at several locations in equipment and machinery, minimizing human error and promoting optimal equipment and machinery functioning (Piechowski et al. 2024). Lubrication systems come in many forms, from manual grease pumps to advanced, automated systems that continuously monitor lubrication requirements in real time.

Additionally, the importance of additives in lubricants cannot be emphasized enough. Additives improve the performance of the base lubricant by offering additional characteristics like enhanced wear resistance, oxidation stability, and corrosion protection (Shahnazar et al. 2016). Additives, like, anti-wear agents, develop protective films on metallic surfaces that help reduce wear in extreme load conditions. The process of these additives plays a crucial role in improving the lubricant performance for various industrial applications.

As technological advancements continue to evolve in the industry and efficiency and sustainability trends persist, lubrication technology will play an ever-larger role. Newer technologies, such as self-lubricating composites, are showing promise for the combination of material science and tribology that allows for the development of components that resist wear by design, rather than by external lubricants (Zhang and Chromik 2022).

Lubrication technology is a broad field that involves the use of various materials, systems, and methods for minimizing friction and improving equipment operation. Engineers and operators need to acknowledge the fundamentals of lubrication as well as the types of lubricants available, as both play a critical role in the effective operation and lifespan of equipment. The future of lubrication technology will bring with it exciting advances, as research continues to press forward and new materials are developed to advance the next generation of lubrication systems that not only make the industry more efficient, but also continue to address the real environmental impact lubricants have on the industry.

8.2 TYPES OF LUBRICANTS

Lubricants, which are substances that help lessen the impact between solid components, are vital components in most machines because they improve performance and stability and increase their lifespan. They are usually divided into categories by physical state, composition, and specific applications. Familiarity with these types is necessary to choose the correct lubricant for diverse industrial processes.

Liquid lubricants are widely used, especially in general industry applications. Mineral oils are the conventional lubricants derived from crude oil. Refineries can be optimized to yield a range of grades, from paraffinic to naphthenic to aromatic oils. Each type has unique properties that render it appropriate for certain conditions. Synthetic oils are lubricants that are engineered for better performance than mineral oil (Rudnick and Bartz 2020). Synthetic oils can be designed to have specific properties like exceptional thermal stability, low volatility, and better lubrication in harsh environments, for example, polyalphaolefins (PAOs) and esters. Natural oils (made of vegetable- and animal-based oils) have become the most famous due to their biodegradable nature and environmental friendliness. Some examples of these are castor oil, palm oil, tallow oil, etc. However, they do not always surpass synthetic options, but are still commonly used in situations where being environmentally friendly is a priority. Since they are also used in machining operations, they help in dissipating the heat while providing lubrication.

Greases are semi-solid lubricants, usually a base oil thickened with a soap or other thickener. They have multiple benefits compared to liquid lubricants. Greases have a consistent stability, making them much easier than liquids to adhere to a surface, and ideal for applications where leakage would not be preferable (Lugt 2016). Greases have a thicker consistency that acts as a barrier to dust and moisture, thus preventing the components from corrosion and wear (Silverstein and Rudnick 2017). Greases are formulated using various thickeners (e.g., lithium, calcium, or sodium soaps) and additives to improve properties like extreme pressure resistance or oxidation stability. Greases are often utilized in bearings, gears, and other machine applications in which a continuous seal of lubricant is difficult to maintain.

The second type of solid lubricant is where solid materials reduce friction not by forming a fluid film but by being present physically. In extreme conditions where it can fail, they are especially advantageous. Commonly in use due to its layered structure, graphite functions as a solid lubricant owing to its resistance to high temperatures as well as high pressures, all while providing low friction (Kumar et al. 2022). Under high-load scenarios, molybdenum disulfide (MoS_2) is a solid lubricant that performs well and is typically found in use among heavy machinery and aerospace applications. Lubricants use other materials too; boron nitride (BN) and polytetrafluoroethylene (PTFE) are also good solid lubricants. They may be used as a coating or blended into other lubricants to optimize performance. They are less common than liquid or solid lubricants but are used in some applications. Air is by far the most common gas lubricant and is widely used in pneumatic systems to cushion moving parts. In specialized environments, other gases may take on similar roles. The right lubricant type also helps in enhancing machine efficiency for optimal performance at all times. Greases make up only a small part of most industrial environments, as the liquid lubricants cover most applications with their flexibility, yet there are also applications where greases are necessary for their benefits of stability and protection against contamination. Under extreme conditions, for example, when extreme temperatures or pressure are needed, solid lubricants can have unrivaled benefits over traditional approaches. Knowing these classifications helps engineers and technicians determine appropriate lubrication methods for their operating environment.

8.3 ROLE OF ADDITIVES IN LUBRICANTS

The introduction of additives greatly improves the functional properties of lubricants in industrial applications. These chemical compounds (organic or inorganic) are blended with base oils to enhance their overall performance, service life, and applications, as shown in Figure 8.2 (Sanchez et al. 2005). Base oil is essential to lubricate the sensitive parts of an engine, but it is an additive that changes the properties of base oil according to the requirements of modern machinery.

Additives are known for improving the inherent characteristics of base oils. Antioxidants, for example, play a critical role in oxidation prevention, a process that results in the degradation of oil quality as time goes by Kiokias et al. (2008). Harmful by-products, including acids and sludge compounds, can be produced through oxidation that may corrode metal interfaces and increase viscosity. This allows the manufacturers to add antioxidants to lubricants that would increase the operational life of the oils at their upper working range, preventing them from degrading. Anti-wear agents are another critical class of additives that create protective films on metal surfaces to reduce wear while a lubricant is in operation (Agarwal et al. 2013). Many anti-wear additives are used, and common examples are zinc dialkyl dithiophosphate (ZDDP) and molybdenum compounds. Their role is to lubricate the metal surfaces to reduce friction between moving parts, avoiding surface damage and enabling longer service life of machinery.

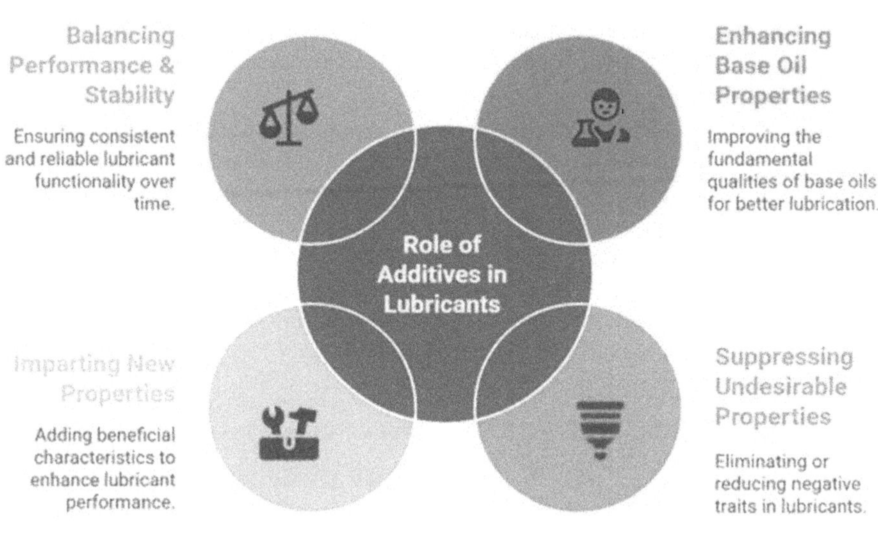

FIGURE 8.2 Optimizing lubricants through strategic additive applications.

To suppress undesirable characteristics of base oils, additives are also used. As an example, some base oils can foam depending on certain operating conditions, leading to inadequate lubrication and increased wear. This problem is reduced by the addition of anti-foam agents, which limit the formation of bubbles by reducing surface tension. Demulsifying agents serve a related purpose, allowing water to be extracted from underlying oil, ensuring that moisture does not reduce the effectiveness of lubrication (Adetunla et al. 2023). In addition, viscosity modifiers play an important role in helping a lubricant retain its viscosity over a wide range of temperatures (Wan et al. 2016). They prevent lubricants from thinning or thickening to enable lubricants to work correctly in both high- and low-temperature conditions. It allows for the possibility of adjusting to different thermal environments, which is critical for the functioning of machines that operate under different conditions.

Additives not only improve and inhibit the characteristics of lubricants, but also provide new properties. Detergents and dispersants, for example, are crucial for keeping engine internals clean, as they prevent deposits and sludge buildup (Bardasz and Lamb 2017). Detergents counteract the acidic by-products formed during combustion, and dispersants keep soot and other contaminants suspended in the oil, rather than allowing them to settle on engine parts. Another innovative category of additives is friction modifiers, used to reduce surface-to-surface friction without developing a complete fluid film. This beneficial adjustment is most effective under boundary lubrication conditions, such as in gears and bearings with high loads.

Formulation of lubricant additives is therefore a complex science, considering many factors in order to achieve optimum performance. Additives make up approximately 0.1% to 30% of the total volume of lubricating oil, depending on the specific application requirements. Different additives must be balanced to ensure that they are compatible and stable in a lubricant system. One of the most important goals is to prevent chemical destabilization of the additives used; the processes must not degrade over time (Xu et al. 2018). It is also essential that the additives be compatible with the base oil to create a uniform blend where all the components can function in their proper manner. The additives are indisputable in advancing the function of the lubricants in numerous industrial applications (Adesoji et al. 2024). These compounds guarantee that lubricants can withstand the strict requirements of modern machines by upgrading existing properties, suppressing undesirable features, and adding new functionalities. However, as industries ramp up technological growth and performance demands, lubricant additive development and optimization will continue to be a key avenue of interest to maintain operational efficacy and equipment life. It is important for engineers and lubrication professionals to understand the different functions that these additives are designed to perform and be responsible for specifying the right lubricants for their applications.

8.4 ADVANCES IN SELF-LUBRICATING COMPOSITES

Nature-inspired self-lubricating composites offer a promising solution for high-performance lubrication in extreme environments. Designed to facilitate effective lubrication without the external use of oils or greases, these novel materials are used in a variety of industries, including aerospace, automotive, and power

FIGURE 8.3 Advances in self-lubricating composites.

generation. Self-lubricating composites have been developed from the need for better performance under harsher conditions, including high temperature, high loads, and corrosive environments, as reflected in Figure 8.3 (Ye et al. 2024).

Self-lubricating composites are often a mixture of a matrix material (metallic, polymeric, or ceramic) and solid lubricants. The addition of solid lubricants, for example, graphite, molybdenum disulfide (MoS_2), and hexagonal boron nitride (h-BN), to these matrices notably enhances their tribological properties (Khalaj et al. 2020). Such solid lubricants are characterized by their lamellar structure, enabling them to easily shear in the loaded condition so as to provide lubricity and mitigate wear between opposing surfaces (Ouyang et al. 2022). The type of matrix material has a direct impact on the composite's overall performance. As an example, metal matrix composites (MMCs) can be good for high strength and thermal stability, and polymer matrix composites (PMCs) can provide lightweight compositions and corrosion resistance (Kangishwar et al. 2023).

This self-lubricating mechanism in these composites is primarily attributed to the solid lubricant's ability to form a protective film on contact surfaces. It avoids direct contact with surfaces, potentially reducing friction and wear and tear between them. In addition, the lubricating properties of some self-lubricating materials can be adjusted according to ambient conditions (Torres et al. 2018). For instance, the solid lubricant acts as a function of temperature in the composite, allowing the formation of effective lubrication under different operational stresses. The influence of nanoparticles on self-lubricating composites has been reported in several recent studies. The addition of TiO_2 nanoparticles into the MMC can improve wear resistance and reduce friction coefficients (Ali and Balaji 2023). This is because lubricant oxide layers are a part of the surface that are formed during operation and aid in lowering the wear rates.

However, self-lubricating composites can provide a significant performance boost in extreme environments where liquid lubricant cannot be used or is ineffective (Kumar and Antonov 2021). As an example, components for aerospace applications, such as bearings and seals, must perform reliably in elevated temperature and vacuum environments. Realizing self-lubrication in high temperatures necessitates the design of a new class of materials with low friction coefficients and low wear rates at very high temperatures (Torres et al. 2018). These composites are, moreover, utilized in automotive usages as just engine mounts, which are capable of managing heat and fluctuating speeds. These self-lubricating composites are not restricted to high-temperature settings, though, making this class of materials well-suited for humid or corrosive environments that are prone to degrading traditional lubricants in a short span of time. Its lubricious nature means that it needs no external lubrication, which again has the effect of lowering maintenance requirements and increasing reliability.

Despite their various advantages, the development and use of self-lubricating composites suffer a number of challenges. A major challenge that must be overcome is to evenly distribute solid lubricants into the matrix material for the material to act consistently as a composite (Zhang et al. 2017). Moreover, new combinations of materials are always being tested for the optimal tribology properties without compromising mechanical integrity (Ogbonna et al. 2023). This will lay the groundwork for future research in more advanced manufacturing technologies like additive manufacturing (3D printing) to allow for more complex geometries and specific properties for their intended application. Moreover, developments in novel solid lubricants and hybrid composite formulations will further facilitate performance improvements of self-lubricating materials (Posmyk and Myalski 2012).

Self-lubricating composites are a significant advancement in lubrication technology, as they provide an efficacious solution for the difficult working conditions. They have a special composition and are able to suit multiple conditions, which makes them priceless in different industries. By developing new materials as research progresses, self-lubricating composites will remain a crucial factor driving machinery efficiency while minimizing maintenance requirements and production costs. This technology is poised for future prospects, leading to more innovations that will best meet the changing needs of industrial applications.

8.5 KEY CHARACTERISTICS OF ADVANCES IN SELF-LUBRICATING COMPOSITES

Self-lubricating composites are a new class of materials that have attracted great interest in various industries because of their capability to provide friction and wear resistance without requiring any external lubricants. These composites are designed for the addition of solid lubricants into a matrix, which could be metallic, polymeric, or ceramic (Posmyk and Myalski 2012). Ample self-lubricating composite applications are made possible by significant parameter advancements.

When it comes to self-lubricating composites, one of the most important properties they should possess is low friction coefficients over a wide range of operating conditions, as evidenced in Figure 8.4. Some of these materials contain solid lubricants, including molybdenum disulfide (MoS_2), graphite, and hexagonal boron nitride (h-BN), that can form a lubricious layer on the surface during operation (Chkhartishvili et al. 2016). Such a layer would greatly minimize friction between contacting surfaces, making it a key component for applications at high load and high speed. For example, several studies demonstrated that composites with h-BN have a

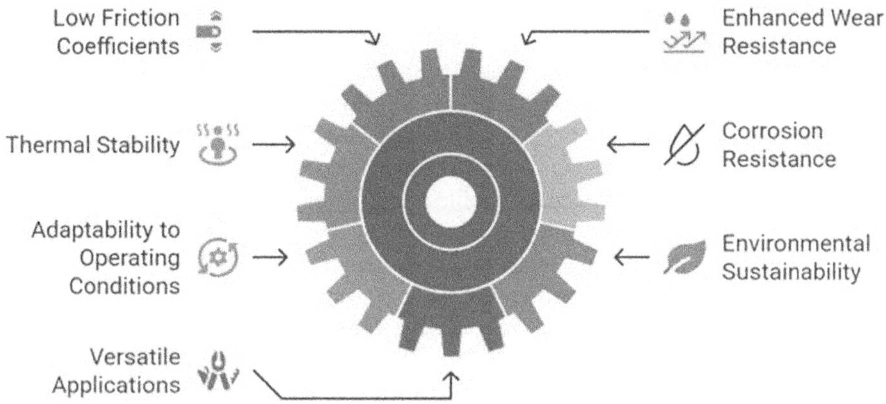

FIGURE 8.4 Key features of self-lubricating composites.

low coefficient of friction (0.1) under a humid environment and thus are good lubricant additives with low wear acceptance. Self-lubricating composites have been engineered to provide higher wear resistance than conventional materials. The solid lubricants embedded in the matrix provide lubrication and maintain the structure of the composite. This combination reduces the material loss while working, resulting in a longer lifetime of machines. According to research conducted by scientists at the National Institute of Standards and Technology, wear rates for self-lubricating composites can be an order of magnitude lower than those of traditional materials (composites and metals) and can be applied to applications where longevity is a higher priority than cost (Ouyang et al. 2022).

Thermal stability is another important feature of self-lubricating composites. Numerous applications necessitate materials capable of reliably performing under high-temperature conditions without material degradation or loss of lubricative properties. Self-lubricating composites are specifically developed to remain mechanically and lubrication-functional at higher temperature ranges. One such formulation can be used in extreme conditions exceeding 900°C, so it can be considered for a variety of application domains, and particularly for automotive and aerospace application sectors where efficient heat management is pivotal (John and Menezes 2021). Many self-lubricating composites have good corrosion resistance, which is, of course, extremely important for components that must work in aggressive environments, such as reduction gears submerged in seawater or bushings in the chemical processing industry. These matrix materials are primarily picked for their built-in corrosion resistance, and solid lubricants are added to protect these materials from further wear due to corrosive agents. This property helps materials retain their function over long periods, even in harsh environments.

Another big benefit is the ability of self-lubricating composites to adapt to changing operational conditions. These materials have the capability to dynamically respond to variations in load, speed, and temperature, helping them retain effective lubrication under diverse conditions. As an example, some micro-dimple structures of self-lubricant composites can improve lubrication behavior by/through load adaptiveness, thereby pre-facilitating contact phenomena under loads. As engineering practices increasingly incorporate sustainability, self-lubricating composites provide a solution to lubrication that is more environmentally viable than traditional oils and greases (Onu and Samuel 2024). They can also contribute to reducing hazardous substances and environmental contamination by minimizing unnecessary external lubricants. In addition, many solid lubricants utilized in these composite materials are developed from non-toxic materials, which increases their sustainable profile even further.

Self-lubricating composites find applications in multiple industries, such as aerospace, automotive, marine, and manufacturing, due to their versatility. They offer a combination of low friction coefficients, wear resistance, thermal stability, and corrosion resistance, making them widely used as bearings, gears, seals, sliding parts, etc. Research is also expected to advance new formulations and combinations of materials that should extend even wider applicability for this field. Recent progress in self-lubrication composites has led to materials with extraordinary properties ideal for high-end applications. With their low friction coefficients, increased wear resistance, thermal stability, and corrosion resistance, they prove to be examples of solutions that address modern engineering complexities. The future of lubrication technology will therefore be shaped by self-lubricating composites, which hold the potential to provide eco-friendly solutions that optimize productivity and minimize system maintenance costs in a wide range of industries (Ramezani et al. 2023). Ongoing exploration of new materials and manufacturing processes will expand their capabilities and foster adoption across additional industries.

8.6 CONCLUSION

From the wide variety of lubricants available to the different additives that enhance their performance, the study of lubrication technology is an intricate whirl of materials science and engineering that plays a crucial role in the smooth running of machines in every sector of the industry. One of the basic properties of lubricants, which we have repeatedly emphasized in this chapter, is to reduce friction and wear between moving entities, resulting in increased performance, improved energy efficiency, and life extension. Note that the constituents in a lubricant are not alone; their effectiveness is greatly enhanced through the careful use of additives. Additives are critical for maximizing lubricant performance by improving desirable characteristics and minimizing undesirable ones. Whether there are anti-wear agents that help create protective films around bearings and other moving parts or antioxidants that help prevent degradation through oxidation, these chemical compounds are critical to lubricant formulation within the lubrication industry; they serve functions beyond merely filling a sump; they help ensure that these products can withstand the rigors of modern industrial applications. Additives are formulated carefully to address specific operational needs, enabling a wide range of solutions across multiple industries to be developed and implemented.

Self-lubricating composites represent a cutting-edge lubrication technology. These materials inherently reduce the need for external lubricants while allowing for better tribological properties critical to higher-performance applications. Self-lubricating composites can provide low friction coefficients and excellent wear resistance while operating at extreme conditions since they have solid lubricants introduced within a matrix. They are also increasingly preferred over alternatives due to their ability to work across different loads and temperatures, enabling them to perform under challenging conditions in industries, including aerospace, automotive, and manufacturing. The industry continues to transform with rising needs for streamlining processes, creating eco-friendly solutions, and introducing reliable measures, making lubrication technology a pivotal aspect in the future industry landscape. However, the continual investigation of new materials and formulations will undoubtedly produce novel solutions that will continue to improve the performance attributes of lubricants and self-lubricating composites. Advanced manufacturing technologies, such as additive manufacturing that incorporates nanotechnology, could enable the large-scale production of next-generation lubricants with unique properties. Additionally, with increasing awareness of environmental impact, the transition to more sustainable lubrication alternatives will propel the innovation of bio-based and eco-friendly lubricants. Such a transition is not just a matter of aligning with the global sustainability goals but also in line with the regulatory bodies, which are promoting green practices for industrial operations.

Lubrication technology is always changing, and its innovation is key to machinery efficiency and longevity. This complex twirl between base oils and additives drives innovations in the field. Moving forward, it is evident that further developments in lubrication technology will be key to attaining operational excellence across countless applications. Adopting these technologies, and knowing how they will affect performance and sustainability, will allow industries to be more ready for the challenges of tomorrow, while helping them get more work done with less environmental impact. As one delves deeper into the possibilities of lubrication technology, they can anticipate improved mechanical performance and increased efficiency in industrial processes around the globe.

REFERENCES

Adesoji, A. A., Onu, P., Afolalu, S. A., Bello, K. A., Lawal, S. L., Monye, S. I., & Ogunniyi, O. J. 2024, April. Tribological analysis of advanced lubricants in automotive engines. In *2024 International Conference on Science, Engineering and Business for Driving Sustainable Development Goals (SEB4SDG)* (pp. 1–11). IEEE.

Adetunla, A., Afolalu, S., Jen, T. C., & Ogundana, A., 2023. The roles of surfactant in tribology applications of recent technology; an overview. In *E3S Web of Conferences*, France (Vol. 391, p. 01021). EDP Sciences.

Agarwal, S., Chhibber, V. K., & Bhatnagar, A. K., 2013. Tribological behavior of diesel fuels and the effect of anti-wear additives. *Fuel*, 106, 21–29.

Ahmed, N. S., & Nassar, A. M., 2013. Lubrication and lubricants. *Tribology Fundamentals and Advancements*, 55–76.

Ali, S. M., & Balaji, D., 2023. Study on tribological performance of nano particle reinforced metal matrix composite. *International Journal of Scientific Research in Mechanical and Materials Engineering*, 7(4), 14–30.

Bardasz, E. A., & Lamb, G. D., 2017. Additives for crankcase lubricant applications. In *Lubricant Additives* (pp. 447–472). CRC Press.

Brandão, J. A., Meheux, M., Ville, F., Seabra, J. H., & Castro, J., 2012. Comparative overview of five gear oils in mixed and boundary film lubrication. *Tribology International*, 47, 50–61.

Chkhartishvili, L., Tabatadze, G., Nackebia, D., Bzhalava, T., & Kalandadze, I., 2016. Hexagonal boron nitride as a solid lubricant additive (an overview). *Nano Stud*, 14(91–98), 2016.

Clauss, F. J. (Ed.), 2012. *Solid Lubricants and Self-Lubricating Solids*. Elsevier.

Czichos, H. 2009. *Tribology: A Systems Approach to the Science and Technology of Friction, Lubrication, and Wear* (Vol. 1). Elsevier.

John, M., & Menezes, P. L., 2021. Self-lubricating materials for extreme condition applications. *Materials*, 14(19), 5588.

Kangishwar, S., Radhika, N., Sheik, A. A., Chavali, A., & Hariharan, S., 2023. A comprehensive review on polymer matrix composites: Material selection, fabrication, and application. *Polymer Bulletin*, 80(1), 47–87.

Kayode, J. F., Lawal, S. L., & Afolalu, S. A., 2023, April. An overview of tribology and its industrial applications. In *2023 International Conference on Science, Engineering and Business for Sustainable Development Goals (SEB-SDG)* (Vol. 1, pp. 1–15). IEEE.

Khalaj, M., Zarabi Golkhatmi, S., Alem, S. A. A., Baghchesaraee, K., Hasanzadeh Azar, M., & Angizi, S., 2020. Recent progress in the study of thermal properties and tribological behaviors of hexagonal boron nitride-reinforced composites. *Journal of Composites Science*, 4(3), 116.

Khonsari, M. M., & Booser, E. R., 2017. *Applied Tribology: Bearing Design and Lubrication*. John Wiley & Sons.

Kiokias, S., Varzakas, T., & Oreopoulou, V., 2008. In vitro activity of vitamins, flavonoids, and natural phenolic antioxidants against the oxidative deterioration of oil-based systems. *Critical Reviews in Food Science and Nutrition*, 48(1), 78–93.

Kumar, R., & Antonov, M., 2021. Self-lubricating materials for extreme temperature tribo-applications. *Materials Today: Proceedings*, 44, 4583–4589.

Kumar, R., Hussainova, I., Rahmani, R., & Antonov, M., 2022. Solid lubrication at high-temperatures: A review. *Materials*, 15(5), 1695.

Lugt, P. M., 2016. Modern advancements in lubricating grease technology. *Tribology International*, 97, 467–477.

Mang, T. & Dresel, W. eds., 2007. *Lubricants and Lubrication*. John Wiley & Sons.

Mortier, R.M. and Orszulik, S.T. eds., 2010. *Chemistry and Technology of Lubricants* (Vol. 107115, pp. 56–75). Springer.

Ogbonna, V. E., Popoola, P. I., Popoola, O. M., & Adeosun, S. O., 2023. A review on recent advances on improving polyimide matrix nanocomposites for mechanical, thermal, and tribological applications: Challenges and recommendations for future improvement. *Journal of Thermoplastic Composite Materials*, 36(2), 836–865.

Onu, P., & Samuel, A., 2024, April. Green Tribology, Sustainable Lubrication Practices and Environmental Impact: A Mini Review. In *2024 International Conference on Science, Engineering and Business for Driving Sustainable Development Goals (SEB4SDG)* (pp. 1–9). IEEE.

Ouyang, J. H., Li, Y. F., Zhang, Y. Z., Wang, Y. M., & Wang, Y. J., 2022. High-temperature solid lubricants and self-lubricating composites: A critical review. *Lubricants*, 10(8), 177.

Piechowski, M., Wyczólkowski, R., & Paszkowski, W., 2024, March. The Concept of a System Supporting the Implementation of an Intelligent Lubrication Strategy Within the Company Using Advanced Information Technologies. In *International Scientific-Technical Conference MANUFACTURING* (pp. 136–150). Cham: Springer Nature Switzerland.

Posmyk, A., & Myalski, J., 2012. Producing of composite materials with aluminium alloy matrix containing solid lubricants. *Solid State Phenomena*, 191, 67–74.

Ramezani, M., Ripin, Z. M., Jiang, C. P., & Pasang, T., 2023. Super lubricity of materials: progress, potential, and challenges. *Materials*, 16(14), 5145.

Rudnick, L. R. (Ed.)., 2020. *Synthetics, Mineral Oils, and Bio-Based Lubricants: Chemistry and Technology*. CRC Press.

Rudnick, L. R., & Bartz, W. J., 2020. Comparison of synthetic, mineral oil, and bio-based lubricant fluids. In *Synthetics, Mineral Oils, and Bio-Based Lubricants* (pp. 359–378). CRC Press.

Sanchez, C., Julián, B., Belleville, P., & Popall, M., 2005. Applications of hybrid organic: Inorganic nanocomposites. *Journal of Materials Chemistry*, 15(35–36), 3559–3592.

Shahnazar, S., Bagheri, S., & Abd Hamid, S. B., 2016. Enhancing lubricant properties by nanoparticle additives. *International Journal of Hydrogen Energy*, 41(4), 3153–3170.

Silverstein, R., & Rudnick, L. R., 2017. Additives for grease applications. *Lubricant Additives*, 431–446.

Spikes, H. A., & Olver, A. V., 2003. Basics of mixed lubrication. *Lubrication Science*, 16(1), 1–28.

Torres, H., Rodríguez Ripoll, M., & Prakash, B., 2018. Tribological behaviour of self-lubricating materials at high temperatures. *International Materials Reviews*, 63(5), 309–340.

Totten, G. E. (Ed.)., 2017. *Friction, Lubrication, and Wear Technology*. ASM international.

Vazirisereshk, M. R., Martini, A., Strubbe, D. A., & Baykara, M. Z., 2019. Solid lubrication with MoS2: A review. *Lubricants*, 7(7), 57.

Wan, S., Tieu, A. K., Xia, Y., Zhu, H., Tran, B. H., & Cui, S., 2016. An overview of inorganic polymer as potential lubricant additive for high temperature tribology. *Tribology International*, 102, 620–635.

Xu, L., Liang, H. W., Yang, Y., & Yu, S. H., 2018. Stability and reactivity: Positive and negative aspects for nanoparticle processing. *Chemical Reviews*, 118(7), 3209–3250.

Ye, W., Shi, Y., Zhou, Q., Xie, M., Wang, H., Bou-Saïd, B., & Liu, W., 2024. *Recent Advances in Self-Lubricating Metal Matrix Nanocomposites Reinforced by Carbonous Materials: A Review*. Nano Materials Science.

Zhang, Y., 2006. Boundary lubrication: An important lubrication in the following time. *Journal of Molecular Liquids*, 128(1–3), 56–59.

Zhang, Y., & Chromik, R. R., 2022. Tribology of self-lubricating metal matrix composites. *Self-lubricating Composites*, 31–71.

Zhang, A., Han, J., Su, B., Li, P., & Meng, J., 2017. Microstructure, mechanical properties and tribological performance of CoCrFeNi high entropy alloy matrix self-lubricating composite. *Materials & Design*, 114, 253–263.

9 Improvements in Tribological Properties of Metal Matrix Composites for Industrial and Automotive Applications

Gurpreet Singh Saini, Narender Singh, Jai Inder Preet Singh, Piyush Gulati, and Ismail Hossain

9.1 INTRODUCTION

Metal matrix composites (MMCs) are high-performance materials produced through reinforcement of a ductile metal matrix (e.g., aluminum, magnesium, copper, or titanium) with hard, wear-resistant particles, fibers, or whiskers. Such reinforcements (most often ceramic, e.g., silicon carbide (SiC), aluminum oxide ($Al2O_3$), boron carbide (B_4C), or graphite) are selected because they can greatly improve the mechanical and tribological properties of the base metal. Increased demand for high-performance, lightweight, and durable materials in industries like transportation, aerospace, and defense, as well as heavy engineering, has stepped up research and development of MMCs (Kumar et al. 2021; Mazahery and Shabani 2010).

Tribological performance, that is, wear resistance, friction, and lubrication behavior, is an important consideration in the efficiency and service life of machine elements. Industrial and automotive applications involving the movement of components are often subjected to harsh system operating conditions: high temperatures, high contact pressures, and continuous friction. In many such applications, conventional metals like steel or pure aluminum are not appropriate due to their high wear rates, surface damage, and the high energy losses due to friction. MMCs eliminate these weaknesses with the provision of tailored surface and bulk properties that will reduce wear, lower friction coefficients, and increase thermal conductivity (Prasad 2001).

Aluminum-based composites have gained special interest among other metal matrices because of their low density, high specific strength, good machinability, and resistance to corrosion. Particle reinforcement, like SiC or Al_2O_3, has been found to

 DOI: 10.1201/9781003620570-9

enhance both the mechanical, for example, hardness and tensile strength, and tribological, for example, wear resistance and friction stability, properties in dry or lubricated conditions (Surappa 2003; Hashim et al. 1999). Likewise, lubricating reinforcements like graphite and MoS_2 are increasingly favored to minimize frictional losses in sliding contacts, further broadening the use of MMCs even more (Sahin 2004). The enhanced tribological performance of MMCs is likely to be of great benefit to the automotive industry. Brake rotors, engine blocks, piston rings, and cylinder liners are starting to find application with MMCs to improve service life, decrease noise and vibration, and improve total energy efficiency. Such composites have superior dimensional stability, seizure and scuff resistance, and lower heat generation during use, which is important in high-speed or loaded automotive systems (Ezatpour et al. 2012; Sharma et al. 2017). More recent advances in processing routes have included stir casting, powder metallurgy (PM), squeeze casting, and, most recently, additive manufacturing (AM), which enables the production of MMCs with better microstructural homogeneity and matrix–reinforcement bonding. Such techniques have also allowed them to exert control over the reinforcement content, particle size, and distribution, which directly influence the tribological performance of the composite (Singh et al. 2018). Moreover, post-processing methods such as heat treatment, surface coating, and cryogenic treatment have demonstrated further opportunities in optimizing the tribological behavior for application (Singh and Goyal 2016).

In this chapter, the author will discuss the major advancements in designing, manufacturing, and testing of MMCs with improved tribological performances. It concentrates on recent developments in materials selection, processing strategies, and performance measures, but it gives specific importance to their use in industrial machinery and car parts. Based on the extensive literature and case study review, the given chapter will help introduce a unified vision of the state-of-the-art MMCs and define the areas that can be improved and investigated in the future.

9.2 LITERATURE REVIEW

Tribological behavior of MMCs has been of great interest over the last several decades, and much effort has been put into enhancing wear resistance, lowering friction, and increasing the service life of parts operating under industrial and automotive conditions. Many scientists have investigated different matrices–reinforcement combinations and processing methods to maximize the tribological performances of MMCs.

9.2.1 Matrix and Reinforcement Materials

The most widely used matrix in MMCs research is aluminum and its alloys because of their low density, good corrosion resistance, and processability. Common reinforcement materials include SiC, $Al_2O_{2,}$ and B_4C to enhance hardness and wear resistance. According to Prasad (2001), Al-SiC composites showed better wear resistance than the unreinforced aluminum in both dry and lubricated sliding. Likewise, it was observed by Mazahery and Shabani (2010) that the SiC content in the Al matrix composites significantly increased the wear rate.

Solid lubricants such as graphite and MoS_2 are commonly used to reduce friction. As it was demonstrated by Sahin (2004), aluminum/graphite composites exhibit enhanced wear resistance and reduced friction coefficients because of the lubricating film formed during the sliding.

The composite type of material is manufactured by joining two or more distinct materials, each having distinctive properties, and coming up with a synergistic performance better than that of the individual components. Composites have two main components: the matrix and the reinforcement. The continuous matrix surrounds, supports, and binds the reinforcement material. Its main functions entail the transfer of stress to the reinforcement and absorption of environmental and mechanical abuse, and it offers bulk shape and toughness to the composite. The matrix also aids in the distribution of loads equally and in crack propagation resistance. Various forms of matrix material—metals, polymers, or ceramics—are used depending on applications, with attractive features that depend upon the application.

Metal matrices with high strengths, stiffness, and thermal resistance are aluminum, magnesium, titanium, and copper that are used in the application of structural MMCs. Aluminum is also common because of its lightweight and non-corrosive properties, and hence it can be used in aerospace and automotive manufacturing industries. Magnesium is even less dense, making it suitable where reducing mass is important. Titanium, which has superior corrosion resistance compared to other metals and good stability in high temperatures, has applications where high performance is required, such as in aerospace and biomedical implants. Composites made with polymer matrices, such as thermosets (e.g., epoxy-based) and thermoplastics (PEEK or polypropylene), may be low-cost, easy to shape, and lightweight which may be needed in civil, marine, and consumer product applications. Ceramic matrices, by contrast, are featured in high-temperature applications such as turbine blades and space structures due to their excellent resistance to heat and oxidation.

The reinforcement stage is stiff and rigid as compared to the matrix and mainly resists the applied load. Additions of reinforcements to the composite have a considerable impact on the mechanical properties of the composite, which include its strength, stiffness, and wear resistance. They may take the structure of continuous fibers, short fibers, particles, whiskers, and wires as per the desired properties and processes undertaken in manufacturing processes. Silicon carbide (SiC) and aluminum oxide (Al_2O_3) are most common in MMCs because of their high hardness, chemical stability, and wear resistance. Another common reinforcement is carbon fibers, which are characterized by high tensile strength, light weight, and the ability to withstand high temperatures, but in turn are prone to oxidation. Composite performance largely depends on the interface between the matrix and the reinforcement. Effective interfacial bonding is required to facilitate effective load transfer, enhance mechanical integrity, and eschew delamination. Weak bonding may cause an early breakdown during loading or in a cyclic load. The thermal expansion coefficient compatibility of the two phases also plays an important role, not only during high-temperature applications but also in preventing internal stresses and microcracking. Interfacial adhesion and stability are usually enhanced through surface treatment, alloying, and coating techniques.

To sum up, the selection of a matrix and reinforcement material, as well as the regulation of the interfacial interaction, is the key to the design of composites with specific properties used in advanced industrial, structural, and functional constructions. Interaction between the matrix and reinforcement ultimately determines the composite's performance, durability, and specific sectors, such as aerospace and automotive, electronics, and the biomedical devices market.

9.2.2 Fabrication Techniques

Even distribution of reinforcements in the matrix is very crucial in dictating the wear behavior of the composite. Stir casting is a very common and economical method of making MMCs. Hashim et al. (1999) emphasized that achieving uniform dispersion of reinforcements and lower porosity in stir-cast composites requires an optimal stirring rate and temperature.

PM provides greater control over the distribution of reinforcement and grain size. It was discovered in the study of Ezatpour et al. (2012) that Al nanocomposites with Al_2O_3 <5 wt.% processed by PM exhibited better hardness and wear resistance owing to refined microstructure and good interfacial bonding.

The making of MMCs is a vital procedure that gives way to their microstructure, interfacial bonding, functions, and mechanical characteristics. Selection of fabrication method depends on the type of reinforcement and matrix, product requirements, expected properties, and cost. In general, the MMCs fabrication techniques are of two types: solid-state processing and liquid-state (casting) processing with new advanced (high in situ) techniques.

The solid-state processing techniques involve working at temperatures that are lower than the melting temperature of the matrix metal. PM is one of the most used ones in this group. In PM, the reinforcement powder and metal powders get mixed evenly, and then, using pressure, they are compacted and subsequently sintered at high temperatures. It offers good control of composition and microstructure, especially when used to add ceramic particles, for example, SiC, Al_2O_3, or B_4C to matrices, for example, aluminum or titanium. PM is, however, expensive and only applies to relatively small components. Diffusion bonding, hot isostatic pressing (HIP), and friction stir processing are other solid-state processes, offering better interfacial bonding and even reinforcement distribution, although they have more complicated production setups.

Liquid-state techniques have been very popular due to ease of fabrication, scalability, and affordability, especially in industries. The most well-liked and cost-efficient liquid-state process is called stir casting or vortex processing. During this process, reinforcements are added to the molten metal by stirring the metal mechanically in order to get a relatively uniform distribution and subsequently cast in molds. Regarding relatively low cost, issues such as wettability of the molten metal and ceramic reinforcement, settling, and clustering tend to compromise the quality of the products. Reinforced casting on the particle dispersion and bonding has been brought about by techniques like the use of ultrasonic-assisted casting and squeeze casting. With squeeze casting, high pressure is applied during solidification to achieve improved grain structure, improved reinforcement–matrix bonding, and lower porosity. An advanced approach is in-situ fabrication, where the reinforcement is synthesized within the matrix during processing through controlled chemical reactions. This method ensures a cleaner matrix–reinforcement interface, better thermodynamic stability, and improved mechanical properties. Common in-situ techniques include chemical reactions involving molten salts or reactive gases introduced into the metal melt. Additionally, spray deposition, thermal spraying, and AM are modern techniques that allow the creation of complex geometries and tailored microstructures, expanding MMC applications in aerospace, automotive, and biomedical sectors.

In conclusion, the selection of a suitable fabrication technique for MMCs is a trade-off between material properties, processing complexity, cost, and application requirements. Stir casting and PM dominate commercial applications due to their relative ease and adaptability, while in-situ and advanced processing methods are paving the way for high-performance, next-generation composite materials.

9.2.3 Tribological Behavior in Different Conditions

The influence of applied load, sliding speed, and temperature on the tribological behavior of MMCs has been studied by a number of researchers. Ramesh et al. (2018) noted that the wear rate of Al-SiC composites rises with the increase in load and sliding velocity, but they remain superior to the unreinforced alloys in the same circumstances. Also, soft or self-lubricating phases have been found to alleviate the temperature and pressure effects due to sliding.

More sophisticated is the in-situ fabrication technique, where reinforcement is created within the matrix during processing by a controlled chemical reaction. The technique provides a cleaner interface between the matrix and reinforcement and greater thermodynamic stability in addition to improved mechanical performance. Some typical in-situ processes entail chemical reactors of molten salts or gaseous reactions with molten metals. Also, modern processes that enable the development of complex geometries and designed microstructures are spray deposition, thermal spraying, and AM, which broaden the usage of MMC in the fields of aerospace, the automotive industry, and the biomedical field.

Finally, the process of choosing an appropriate method of fabrication of MMC involves economic issues of material and the complexity of the method of processing, as well as costs and needs in the field of application. The leading commercial applications, predominantly owing to their relatively simple nature and flexibility, are stir casting and PM, whereas in-situ and advanced processing are the most rapidly developing—but highly focused and specific—methods for the development of high-performance, new-generation composite materials. Temperature plays a major role in the tribology of MMCs, as it has a great impact on the matrix and reinforcement. Most metal matrices undergo thermal softening at high temperatures, which results in loss of hardness and load-carrying capacity. This might lead to additional wear, which is caused by plastic deformation, adhesion, and the lack of reinforcement support. As an example, the wear rates of aluminum-based MMCs reinforced with SiC particles can increase significantly at temperatures above 300°C. Moreover, elevated temperatures can reduce the strength of interfacial bonding between the reinforcement and matrix, resulting in debonding, pull-out of particles, or development of microcracks at the interface. Nevertheless, particular MMCs may demonstrate oxidation wear resistance, owing to the formation of protective oxide scales, particularly where oxides are hard and adherent. The presence of these scales does the job of a solid lubricant or a barrier to further wear, in case they do not break apart. The temperature effect is positive or destructive depending on the kind of reinforcement and matrix, thermal expansion misfit, and chemical stability.

Altogether, the tribological behavior of MMCs is a complicated interaction of mechanical and thermal effects. At high loads and sliding speeds, wear often becomes a problem, but in the case of hard reinforcements, this, to a certain degree, is sorted out in the sharing of the load and resisting the deformation. Wear performance is generally dependent on elevated temperatures, leading to degradation by softening and interfacial degradation, but wear may be temporarily cushioned against by the formation of oxide layers in certain circumstances. To streamline the performance of MMC, there is close consideration to matrix–reinforcement combinations along with a knowledge of the wear process in different tribo-conditions.

9.2.4 Hybrid and Nanocomposites

Hybrid composites, which contain two or more kinds of reinforcements (e.g., SiC + graphite), are of interest because they combine hardness and lubrication. Sanjeev Kumar et al. (2019) revealed that hybrid Al-based MMCs containing ceramic and lubricating fillers display a well-rounded mechanical and tribological performance that can be applied in dynamic automotive components.

New progress has also resulted in the creation of nano-reinforced MMCs. Nanoparticle size, or nano-sized particles, offers increased surface area, and this leads to improved load transfer and grain refinement. Tested nanocomposites have exhibited reduced wear rates and improved surface integrity in comparison to their micro-sized reinforced counterparts (Ezatpour et al. 2012).

Advanced classes of engineered materials include hybrid composites and nanocomposites because the alternatives are all characterized by enhanced performance, combining several reinforcement approaches or using materials at the nano-scale. The composites have become one of the better prospects to the traditional MMCs and have shown improvement in terms of mechanical, thermal, and tribological characteristics in regard to specialized applications in the aerospace, automotive, defense, and biomedical markets.

Hybrid composites consist of a mix of two or more distinct forms of reinforcement in the same matrix system. The reinforcements may be fibers, particles, whiskers, or a combination of them. The motivation behind hybridization is to synergistically offer the benefits of multiple reinforcements so as to beat the weaknesses of single-reinforcement systems. As an example, the addition of both hard ceramic particles (such as SiC or B_4C) and ductile fibers (such as graphite or carbon fibers) into an aluminum matrix may yield a composite material with high strength, excellent wear resistance, and enhanced fracture toughness. Stiffness and wear resistance can be increased through the addition of ceramic particles, whereas lubrication and friction can be decreased with the use of soft reinforcements such as graphite. The distribution, orientation, and volume fraction of the hybrid reinforcements are very crucial in determining the final property. Hybrid MMCs are useful where there are conflicting requirements in the performance requirements of a product, which include wear resistance, lightweight design, and strength.

Nanocomposites, on the other hand, entail dispersion of the nanoscale reinforcements, which are usually less than 100 nm in size within the matrix material. They are nanoparticles (e.g., Al_2O_3, TiO_2, and SiC), nanofibers, carbon nanotubes (CNTs), and graphene nanoplatelets. Nano reinforcements can only provide a very high proportion between the surface area and the volume; hence, the extremely good load transfer characteristics, resistance to dislocation movement, and also a high barrier to heat and wear. Even at low volume fractions of reinforcement, a great increase in mechanical strength, hardness, thermal stability, and tribological behavior is observed in nanocomposites. As an example, nano-sized addition of Al_2O_3 (12 wt.%) in an aluminum matrix may lead to a dramatic enhancement in hardness and decrease wear rate. Nevertheless, issues, including agglomeration of nanoparticles, uniform dispersion, and bonding at the interface, are the problems that need to be resolved to achieve their potential to the fullest. Such methods as ultrasonic dispersion, ball milling, and surfactant/in-situ processing are used to provide homogeneous distribution and adherent interfacial connections.

Both hybrid and nanocomposites have increased tribological performance in a significant way. The hybrid composites can also be customized in such a way that they behave well under predetermined loading, sliding, and thermal conditions through a manipulation of reinforcement combination. For example, in Al/SiC/graphite composites, SiC resists wear, with graphite serving as a solid lubricant to reduce friction. The fine reinforcement size and high mechanical quality of the interlocking leads nanocomposites to exhibit lower wear rates, improved resistance against adhesive and abrasive wear, and improved stability of friction when compared to micro-composites. Besides, sliding and its associated tribofilm formation on a nano-scale further enhance surface protection and wear.

To sum up, hybrid and nanocomposites already occupy the leading positions in materials engineering, providing the capability of designer materials with specific properties that may be tuned to handle complicated service conditions. Although hybrid composites are concerned with the increment of the property of the composite via the utilization of multi-reinforcement solutions, nanocomposites exploit the phenomenon of nanoscale size to obtain even better results and behaviors. Nevertheless, fabrication and dispersion difficulties remain; research and advances in the process nonetheless keep expanding the envelope of what these high-tech composites may accomplish in industrial and high-performance applications.

9.2.5 Heat Treatment and Surface Treatments

It has also been reported that surface treatment of MMCs, for example, laser cladding, thermal spraying, and cryogenic processing, further improves wear resistance. Cryogenic treatment, as an example, enhances hardness and polishes the matrix microstructure, hence lowering the wear rate of composite materials. It is also due to heat treatment that the distribution of alloying elements is modified and the strength of the matrix is increased, which also positively affects tribological performance (Sarkar et al. 2017).

Based on the literature, it can be concluded that the tribological performance of MMCs can be substantially enhanced by a judicious choice of reinforcements, proper processing methods, and post-fabrication treatments. Nevertheless, problems of homogeneous dispersion, economically large-scale synthesis, and wear mechanisms at micro- and nano scales still exist. Hybrid, functionally graded, and nanostructured composites should be investigated further as automotive and industrial applications put greater demands on materials that have longer life and less maintenance.

9.3 WEAR MECHANISMS IN MMCS

Wear is a certain surface phenomenon that leads to material removal and component deterioration during service. Automobile and industrial machinery components (brake rotors, pistons, cylinder liners, bearings, etc.) are subject to continuous wear. Advanced composites are replacing conventional materials (cast iron or unmodified aluminum alloys), which have proven inadequate in conditions of either high loads or high temperature. However, their wear behavior is quite complex, and it depends on many factors, which are interrelated.

Elementary Wear Processes in MMCs the main wear processes are

9.3.1 Abrasive Wear

Abrasive wear occurs when hard asperities or abrasive particles slide over a softer material surface and wear mainly through micro-cutting, micro-plowing, or micro-fracture. It is a mechanical degradation process sustained by the relative movement and the hardness disparity between the interacting surfaces or trapped particles. The abrasive wear is essentially dependent upon the fact that the abrasive medium (either free particles or hard surface asperities) should be much harder than the material being worn.

Abrasive wear occurs in two distinct ways: two-body abrasion and three-body abrasion. Two-body abrasion is when hard protuberances or asperities on one surface move directly against another surface, and serve as a cutting tool. By comparison, three-body abrasion includes free particles that may roll or slide between two surfaces, as might occur in contaminated lubricating systems environments, or in sliding contact with trapped debris. Both processes result in material removal; however, the form and extent of wear vary according to the interplay between the surfaces and the abrasives.

Abrasive wear mechanisms are functions of a number of factors, which include the hardness and toughness of the material; shape, the size, and distribution of the abrasive particles; the load applied; the sliding velocity; and the environmental conditions like lubrication and temperature. Micro-cutting occurs when the abrasive particles plow through the material, cutting it into small chips. Micro-plowing also causes plastic deformation that does not remove material but may cause surface fatigue and subsequent detachment. Micro-fracture occurs more in brittle material, where the removal of fragments and cracking is predominant.

To estimate abrasive wear quantitatively, the wear volume can be predicted by the Archard wear law, which correlates the wear volume to normal load, sliding distance, and the hardness of the material. Study of worn surfaces and the predominant wear

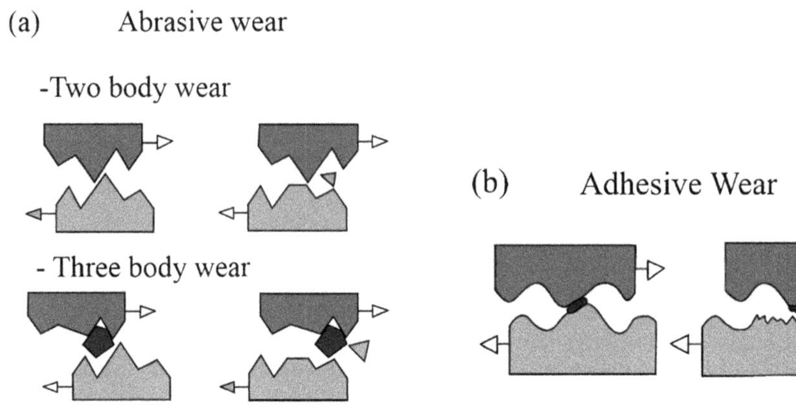

FIGURE 9.1 Schematic diagram showing abrasive wear (Jaydeep Vishwakarma et al. 2022).

modes: Worn surfaces and the prevailing wear modes are typically studied by more sophisticated surface characterization techniques like scanning electron microscopy (SEM), atomic force microscopy (AFM), and profilometry by researchers. Further, pin-on-disk, ball-on-flat, or rubber wheel abrasion tests are common tribological tests to study abrasion resistance under controlled conditions (Figure 9.1).

The important aspect of abrasive wear is used in a wide range of industrial activities, such as machining, mining, transport, and biomedical implants. For example, in MMCs, abrasion resistance may be enhanced by the hard reinforcement particles; however, inappropriate particle distribution or poor interfacial bonding may have the opposite effect and even promote higher wear rates. Therefore, surface engineering and optimization of material microstructure are vital in the reduction of abrasive wear. More complex coatings, surface treatments, and self-lubricating materials remain an important area of active research into abrasive wear to increase the life and reliability of components working under severe conditions.

9.3.2 Adhesive Wear

Adhesive wear occurs quite frequently and is observed when two solid surfaces slide past each other under a load, and the material is transferred between the two surfaces at the points of contact by the strong adhesive forces. Relative motion between two surfaces, on a macroscopic scale, appears smooth; however, on a microscopic scale, the surfaces are really composed of microscopic roughness referred to as asperities. When the surfaces come into contact, these asperities touch each other, and under the pressure and heat, become micro-welded junctions through adhesion. These junctions are broken during sliding by the shearing forces that develop at the contact points. Such breakage may leave pieces of one surface attached to—and/or torn off—the other, creating wear particles or debris. The process causes damage to the surface in terms of scoring, grooving, or the creation of pits and scratches. Adhesive wear becomes worse in cases where lubrication is deficient, when soft metals are involved, under heavy loads, or when surfaces are clean and free of oxides. This type of wear

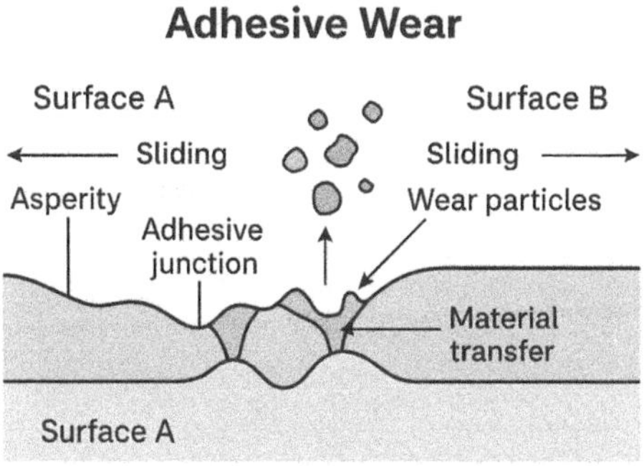

FIGURE 9.2 Schematic diagram showing adhesive wear.

may eventually lead to a performance and service life degradation of the mechanical elements (gears, bearings, and piston-cylinder assemblies).

A schematic illustration of adhesive wear in Figure 9.2 includes two surfaces in sliding bearing asperities on each side, which are microscopic in scale. Little adhesive contacts are created by pressure at the interface. With further sliding, such junctions fail, and material can be either transferred to the other surface or lost as wear particles. Surface A and surface B would be shown in the diagram with contact asperities: arrows show the direction of sliding, adhesive contact junctions, and material transfer or voids left after breakage. This illustration is useful in the conception of the adhesive wear that happens at the micro-contact level during the surface interaction.

9.3.3 Oxidation Wear

Oxidation wear is caused by the creation and removal of oxide layers on metal surfaces during relative motion. During sliding contact between two metallic surfaces, particularly when this happens at high temperatures, the heat produced at the points of contact may cause oxidation of the surface layers. Such oxide films tend to be harder and more brittle in comparison to the base metal. During continuing sliding, these brittle oxides may develop cracks, and pieces may break off, forming wear particles. As opposed to adhesive or abrasive wear, the material removal in this case is caused by the chemical reaction with oxygen and not by the pure mechanical impact. But oxide particles that break loose may subsequently serve as third-body abrasives and still further increase the rate of wear in an oxidative–abrasive wear process.

An oxidation wear diagram has two metallic surfaces, for example, Surface A and Surface B, in contact and sliding past one another. Oxide layers are formed on the interface under the influence of high temperatures and oxygen exposure. During the sliding process, pieces of the oxide film peel off, and fragments of wear debris are formed. Such particles of debris could be visualized as tiny flakes or chips that might either fall off or get retained in between the two surfaces, resulting in extra mechanical wear. The diagram is usually used to show heat zones, oxide layers, direction of motion, and cracked particles to indicate the entire cycle of oxidation wear (Figure 9.3).

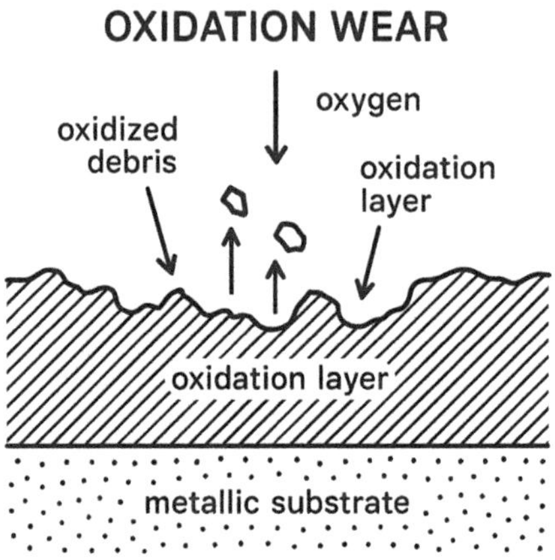

FIGURE 9.3 Schematic diagram showing oxidation wear.

9.3.4 Delamination Wear

Delamination wear is a severe form of wear mechanism that mostly occurs in ductile materials like metals and MMCs when they are subjected to repeated sliding or rolling contacts. It is formed underneath the surface through the buildup of shear stresses and plastic deformation by constant sliding. This causes stresses that result in the creation of subsurface cracks, which extend parallel to the surface. With time, the cracks join together, leaving thin layers or sheets of the material to peel off, a phenomenon called delamination. The worn flakes can be big and flat pieces of wear debris, and this helps to differentiate this type of wear mechanism from abrasive or adhesive wear, which tends to produce smaller particles. Delamination wear may shorten the life of components and is often seen in gears, bearings, and other components that experience high loads and cyclic stresses. Wear factors affecting delamination include material microstructure, hardness, load, sliding velocity, and lubrication conditions. Prevention measures include enhancement of surface hardness, coating, maximization of material toughness, and adequate lubrication.

Delamination wear is a sequential material wear process that is initiated and propagated by subsurface cracks that later culminate in peeling or flaking off of material layers. Delamination initiates below the surface, unlike other forms of surface-originated wear, such as abrasion or adhesion. Delamination initiates below the surface, typically at a depth equal to the maximum shear stress under cyclic loading. This is a particularly serious wear mechanism in situations where parts experience recurring sliding or rolling movements, for example, bearings, gears, shafts, and machine tool interfaces.

Wear debris associated with delamination is generally larger, flat, and in most cases takes the shape of small metal sheets. Such flakes may also cause third-body abrasion unless they are cleared out of the area of contact. Once started, the process can speed up, causing quick failure of components. Delamination can be promoted by factors like material purity, inclusions, microstructural discontinuities, and inadequate lubrication.

9.4 SURFACE ENGINEERING AND COATINGS FOR METAL MATRIX COMPOSITES (MMCS)

MMCs are a high-tech group of materials that consist of a metal or alloy matrix reinforced by particles, fibers, or whiskers of a different material. Common matrices include aluminum, magnesium, titanium, and copper; common reinforcements include ceramic particles, for example, SiC and Al_2O_3, carbon fiber, and boron carbide. Superior mechanical property, high wear resistance, good thermal stability, and light weight in nature are some of the properties that make MMCs find application in automotive, aerospace, defense, and industrial applications.

However, imperial ecosystems notwithstanding, surface properties of MMCs often require enhancement in order to become commensurate with the harshness of the operating environment, that is, elevated wear, corrosion, heat, or friction. To improve these surface-related properties, and to raise component durability and their overall performance, coating and surface engineering are of importance.

Coatings and surface engineering are a very important field of materials science and engineering. It is the field that deals with the alteration of the surface energy of materials with a view to improving performance, life span, and functionality. Surface engineering is a broad term defining a group of techniques and processes that aim to enhance the properties of materials such as wear resistance, corrosion resistance, and cosmetism. Surface engineering coatings are a type of surface engineering, in which a layer of material is deposited over a substrate to produce the required behavior on the surface. The role of surface engineering and coatings cannot be overstated, as they significantly influence component lifespan, reduce the cost of maintenance, and improve overall efficiency in most industries.

9.4.1 Thermal Spray Coating

MMCs are widely used in aerospace, automotive, and industrial areas due to their excellent mechanical, thermal, and wear-resistant abilities. However, their surface can be enhanced even in hostile environments to augment extreme wear, corrosion, and thermal conditions. One of the most successful surface engineering processes is thermal spray coating, which may result in a tremendous enhancement of surface hardness, wear resistance, and corrosion behavior of MMCs with an insignificant or no bulk property change. Thermal spraying is a group of coating processes in which finely divided metallic or non-metallic materials are melted or partially melted and deposited, under high temperature, on a prepared surface to produce a coating. This process is particularly useful with MMCs, as it enhances surface properties like wear resistance, corrosion protection, and thermal insulation without affecting bulk material.

A typical mechanism of thermal spraying is described graphically in Figure 9.4. It generally involves feeding a coating material (generally a powder or a wire) into a spray skillet, where it is melted by a high-temperature flame or plasma generated by a gaseous supply. This hot material is then shot against a substrate at high speed, forming a protective layer coating when it hits and solidifies. This process enhances the surface characteristics of materials, which include resistance to wear, corrosion,

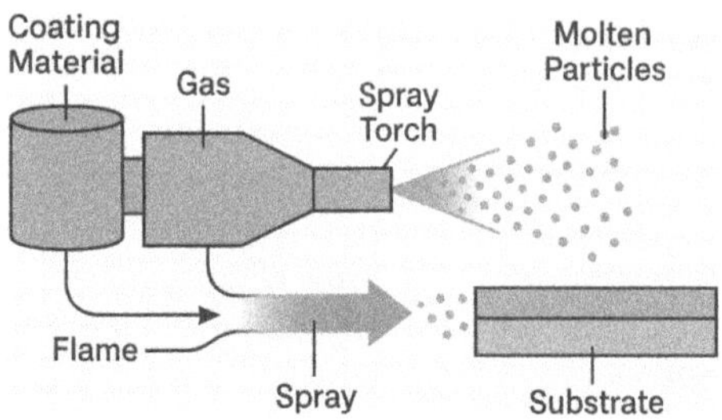

FIGURE 9.4 Schematic diagram of spray coating system.

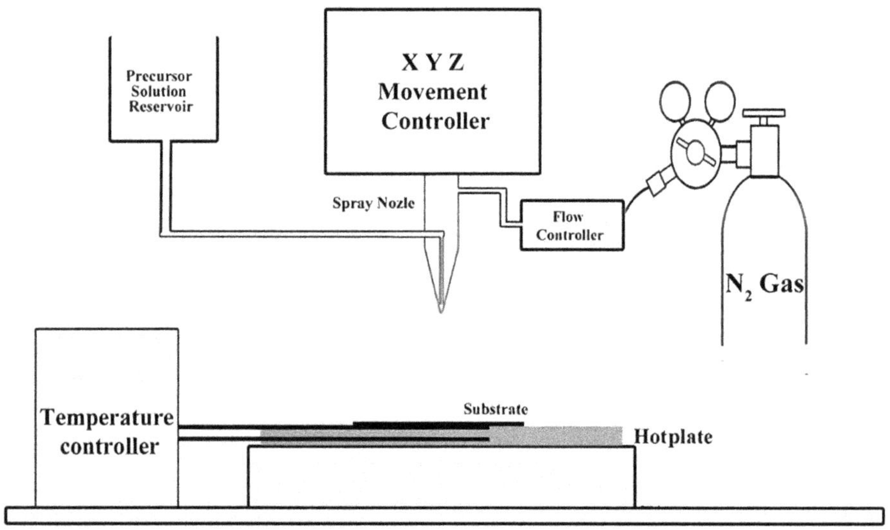

FIGURE 9.5 Schematic diagram of spray coating system (Prabeesh et al. 2021).

and thermal stability of the materials without altering the bulk characteristics of the material beneath. Therefore, it is of great utility to the novel materials, and this even covers the utilization of metals in MMCs.

Figure 9.5 shows an example of typical spray pyrolysis as a method of thin film deposition of materials which can be applied to coat substrates and make them functional. The reservoir problem presents a precursor solution that is sprayed into a spray nozzle that is finely controlled along the three dimensions with the use of an XYZ movement controller. The precursor gets atomized into a fine mist in the presence of nitrogen gas (N_2) that is controlled by the flow controller. This mist is speckled onto a hot substrate plated on a hotplate, and the temperature of this hotplate is controlled by a temperature controller. When the precursor droplets come into contact with the hot substrate, they are broken down by heat and spread out in the thin film. The method finds extensive application in depositing coatings such as oxides and semiconductors, among others, in electronics and materials studies.

9.4.2 Physical and Chemical Vapor Deposition (PVD/CVD)

Coating technologies are useful in the improvement of wear, corrosion, and thermal resistance of materials in modern surface engineering. Two of the most popular of these are physical vapor deposition (PVD) and chemical vapor deposition (CVD) methods of depositing thin high-performance coatings on metal, ceramic, and polymer substrates. Such techniques are of particular importance in the area of MMCs, where the inherent mechanical performance of the base material is augmented by the engineered surface performance of the coatings. PVD is a coating process that is done in a vacuum where the material to be deposited is physically changed into a vapor phase by means of evaporation, sputtering, or an arc discharge. This vapor is condensed on the substrate, wherein a thin, dense film is created. PVD has a number of techniques, which include thermal evaporation, electron beam evaporation, sputter deposition, and arc vapor deposition. The methods have differences related to how the source material is vaporized. To take a few examples, in sputter deposition, ions (typically argon) are fired at a target material, knocking off atoms, which fly to the substrate and deposit there. The properties of PVD coatings include great hardness, reasonable adhesion, and high purity. Titanium nitride (TiN), chromium nitride (CrN), and aluminum oxide (Al_2O_3) are common materials deposited by PVD. PVD finds extensive application in cutting tools, aerospace, decorative applications, and microelectronics because it can produce hard, wear-resistant coatings with good thermal stability.

Figure 9.6 represents the general working principle of PVD and CVD processes that are used in the vacuum chamber, which involves vaporization or chemical reaction of the target material, resulting in a thin film deposited on the substrate. In PVD, ions or atoms are sputtered or evaporated off the target and deposited onto a substrate to create a coating. In CVD, the reactive gases are added to the chamber; then they enter into a chemical reaction at the hot surfaces on the substrate, forming a solid thin film. The environment of a vacuum assures the control of the gas system, equal deposition, and high-purity depositions that are suitable in advanced applications such as working in electronics, optics, and MMCs.

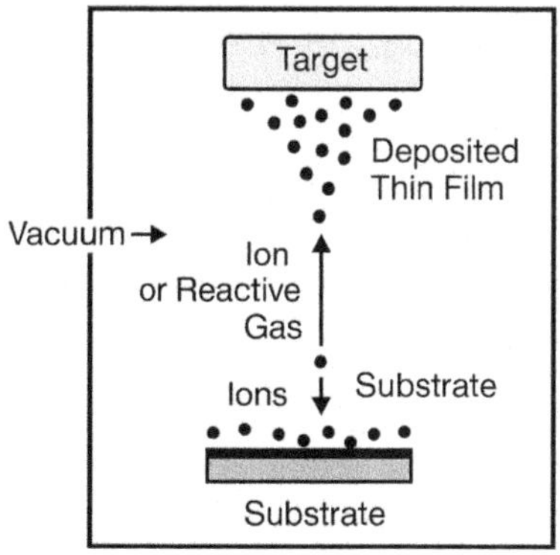

FIGURE 9.6 Schematic diagram of physical and chemical vapor deposition (PVD/CVD).

CVD is a chemical reaction in which volatile precursor gases are reacted on a heated substrate to create a solid coating and products. The reaction is typically carried out at high temperature (usually between 300°C and 1100°C), but low-temperature versions of the process, including plasma-enhanced CVD (PECVD), exist to allow temperature-sensitive materials. The coating in CVD occurs on all surfaces equally, which are in contact with the reacting gases, whether internal or complex shapes; hence, it has an advantage over PVD regarding conformality. Silicon carbide (SiC), titanium carbide (TiC), and diamond-like carbon (DLC) are common coatings deposited by CVD and offer excellent thermal and chemical resistance. A key benefit of CVD is that it can produce dense, uniform, and adherent films, including diffusion barriers, wear-resistant coatings, and corrosion protection. It is widely applied in the semiconductor industry, aerospace, and tool making. The temperature of the processing, however, may be a constraint in using it with temperature-sensitive substrates. Also, some precursor gases used can be toxic and corrosive (e.g., silane, ammonia, and hydrogen chloride) and require special care and disposal, making CVD systems more expensive and complicated.

Although both techniques are used to deposit high-quality coatings, the choice between PVD and CVD depends on the application. PVD is preferred when low-temperature processing is required or when coatings must be extremely pure and hard. On the other hand, CVD is chosen for applications requiring uniform coatings on complex geometries or improved chemical stability.

Both PVD and CVD have transformed surface engineering by facilitating the creation of sophisticated functional coatings with customized characteristics. As the need for high-performance materials grows in industries like aerospace, automotive, electronics, and biomedical engineering, additional advancements in hybrid methods (such as CVD-PVD combinations and atomic layer deposition) are anticipated to overcome the constraints of each individual process and expand their use across various industrial sectors.

9.4.3 Electrochemical Surface Treatments

Electrochemical surface treatments play a wide role in materials engineering, where an effort is made to enhance the surface qualities of metals and alloys, where qualities are to be enhanced in increasing corrosion resistance, wear resistance, electrical conductivity, and aesthetic appearance. Those methods are based on electrochemical reactions that usually concern the electron transfer within the electrolyte solution, as enough of the bulk material to be modified can be avoided. Some of the most well-known ways of electrochemical surface treatment are anodizing, electroplating, electropolishing, and conversion coatings. All of these methods have different advantages and uses depending upon the material and performance that surface requires. Anodization: A controlled electrochemical oxidation process, usually of aluminum and aluminum alloys. During the anodizing process, the metal is the anode in an electrolytic cell and causes a thick, stable oxide film to be formed on its surface. The layer of this oxide is porous and can be subjected to further treatments like dyeing, and has outstanding corrosion and wear resistance. Each of these factors affecting the thickness and properties of an anodized layer, including the voltage, the composition of the electrolyte, and the duration of processing, can be modified accurately to create fine control over the thickness and properties of the anodized layer. The aerospace, automobile, and architectural industries commonly utilize the process of anodizing. Electroplating, on the other hand, involves placing a thin coating of metal on the surface of a workpiece (the cathode) using a metal salt solution by using electric current. The method is very common in improving the surface hardness, reducing friction, and improving corrosion resistance and decorative finishing. Several typical metal elements released into electroplating include chromium, nickel, zinc, and gold. Factors affecting the quality of the electroplated layer include the following: composition of the bath, pH, temperature, and the current density, as well as the plating time. Technological improvement in the field of electroplating also led to the production of composite coating and multilayer systems to maximize the functional performances.

Electropolishing is an anodic treatment that smooths and brightens the metal surface by dissolving only microscopic surface peaks. This process works specifically well on stainless steel and other corrosion-resistant alloys, achieving perfection in their passivity and cleanliness. Electropolishing minimizes the roughness of the surface, which is very essential in medical devices, food processing equipment, and semiconductor components where surface finish and cleanliness are essential.

Another type of electrochemical treatment is conversion coatings, where a layer is created as a result of the chemical reaction between the coating material and the metal substrate. This can include an oxide, chromate, and phosphate coating. Such coatings are corrosion barriers, besides enhancing the adhesion of paint. An example is the chromate conversion coating, commonly applied to aluminum and zinc alloys, but due to environmental issues caused by hexavalent chromium, alternatives that are more environmentally friendly have been created.

Innovation in the electrochemical surface treatment in the recent past has been directed at nanoscale control of the surface structures, hybrid coating integration, or utilization of eco-friendly electrolytes. Other methods, such as plasma electrolytic oxidation (PEO), integrate the properties of anodizing and plasma processes to form ceramic-like protective layers that are more resistant to wear and temperature. Also, the advent of pulse current and reverse pulse plating has enhanced the quality and uniformity of electroplated layers, even on complex geometries.

Finally, electrochemical surface treatment still continues to be an irreplaceable resource in the contemporary surface engineering world, offering customized solutions to prolong the component life and performance. These treatments are constantly being developed to match the advanced industrial needs, and thus, continuous research is being done to achieve better efficiency, improvements in sustainability, and functionality.

9.4.4 Laser Surface Engineering

Laser surface engineering (LSE) is a new theory of processing material composites where very intense and regulated energy of the laser is applied to alter the surface characteristics of the material without a single iteration of the substance's bulk. This has incomparable benefits in the area of accuracy, repeatability, and versatility, and is therefore suited to use in cases demanding local surface improvement. It is also commonly applied in aerospace, automotive, biomedical, and tooling to enhance wear resistance, corrosion protection, hardness, fatigue life, and other functional properties of metallic and non-metallic materials.

The underlying principle of LSE is that a concentrated laser beam is applied to the surface of the material, and due to its intense heat energy, it becomes heated up, melts, and solidifies or vaporizes according to the process parameters. Interaction of the laser and material may cause numerous changes in the surface, such as laser hardening, laser cladding, laser alloying, laser texturing, and laser shock peening. Modern fabricating methods expose the piece to accurate thermal input and small regions of heat-affected material necessary to preserve the integrity of components with more intricate configurations or delicate interior structures.

Laser hardening is usually applied to ferrous alloys because of the specific optical properties of such alloys. Laser hardening can be completed locally at a specific surface at austenitizing temperatures and is subsequently rapidly cooled by the surrounding cooler bulk mass. This will make a hardened martensitic layer, which is an excellent improvement in wear resistance and fatigue. It does not require quenching media, which makes it ideal for precision components; also, it does not result in distortion.

Laser cladding is a process in which a substrate is covered with a metallic coating material that is melted with the help of a laser beam. These create a layer which is bonded metallurgically and has advanced wear, corrosion, and thermal properties. Laser cladding is being increasingly applied to repair costly parts like turbine blades and molds or to process the functionally graded materials by preconditioning surfaces.

Laser alloying alters the surface composition; the surface is melted and alloyed with another element previously deposited or a pre-alloying element. This is an improvement technique on certain surface properties where new phases or microstructures are produced that are not present in the basic raw material. As an example, combining the aluminum alloy with ceramic forming elements, such as Si or Ti, may have huge hardness and wear-resistance improvements.

Laser texturing and laser shock peening are non-thermal laser-based processes that have tribological or mechanical/property enhancement goals. Laser texturing produces directional micro- or nano-texturing patterning of the surface in order to modify friction or lubrication behavior, or hydrophobicity. This is useful in biomedical implants, cutting tools, and microfluidics. In contrast, laser shock peening applies high-intensity laser pulses and accomplishes this by developing compressive residual stresses, which improve fatigue strength and stress corrosion cracking resistance.

Hot events in LSE have been concerned with hybrid methodologies, automation of processes, and the combination of real-time monitoring and closed-loop control systems in order to enhance reliability and performance. With ultrafast and pulsed lasers (like femtosecond lasers or picosecond lasers), novel opportunities for precise micro- and nano-structural changes have developed that have low thermal damage. Moreover, LSE coupled with the AM process is emerging as an effective surface functionalization method for 3D printed parts.

Figure 9.7 is an example of an LSE setup in which a CO_2 laser system is used, and a laser beam (2) is produced by the CO_2 laser source (1) and is reflected by a mirror (3) to a focusing lens (4). The beam is focused by the lens to form a spot of high energy

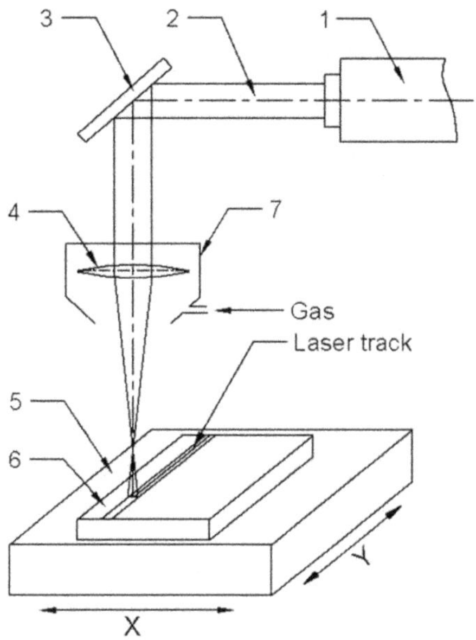

1. CO2 Laser; 2. Laser Beam; 3. Mirror; 4. Lens;
5. Worktable; 6. Workpiece; 7. Cap with Gas Jet Nozzle

FIGURE 9.7 Schematic diagram of laser surface heat treatment (Amine et al. 2014).

that acts upon the surface of the workpiece (6), which is then placed on a sliding worktable (5). Oxidation and control of the thermal environment of the laser interaction zone are achieved by gas jets from a cap (7) through which a protective gas (usually an inert gas such as argon) is directed. The laser beam traces a path along the surface as directed by the X-Y movement of the table, and this produces a "laser track" that modifies the microstructure or composition of the surface to provide the desired increased properties of the surface, for example, wear or corrosion resistance.

Summarizing on LSE, it is a very controllable, versatile, and effective way of customizing surfaces to fit the requirements of a hard application. Its performance through processing of broad materials with both high spatial resolution and limited environmental effects makes it a massive technology in contemporary surface modification as well as in advanced manufacturing.

9.5 CHALLENGES AND CONSIDERATIONS IN SURFACE ENGINEERING AND COATINGS FOR METAL MATRIX COMPOSITES (MMCS)

MMCs' surface-engineering and coating have high potential value as far as wear, corrosion, and thermal resistance go. Nevertheless, multiple technical issues are involved in the dimensions of MMCs as a complex material materialized by a metal base and reinforced with hard ceramics, which should be considered to maintain their efficient and durable performance in the involvement of surfaces by any treatment. The challenges are embedded in the actual differences that exist between the coating materials and the composite substrate on physical and chemical bases, as well as the interaction between the matrix and reinforcement phases in the case of processing at the surface.

A major task is to firmly bond the coating and the MMC substrate. The bonding nature at the interface of MMCs may greatly differ due to the disparate nature of MMCs with respect to the bonding nature interacting with the metallic matrix and the ceramic reinforcement. Such a variation may cause specific weak areas that can start delaminating or cracking, especially when the construction is subjected to mechanical or thermal stress. Furthermore, differences in thermal expansion coefficients between the coatings and the composite substrate may induce residual stresses during the deposition process or in service life, contributing to the problem of the coating failure.

Roughness and topography on the surface are also serious challenges during the coating of MMCs. The presence of hard reinforcement particles like SiC, Al_2O_3, or TiC in the matrix may lead to uneven surfaces or protrusions that can influence the uniformity of coatings. These topographical variations may result in in-fill, pinholes, or micro-cracks in thin films or deposited layers. An example is in the process of thermal spraying, where the hard particles may deflect the molten droplets, and the density and the strength of the bond of the coating would be influenced. This renders surface treatment (i.e., grit blasting or polishing) a necessary and yet challenging exercise to achieve a fine equilibrium between roughness and compatibility between the coating and the surface.

Other important issues are possible chemical reactions between the coating material and the MMC components. However, in thermal-processing applications, such as LSE, thermal spraying, or chemical vapor deposition, in some cases, undesirable interfacial reactions can exist. These responses may deteriorate the reinforcing step, change the interface between the matrix and reinforcement, or create a brittle intermetallic compound that impairs the mechanical quality of the coating-substrate assembly. And, to take an example, nickel- or copper-based coatings in aluminum-based MMCs can give brittle Al-Ni or Al-Cu phases that can degrade both corrosion resistance and fatigue life.

Thermal management forms an important aspect in the process of surface treatment of MMCs. Most surface engineering processes have localized or bulk heating that may result in alteration in the MMC substrate microstructure. Thermal gradients and phase changes may result from differential heating, changing properties of the matrix, or distorting it. Moreover, heat transfer through the composite might be completely different depending on the reinforcement content and influence heat removal in the laser cladding or plasma spraying. Such effects require exact control of such processing parameters as power input, scanning rate, and cooling rate.

Compatibility with the environment or the process should also be taken into consideration. Certain surface treatments use toxic precursors (such as in CVD processes) or necessitate inert conditions, which can add cost to operations and complicate them. Moreover, the coatings made on normal metals are not likely to play a similar role to the MMCs, and we will have to customize coatings and application approach to MMCs. The feasibility of coating processes and economic viability are also considerations, especially in those industries where mass production is important and uniformity of quality matters.

Finally, as revealed, though surface engineering is a major science that has increased the functionality of MMCs, it nonetheless comes with technical challenges. It is essential to attain high adhesion strength, and establish thermal and chemical interactions, as well as surface morphology and compatibility with the environment. Proper comprehension of the microstructure and behavior of the MMC in the processing environment is critical to identifying and optimizing coating processes. Further studies in designs of specific coatings, novel deposition techniques, and surface analysis would slowly defeat these and allow the full potential of MMCs to be realized in advanced engineering solutions.

9.6 TRIBOLOGICAL BEHAVIOR OF METAL MATRIX COMPOSITES (MMCS) UNDER LUBRICATED CONDITIONS

MMCs have been subjected to much attention in connection to tribological applications, given their pronounced wear resistance, strong strength-to-weight ratio, and outstanding thermal stability. MMCs with under-lubricated conditions during operation exhibit a substantial effect on tribologically attributed properties due to the kind of matrix material utilized, reinforcement character and dispersion, operating parameters, and the nature of the lubricant utilized. The essential role of lubrication is clear, and it involves a low-friction wear reduction or reduction of friction by forming a protective film that separates the contacting surfaces, avoiding direct contact of the metals. The lubricated performance is also increased by the existence of hard reinforcements (e.g., SiC, Al_2O_3, or graphite) that may result in mechanical support, or even chemical interaction, of the lubricant, and thus be able to change the tribo-interface in MMCs.

Lubricants lessen the severity of the wear processes like abrasive, adhesive, and oxidative wear processes, which are usually seen under dry conditions. In both boundary and mixed lubrication conditions, the tribological performance of MMCs is enhanced since a film of lubricating layer is created, which reduces warming or oxidation and provides room for load-bearing capacity. The performance of this film, however, will be a factor of such properties as surface roughness, sliding velocity, and contact pressure, as well as the capacity of the reinforcement to hold the lubricant or have contact with it. The graphite-reinforced MMCs, for example, can exhibit better lubricated wear performances because of the self-lubricating property of graphite particles, which create a transfer layer at the interface and increase film stability.

Under the hydrodynamic aspect of lubrication, the friction coefficient in oil-lubricated environments is relatively dictated by the property of the lubricant, and as such, the reinforcement effect is less pronounced. The contact between the additives in the lubricant and the composite surface, however, becomes critical during boundary lubrication. The wear can be further minimized by reacting the surface with additives; a tribo-chemical film is formed, such as zinc dialkyldithiophosphate (ZDDP). Such tribofilms are usually anchored by MMCs that contain ceramic reinforcements, resulting in greater longevity of the surfaces in contact.

In addition, the thermal conductivity of the MMCs also influences the tribological responses during lubricated surfaces. Fine thermal conductivity aids the natural removal of frictional heat, reducing lubricant degradation, and preserving its viscosity. This is especially critical in high-speed applications, whereby temperature rise may cause the lubricant to fail. The reinforced MMCs, and those with carbon-based materials or possessing ceramics with high conducting tendencies, are generally able to exhibit better performance in this respect.

Conclusively, MMCs have lower tribological properties at lubricated conditions than conventional alloys when they are properly designed with regard to the type and distribution of reinforcements. The interaction between the lubricant and the surface made up of composites is of crucial importance to the frictional behavior and wear rate. Therefore, to achieve enhanced tribological behavior in lubricated conditions, an integrated strategy taking into account the design aspect of materials and the engineering of lubricants is indeed required. Further investigations in the area are essential to develop MMCs with tailored properties advanced for tribological applications, including automobile engine parts, airplane systems, and high-speed equipment.

9.7 TRIBOLOGICAL BEHAVIOR OF METAL MATRIX COMPOSITES (MMCS) UNDER DRY CONDITIONS

The tribological response of MMCs to dry sliding conditions represents an area of primary interest in view of extreme service conditions where they are expected to be utilized such as in the aerospace, automobile, and defense industries. The performance of MMCs under dry conditions, that is, when no external film is used, is largely determined by inherent qualities of the matrix, the type and grain size distribution, content of reinforcements, and the operating parameters, such as the sliding velocity, the applied load, and the temperature. Direct contact of the asperities results in increased wear rates and friction and necessitates the use of materials that have high wear resistance in nature, as well as being stable at a high temperature in the absence of lubricants.

MMCs also have enhanced wear resistance in dry sliding as compared to monolithic metals; this is because MMCs form strong bonds through hard ceramic reinforcements (silicon carbide (SiC), aluminum oxide (Al_2O_3), boron carbide (B_4C), and titanium carbide (TiC)). These reinforcements are useful in supporting a high percentage of the total amount of loading applied and opposing plastic deformation, thus aiding in the reduction in actual area of contact. Consequently, MMCs have a lower wear rate compared to unreinforced ones. The efficiency of this mechanism, however, is greatly dependent on the interfacial bonding of the reinforcement and the matrix. The low interfacial strength may cause the pull-out of particles, which causes a higher amount of third-body abrasion.

During mechanical testing under dry conditions, wear mechanisms of MMCs could be different—such as abrasive, adhesive, oxidative, and delamination wear—depending on the conditions and the composition of the test materials. Abrasive wear may prevail where there is plowing or cutting by hard particles or asperities into the weaker matrix. This is resisted using reinforcements such as SiC and Al_2O_3, which serve as a block to the cutting action. But at greater loads or sliding velocities, there may be thermal softening of the matrix with resultant adhesion or delamination. In some cases, oxidation wear even becomes significant, where frictional heat causes oxidation of the surfaces, forming oxide imprints, which periodically might help in the formation of a protective film, but these can also crack and contribute to debris formation.

The other important issue under dry sliding is the production and significance of wear debris. In most MMCs, the debris may further cause wear by serving as abrasive third bodies or, in contrast, create a compacted tribo-layer that stabilizes the surfaces and provides protection. Formation of this layer is influenced by composition, temperature, and reinforcement morphology. As an example, graphite- or MoS_2-reinforced MMCs have self-lubricated behavior where the solid lubricant phases can creep across the contact interface to de-stress the interface and create a continuously renewed lubricated layer, thereby inhibiting wear and friction both with and without fluids.

Dry sliding also exhibits the pronounced thermal effects. The interface heating can cause matrix softening and/or stresses caused by thermal expansion mismatch or even degradation of the composite structure. Hence, the thermal conductivity of the MMC is also a crucial factor, and the materials that have higher conductivity can eliminate heat efficiently, and there is a lesser possibility that the parts will implode due to localized melting or the failure of structures.

Overall, the tribological behavior of MMCs under dry sliding conditions is characterized by a complicated interaction of different phenomena due to the joint impact of mechanisms of mechanical and thermal phenomena on tribological behavior. Choosing the reinforcement and its design is important in customizing MMCs to fit applications that are wear-resistant. They perform very well when compared to other traditional metals when the environment is dry, but it is critical to optimize some of the parameters like the bonding of reinforcement with the matrix, the volume of the reinforcement, and also the parameters of operation in order to tap their potential to the full. Further development of nano-reinforced MMCs and functionally graded composites has the potential to further improve even in the dry wear conditions.

9.8 THERMAL AND MECHANICAL EFFECTS ON TRIBOLOGICAL PROPERTIES

9.8.1 Thermal Effects on Tribological Properties

The effect of temperature is a key determining factor in the influence of tribological properties of materials. In high temperatures, friction and wear can be significantly impacted by material softening, oxidation, phase transitions, and interfacial property variations. In MMCs, the matrix and reinforcement materials may differ in their thermal expansion coefficients and may create residual stresses or debonding at the interface when subjected to thermal cycles. As an example, aluminum-scaffolded MMCs strengthened with SiC or Al_2O_3 deteriorate in wear resistance over 300°C since the matrix softens and the bearing capacity decreases. Also, the formation of oxide films at elevated temperatures may not only protect the surface (forming a lubricating tribo-film) but may also act as a third-body abrasive (depending on how the oxide forms mechanically and its adhesion).

Moreover, thermal conductivity has an effect on heat loss in the process of sliding. Low thermal conductivity results in localized heating, which accelerates wear and can even lead to thermal fatigue or cracking. To illustrate, titanium-based composites having a low level of thermal conductivity are highly susceptible to adhesive wear at high speeds during sliding. By comparison, other composites that contain graphite as a reinforcement material are more thermally stable and have a low friction coefficient even under equivalent circumstances.

In extreme temperatures, metallic matrices are likely to lose the strength that they have as a result of thermal softening. This leads to decreased load-bearing capacity in sliding or rolling contact, hence raising the wear rate. As an example, at temperatures above 300 °C, the wear of aluminum-based MMC reinforced with SiC drastically increases because the material softens and provides little support to reinforcements.

Alpas and Zhang (1992) observed that at or above certain temperatures, the softening of the aluminum matrix is a result that allows the aluminum material to experience more plastic deformations during the sliding process, therefore contributing to the removal of the material through adhesion and delamination. The result of this is that the worn surface has a smeared morphology, as experienced with soft metals during sliding.

MMCs are matrix metallic materials, which are reinforced by ceramics. These materials tend to have varying coefficients of thermal expansion (CTE). The outcome of such a mismatch is the formation of residual stresses at the interface under thermal loading or cycling. At higher temperatures, interfacial debonding or microcrack formation may be reached, the reinforcement-matrix bonding is weakened, and the wearability of the composite will increase.

To illustrate, some composites of Al-SiC have thermal expansion that is almost threefold that of SiC, so at high temperatures, there exist stresses that lead to tension on the SiC particles. This causes interfacial decohesion with a consequent reduction in the efficiency of load transfer and an increase in matrix wear (Ravichandran and Karthikeyan 2001).

Oxide films on the surface of the material can be formed when sliding contacts are working at high temperatures. Depending on their nature and stability, these oxides may be used as solid lubricants or third-body abrasives. As an example, friction can be minimized when aluminum oxide (Al_2O3) precipitates during the high-temperature conditions because of its high hardness and thermal stability. But where the oxide layer is tough or weakly attached, it is also possible to spall off and add to wear by abrasion processes.

Composites research on titanium- and nickel-based composites also indicates the formation of tribo-films such as TiO_2 and NiO, alleviating metal-to-metal interaction, thereby enhancing resistance at high-temperature contact (Sundararajan and Venkataraman 2000).

Other MMCs change phases or dissolve when exposed to high temperatures. The effect of this microstructure alteration is on hardness, toughness, and wear resistance. To take an extreme case, in Al-based composites with precipitate reinforcement by high-temperature treatment, the particulate can lead to precipitate coarsening or even dissolution during prolonged sliding at a high temperature, reducing the matrix hardness and enhancing wear (Hemanth 2009). Conductivity and heat dissipation capabilities of a material directly affect the surface temperature of the material during sliding. The high-thermal-conductivity materials (e.g., graphite- or copper-reinforced MMCs) reduce the surface temperatures and thus aid in avoiding thermal softening and localized microstructural damage. For example, even when irradiated over an extended time, the friction coefficient of graphite-reinforced Al composites is more stable than in the case of plain alloys because graphite is considered a lubricant and heat dissipator (Rohatgi et al. 2001a).

9.8.2 Mechanical Effects on Tribological Properties

Friction and wear performance are directly connected to mechanical parameters, especially to load, sliding velocity, contact pressure, and hardness. Normally, the increase in the normal load applied tends to result in an increase in the real area of contact, which boosts adhesive interactions and thus wear. But, in MMCs, inclusion of hard reinforcements (such as SiC, B_4C, or AlN) leads to a dramatic difference in the mechanical behavior of the contact, hence affecting the deformation in the contact area and enhancing load-carrying capacity. These effects usually increase wear resistance as the reinforcement content increases, though a very high reinforcement content level may cause fracture of reinforcements or debonding of matrix–reinforcement interfaces, causing greater wear.

The dynamic effects that are brought up by sliding speed include surface oxidation and frictional heating. MMCs tend to exhibit stable wear behavior at moderate speeds because of the balance between heat generation and heat dissipation. But at extremely high speeds, the wear mechanisms of thermal softening and delamination take over, especially in ductile matrix composites. Hardness is another important factor: the harder the material, the less it should wear, as is the case with ceramic-reinforced aluminum composite. Nevertheless, highly stiff reinforcements may also result in counterface wear, hence being less preferable to use in soft mating surface applications.

Contact pressure directly depends on the normal load used; it is responsible for altering the actual contact area of two surfaces. The plastic deformation of asperities on the surface, which leads to adhesion and material transfer, increases with the load and leads to increased wear and the friction force. In MMCs, the extent to which a particular type of response to the load depends on the nature, the volume fraction, and the distribution of the reinforcement phase.

Researchers in Al-SiC composites experienced a switch between mild oxidative wear and serious delamination wear once the applied load exceeded a particular limit. It shows that wear at low loads is dominated by microplowing and a light form of abrasion, whereas microplowing and slight particle fracture dominate at higher loads (Alpas and Zhang 1993).

Frictional heating makes sliding speed an important factor that affects the temperature at the contact interface. The interaction is more mechanical (abrasion or plowing) at low speed, and at high speed, effects of heating are mixed with those of mechanical action, which enhances faster softening of the matrix, oxidation, and fatigue. An investigation of graphite-reinforced Al MMCs showed that the friction coefficient became lower with the increasing speed as a result of the creation of the lubricious tribo-film, and the wear rate either declined (because of improved lubrication) or increased (because of softening) within the working range (Zhang et al. 2007). One of the major mechanical properties that influences wear resistance is the hardness. Ceramic-reinforced MMCs incorporated with particles such as SiC, Al_2O_3, or B_4C usually have enhanced wear resistance in the form of greater surface hardness. Ravikiran and Surappa (1997) argue that the wear rate of Al-SiC composites slows down as the amount of SiC increases, in large part due to the hardness of the composites and their ability to share a load. But if over-reinforcement takes place, it could result in brittleness as well as an elevated case of particle pull-out during loading.

Morphology (shape and size) and distribution (uniformity and dispersion) of the reinforcement particles occupying an MMC affect its tribological behavior, with respect to the wear resistance, phonon behavior, and load-bearing capacity, to a great extent. Such microstructural attributes have a direct influence on the way the composite reacts to both mechanical and thermal stresses when in the state of sliding or in an abrasive contact. MMC reinforcement particles are freely varied in shape and include angular to spherical shapes and flake-like to fibrous shapes, and these shapes are critical to load transfer and the wear process. Wear resistance is improved by angular and irregular shapes of particles, for example, crushed SiC or B_4C, which increase the mechanical interlocking between the particles and the matrix. Nevertheless, on the one hand, sharp edges of such particles may become stress enhancers and facilitate the initiation and propagation of cracks, particularly when they occur in cyclic or high-load conditions (Sahin 2003). On the other hand, spherical or rounded particles have a tendency to eliminate stress concentration, giving rise to improved fatigue characteristics as well as ductility of the composite. They may, however, exhibit lower anchoring in the matrix, hence less wear resistance in harsh abrasive conditions. Tribological performance is affected twice by particle size. The smaller particles present increased load-transfer surface area and dispersion and improved hardness, wear rate, and more even deformation behavior. They also contribute to the development of a compact tribo-layer during sliding, which will be a protective layer against any wear. But on rare occasions, ultra-fine or nano-sized reinforcements may agglomerate during processing, forming local inhomogeneities and voids that are detrimental to wear performance. One should also be concerned with the size of larger particles, which are easier to dispense when casting or processing a powder, debond under the load, or behave as abrasives themselves, causing more wear of the counterface (Torralba et al. 2003). The tribological behavior has to be consistent, and thus, a uniform distribution of reinforcement particles is imperative. Even dispersion assists in a stable, steady distribution of load on the surface of contact, minimizes the occurrence of local deformation, and increases the creation of a stable transfer film. It also thwarts concentrations of overheating or premature wear points when the sliding or abrasive wear takes place. Particle clustering might be caused, however, by non-uniform distribution. Such clusters form brittle areas liable to cracking and debonding that greatly impair wear resistance. Clustering can also cause an uneven removal of material and uncontrollable variation in the coefficient of friction (Kok 2006). Tribological properties are also influenced by orientation in discontinuously reinforced MMCs, the reinforcements of which are fibers or flakes (e.g., short fibers or graphite flakes). The sliding can be simulated by reinforcements that are oriented along the direction of the sliding, and this can help in decreasing the friction either by the use of solid lubricants or by reducing the roughness of a contact. Nevertheless, the transverse or random orientation is superior in terms of wear resistance, as it intercepts more wear tracks and causes resistance to the matrix deformation (Tjong and Ma 2000).

In case of non-intensive bonding of particles or poor distribution, there exists clustering, which would generate high-stress areas. These clusters are prone to debonding or being pulled out during sliding to create voids, which initiate wear. The loose particles can also become third-body abrasives, which raise the wear rate and the surface roughness. Various studies establish that one of the main reasons that cause wear impairment in inadequately made MMCs, particularly those that have been made through the use of stir casting with poor stirring or wetting performance, is particle pull-out (Surappa 2003).

The nature of tribology is even more complicated when thermal and mechanical stresses are present together (i.e., in high-load and high-speed sliding). Such environments in many composites bring about a shift from mildly oxidative wear to rather harsh adhesive or delamination wear. As an example, Al-SiC studies have revealed such behavior: the wear rate and coefficient of friction rise steeply at a critical temperature–load threshold, mainly due to matrix softening and interface decohesion. Additionally, through repeated application of the thermal cycling, the wear performance may be compromised due to the fatigue that forms at the interface. Composites and reinforcement (such as TiC or MoS_2) ideally should be designed to be thermally stable to relieve such impairments; however, employing solid lubricants like graphite and WS_2 can also have a beneficial influence. Those with hard and lubricious reinforcements blended (hybrid composites) offer potential in experiencing lower wear and friction even when exposed to severe thermo-mechanical conditions. These techniques are particularly useful in components of aerospace, automotive, and braking systems because high temperatures and stresses are present concurrently.

9.9 CHALLENGES IN TRIBOLOGY OF MMCS

MMCs are created materials that have the same hardness and ductility properties as metals and the strength and wearability of ceramic reinforcements. Although they have much promise as applications to the automotive industry, aerospace industry, military, and tooling industry, the tribological issues affecting their performance in these areas with different environmental conditions have yet to achieve desirable solutions. MMCs' tribology, which focuses on their friction, wear, and lubrication behavior, is inevitably complex because these composites are heterogeneous in nature and the environments under which they operate can be intense.

Non-uniform distribution of the reinforcement particles, for example, silicon carbide (SiC), aluminum oxide (Al_2O_3), or graphite, is one of the most important issues in the tribology of MMCs. This non-uniformity during the manufacturing process may result in the agglomeration or clustering of reinforcements and result in the localized stress concentration, difference in wear, and even micro-cracking under cyclic loads. This is particularly important when a smooth surface characteristic is called upon, as in the case when using brake rotors or lining in an engine's cylinder (Prasad & Asthana 2004). Even where the reinforcement and matrix materials have different densities and thermal expansion coefficients, the problem of porosity or poor interfacial bonding can normally be expected (Surappa 2003).

The other serious impending issue lies in the interface interactions between the reinforcement phase and the matrix. The strength and stability of this interface are critical to wear resistance and load-bearing properties of MMCs. Increased temperatures experienced at high temperatures can cause a chemical reaction between the matrix and reinforcement to form brittle intermetallic compounds, which will adversely affect the tribological properties of the composite. As an example, Al-SiC composites are also potentially unable to resist interfacial degradation, because when the Al_4C_3 is formed, the result is hygroscopic, and it erodes mechanical integrity (Zhang et al. 2021). Debonding along the interface in the case of loss may also result in particles of reinforcement being pulled out, causing a greater coefficient of friction and roughness.

Another complexity exists because of temperature effects in tribological interactions. Thermal softening of the metallic matrix of MMCs in high-temperature applications, such as blade materials in turbines or rings in pistons, can also decrease hardness and result in wear. Simultaneously, reinforcements might not stretch or carry heat identically to the matrix, creating internal stress and microstructure wear and tear. When applied to dry sliding conditions, such thermal effects may cause oxidative or delamination types of wear, either of which has an abysmal effect on the life of a component (Davis 2001; Bhushan 2013).

There are also limitations in the production and finishing of MMCs and their surface finishing. MMCs are not as suitable for traditional machining methods because of the existence of hard particles and hence their consequent tool wear and poor finishing. Additionally, the mass production of MMCs faces a considerable obstacle, being rather expensive and energy-demanding (requiring the advanced method of PM or squeeze casting, etc.) (Rohatgi et al. 2001b). Despite their potential in improving the tribological performance, surface treatments such as laser cladding, thermal spray, and PVD/CVD coatings have been recorded to face adhesion problems or residual thermal stresses, which result in a performance setback in the long run.

Still an unexplored but essential area is the lubrication and environmental compatibility of MMCs. MMCs have been applied in partially or dry lubricated applications in most instances. Due to the contact of lubricants with MMC surfaces at high temperatures, a tribochemical reaction or lubricant degradation may occur. Using an example, likely, traditional oils will not be able to offer adequate protection to MMCs made of aluminum when placed under extreme pressures, which results in scuffing or seizure (Prasad & Asthana 2004). It also happens when certain types of reinforcements are chemically incompatible with lubricant additives and speed up wear or corrosion.

Finally, scaling up from laboratory testing to real-world industrial applications presents challenges in standardization and predictability. Many tribological evaluations are conducted under idealized conditions that do not replicate real service environments involving variable loads, speeds, and temperatures. As a result, there is a lack of standardized test protocols and long-term field data, making it difficult for designers and engineers to confidently implement MMCs in critical components (Zhang et al. 2021). The cost-performance trade-off is also a major concern; while MMCs offer superior wear resistance, their higher processing and machining costs must be justified by improved life-cycle performance and reduced maintenance.

In conclusion, while MMCs exhibit excellent potential for wear-critical industrial applications, their widespread adoption is hampered by challenges related to material design, manufacturing complexity, interfacial stability, thermal degradation, lubrication incompatibility, and performance predictability. Addressing these issues requires interdisciplinary research focused on advanced processing methods, robust modeling of tribological behavior, innovative lubrication strategies, and real-time monitoring of wear phenomena under operational conditions.

9.10 FUTURE TRENDS OF TRIBOLOGY OF MMCS

The changing face of materials science and the growing needs of the auto, aircraft, and industrial industries for high-performance, wear-resistant materials keep fueling the progress of tribology in MMCs. Gazing into the future, the current trends in this area can be estimated mainly as the innovations in material design, smart tribological systems, nanotechnology, and eco-friendly manufacture. The goal of these developments is to increase the performance, reliability, and environmental compatibility while reducing costs.

The development of nano-reinforced composites qualifies as one of the most noticeable future trends in MMC tribology. Some nanoparticles, which include graphene, CNTs, and nano-silicon carbide, provide a better surface area-to-volume ratio, good enough to enhance load-bearing capacity, minimizing friction and wear. Compared to the conventional micro-level reinforcements, nanomaterials have the potential to increase hardness, thermal stability, and the formation of the tribo-films in a sliding system (Ravindran et al. 2019). Nevertheless, even dispersion and interfacial bonding of such nanoreinforcements continue to pose processing problems, which are currently actively being sought with emerging methods such as ultrasonic-assisted casting and surface functionalization.

The other important trend is that of incorporating self-lubricating materials into MMCs. Dry/semi-lubricated operation in harsh environments. There exists a potential to run without lubricant or with a minimal amount of lubricant possessing a metallic matrix, for example, with solid lubricant graphite, molybdenum disulfide (MoS_2), or hexagonal boron nitride (h-BN) incorporated into it. These materials have the potential of forming lubricious tribo-layers during sliding in order to minimize friction and wear without external lubrication. There is new research on dual-reinforcement approaches, mixing hard ceramic and soft solid lubricant paste powder to fine-tune both mechanical and tribological properties (Alidokht et al. 2011).

There is also the aspect of smart tribological systems, which is a revolutionary trend toward MMC application. These are composites that have adaptive/responsive behavior to external stimuli, like load, temperature, or speed. An example is that functionally graded MMCs could deliver spatial hardness and wear resistance to suit the stress profile in a part. Self-healing composites, that is, surfaces that can renew by regrowing a wear layer through embedded phase transition or micro-capsule release of lubricant, are also being pursued by researchers. Such materials are especially appealing to aerospace and biomedical uses (where maintenance cannot be done, or is hard to do, in operation).

Coating surfaces and engineering are also likely to change a lot when it comes to MMCs. There is potential in hybrid surface treatments integrating PVD, laser cladding, and thermal spraying with high-tech surface coatings such as DLC, nanocomposite, or ceramic in order to increase surface hardness and wear resistance. The cost-effective enhancements in high-friction components like cutting tools, valve seats, and piston liners are achievable not by changing the bulk material but by tailoring the surface properties (Kumar et al. 2021).

Computational perspective: In regard to the future of tribology in MMCs, multi-scale modeling and machine learning are of great benefit. There are plans to develop advanced simulation programs combining molecular dynamics (MD), finite element methods (FEM), and tribomechanical models to have more accurate predictions of wear and friction over various lengths and time scales. To find and optimize new MMCs with specific tribological qualifications more cheaply, machine learning algorithms are being taught over huge superstructures of material properties and wear performance

The future of MMCs is also shaped by sustainability and environmental considerations. Recyclable matrices (e.g., magnesium and aluminum) and bio-based or waste-derived reinforcements are becoming increasingly interesting exploits, particularly with regard to environmentally friendly uses in tribology. In one example, ashes from agricultural waste or even industrial by-products are under consideration as low-cost and environmentally friendly reinforcement having reasonable tribological characteristics. Moreover, AM technologies are gaining more and more popularity to produce MMCs with desired architecture, minimal production breakdown, and specific surface properties.

To sum up, the tribology in MMCs is multi-dimensional and innovation-based in the future. As the fields of nanotechnology, smart materials, and surface engineering are joined together with data-driven design, MMCs have an opportunity to provide high-performing and tailored solutions to the tribology problems of present industries. To ensure that these trends could be adapted to become broadly applicable within the industry, it will be important to overcome the barriers that currently exist in terms of processing, scalability, and long-term performance validation.

REFERENCES

Alidokht, S. A., A. Abdollah-Zadeh, and S. Soleymani. 2011. "Tribological Behavior of A356/10%SiC Aluminum Matrix Composites Fabricated by Stir Casting." *Wear* 271(11–12): 1650–1659. https://doi.org/10.1016/j.wear.2010.12.055

Alpas, A. T., and J. Zhang. 1992. "Effect of Temperature on the Sliding Wear Performance of Aluminum Alloys and Composites." *Wear* 155(1): 83–104.

Alpas, A. T., and J. Zhang. 1993. "Wear Mechanisms in Aluminum Matrix Composites." *Wear* 161(1–2): 125–140.

Amine, T., J. W. Newkirk, H. D. F. El-Sheikh, et al. 2014. "Microstructural and Hardness Investigation of Tool Steel D2 Processed by Laser Surface Melting and Alloying." *International Journal of Advanced Manufacturing Technology* 73: 1427–1435. https://doi.org/10.1007/s00170-014-5882-8

Bhushan, B. 2013. *Introduction to Tribology*. 2nd ed. John Wiley & Sons.

Davis, J. R., ed. 2001. *ASM Specialty Handbook: Aluminum and Aluminum Alloys*. ASM International.

Ezatpour, H. R., M. H. Sabzevar, and S. H. Seyed Ebrahimi. 2012. "Microstructure and Wear Behavior of Al–Al_2O_3 Nanocomposites Fabricated by Powder Metallurgy." *Journal of Materials Science & Technology* 28(3): 252–257.

Hashim, J., L. Looney, and M. S. J. Hashmi. 1999. "Metal Matrix Composites: Production by the Stir Casting Method." *Journal of Materials Processing Technology* 92–93: 1–7.

Hemanth, J. 2009. "Investigation on Wear Characteristics of Permanent Mold Cast MMC and Its Hybrid Composites." *Wear* 266(11–12): 1168–1177.

Kok, M. 2006. "Production and Mechanical Properties of Al_2O_3 Particle-Reinforced 2024 Aluminium Alloy Composites." *Journal of Materials Processing Technology* 171(2): 262–267.

Kumar, S., R. Kumar, and A. K. Jha. 2021. "Recent Developments and Future Trends of Surface Coatings for Enhanced Tribological Performance: A Review." *Surface & Coatings Technology* 421: 127390. https://doi.org/10.1016/j.surfcoat.2021.127390

Kumar, S., et al. 2019. "Development and Characterization of Hybrid Aluminum-Based Metal Matrix Composites for Automotive Applications." *Materials Today: Proceedings* 18: 3812–3818.

Mazahery, A., and M. O. Shabani. 2010. "Influence of Reinforcement Type on Tribological Behavior of Aluminum Matrix Composites." *Tribology International* 43(1–2): 455–460.

Prabeesh, P., V. G. Sajeesh, I. P. Selvam, et al. 2021. "Influence of Thiourea in the Precursor Solution on the Structural, Optical and Electrical Properties of CZTS Thin Films Deposited via Spray Coating Technique." *Journal of Materials Science: Materials in Electronics* 32: 4146–4156. https://doi.org/10.1007/s10854-020-05156-y

Prasad, B. K. 2001. "Wear Characteristics of Aluminum Alloy–SiC Composites under Various Conditions." *Wear* 248(1–2): 73–78.

Prasad, S. V., and R. Asthana. 2004. "Aluminum Metal-Matrix Composites for Automotive Applications: Tribological Considerations." *Tribology Letters* 17(3): 445–453. https://doi.org/10.1023/B:TRIL.0000044492.91991.f3

Ramesh, S., T. R. Ramesh, and S. J. Rajesh. 2018. "Review on Tribological Performance of Metal Matrix Composites." *Materials Today: Proceedings* 5(1): 2743–2751.

Ravichandran, M., and R. Karthikeyan. 2001. "Influence of Thermal and Mechanical Loads on the Wear Behaviour of Al-SiC Composites." *Materials Science and Engineering: A* 315(1–2): 104–113.

Ravikiran, K., and M. K. Surappa. 1997. "Effect of Sliding Speed on the Wear Behavior of A356–SiCp Composites." *Wear* 206(1–2): 33–38.

Ravindran, P., K. Manisekar, and N. Murugan. 2019. "Nanoparticle-Reinforced Metal Matrix Composites – A Review." *Materials Today: Proceedings* 18: 897–906. https://doi.org/10.1016/j.matpr.2019.06.105

Rohatgi, P. K., R. Asthana, and S. Das. 2001a. "Solidification, Structure, and Tribological Behavior of Cast Metal-Ceramic Particle Composites." *International Materials Reviews* 37(1): 1–29.

Rohatgi, P. K., et al. 2001b. "Metal-Matrix Composites for Automotive Applications." *JOM* 43(4): 10–15.

Sahin, Y. 2003. "Preparation and Some Properties of SiC Particle Reinforced Aluminium Alloy Composites." *Materials & Design* 24(8): 671–679.

Sahin, Y. 2004. "The Effects of Reinforcement Content and Matrix Particle Size on Wear Resistance of Aluminum–Graphite Composites." *Materials & Design* 25(5): 371–376.

Sarkar, S., et al. 2017. "Effect of Heat Treatment on Mechanical and Tribological Properties of Aluminum-Based MMCs." *Materials Research Express* 4(10).

Sharma, N., R. Khanna, G. Singh, and V. Kumar. 2017. "Fabrication of 6061 Aluminium Alloy Reinforced with Si_3N_4/N-Gr and Its Wear Performance Optimization Using Integrated RSM-GA." *Particulate Science and Technology: An International Journal* 35 (6): 731–741.

Singh, G., and S. Goyal. 2016. "Dry Sliding Wear Behaviour of AA6082-T6/SiC/B_4C Hybrid Metal Matrix Composites Using Response Surface Methodology." *Proceedings of the Institution of Mechanical Engineers, Part L: Journal of Materials: Design and Applications* 232 (11): 952–964.

Singh, G., S. Goyal, G. Miranda, and N. Sharma. 2018. "Parametric Study of the Dry Sliding Wear Behaviour of AA6082-T6/SiC and AA6082-T6/B_4C Composites Using RSM." *Journal of Mechanical Science and Technology* 32 (2): 579–592.

Sundararajan, G., and B. Venkataraman. 2000. "The Role of Oxide Layers in Controlling the Tribological Behavior of Aluminum Matrix Composites." *Wear* 245(1–2): 22–38.

Surappa, M. K. 2003. "Aluminium Matrix Composites: Challenges and Opportunities." *Sadhana* 28(1–2): 319–334. https://doi.org/10.1007/BF02717141.

Tjong, S. C., and Z. Y. Ma. 2000. "Microstructural and Mechanical Characteristics of In Situ Metal Matrix Composites." *Materials Science and Engineering: R* 29(3–4): 49–113.

Torralba, J. M., F. Velasco, and J. Rodríguez. 2003. "Wear Behavior of Particulate-Reinforced Metal Matrix Composites for Automobile Applications." *Journal of Materials Processing Technology* 133(1–2): 538–544.

Vishwakarma, J., K. Chattopadhyay, and N. C. Santhi Srinivas. 2022. "Effect of Build Orientation on Wear and Erosion Behavior of Maraging Steel Processed by Powder Bed Fusion Using Laser Beam (PBF-LB)." *Surface Topography: Metrology and Properties* 10: 045006

Zhang, D., Z. Chen, and J. Liu. 2007. "Effect of Sliding Speed on the Tribological Behavior of SiC Particle Reinforced Aluminum Matrix Composites." *Wear* 262(11–12): 1353–1359.

Zhang, S., L. Wang, and M. Chen. 2021. "Tribological Behavior of Advanced Metal Matrix Composites: A Review." *Journal of Materials Science & Technology* 60: 97–117.

10 Advanced ML Algorithms for Nanotechnology

Shaman Gupta, Neeraj Sharma, and Lagouge Tartibu

10.1 INTRODUCTION

Nanotechnology, the process of deliberately changing the structure of matter at the microscopic scale, has been applied in medicine, electronics, material science, and power (Contera, 2019). As a branch of technology that has a dimension of less than 100 nanometers, nanotechnology makes it possible to produce compounds and devices that perform and function differently at a larger scale (Madkour, 2019a). Its possible uses include drug delivery nano-platforms, superb energy storage devices, and superior and reusable building materials, which are lighter, stronger, and more flexible than the typical types.

According to Dananjaya et al. (2024), the enhancement of machine learning (ML) with nanotechnology is the catalyst for subsequent progress. ML algorithms are particularly good at making predictions or conclusions. They are based on large and difficult-to-manage data sets. It is a useful feature when it comes to solving problems at the nanoscale (Dananjaya et al., 2024). They can predict material properties, thereby lowering the experimental costs. In the design and optimization of nanomaterials and devices, automation with the aid of ML speeds up the process, while deep learning reveals information that would be difficult for the human eye to discern on the microscopy images and simulation molecular data.

This book discusses the coupling of ML and nanotechnology and how new algorithms address issues such as lack of data, nano-scale modeling, and high-resolution measurements. Credited for bringing together these domains, this book intends to assist researchers, students, and professionals to take advantage of ML to advance innovative possibilities in nano-systems.

10.2 CORE MACHINE LEARNING ALGORITHMS FOR NANOTECHNOLOGY

Applying principles of ML on nanoscale problems provides significant advantages for designing, predicting, and processing variants at the nanoscale level.

DOI: 10.1201/9781003620570-10

10.2.1 Supervised Learning in Nano-Applications

Supervised learning is concerned with feeding the algorithms with data that is already tagged to predict something or put an observation into a certain class. In nanotechnology, it is widely used in the following applications:

10.2.1.1 Predicting Material Properties

Nano-materials are characterized by often counterintuitive mechanical, thermal, and electrical characteristics (Pizzagalli et al., 2024). These properties can be predicted through supervised models such as the support vector machines (SVMs) and random forests by using features like chemical composition, structure configuration, and synthesis parameters. For example, the random forest algorithm precisely estimated the thermal conductivity of carbon nanotubes over 90% and provided the input of chirality and defect density data (Malek et al., 2024). These predictions cut the need to rely heavily on experimental tests, most of which are both time- and capital-intensive.

10.2.1.2 Defect Detection in Nano-manufacturing

As cited by Mutha et al. (2024), manufacturing at the nano-scale needed accuracy, and even small flaws in the structure would significantly impact the product's performance. In this context, supervised learning algorithms are used to extract defects from microscopy images or other sensor data. For instance, according to Zheng and Gu (2020), to determine the manufacturers' defects in graphene sheets by supervised learning, convolutional neural networks (CNNs) were implemented to achieve an accuracy of above 95%. A basic version is seen in Figure 10.1. It minimizes the need for intervention by quality control personnel and equally hails a decrease in material wastage (Cho et al., 2021).

FIGURE 10.1 Defect finding.

10.2.2 Unsupervised Learning for Nano-material Discovery

Unsupervised learning arrangements allow a machine to learn from unlabeled data with the intent of discovering patterns. It is most useful for nano-material identification. It finds uses in the characterization of materials. In these cases, data is generally scarce and complex.

10.2.2.1 Clustering and Classification of Nano-compounds

K-means and hierarchical clustering aid in the categorization. They further aid in the finding of materials. They do so by grouping and classifying nanocompounds. They are based on their attributes (Kaushik and Mathur, 2014). For instance, the k-means clustering algorithm was used for a nano-catalyst classification. It is regarding the reactivity and stability of samples to find candidates for sustainable chemical reactions (Kaushik and Mathur, 2014). The technique suggested that three clear classes of catalysts existed. They have high, moderate, and minimal catalytic activity.

10.2.2.2 Dimensionality Reduction for Nano-data Visualization

Databases in nano-science are mostly high-dimensional data. They arise from spectroscopic measurements or simulations. Data aggregation trends ease this information for visualization and analysis (Sarmina et al., 2023). Some trends are principal component analysis (PCA) and t-distributed stochastic neighbor embedding (t-SNE). For instance, its application includes preparing the spectroscopic data of nano-coatings by using PCA for dimensionality reduction with over 100 features condensed to three features, which contain only a 95% loss of variance (Madkour, 2019b). In this case, reaction profiles augment visualization that allows for the identification of trends associated with anti-corrosion performance.

10.2.3 Deep Learning for Nano-imaging and Design

Deep learning is a branch of ML. It is relatively good at analyzing data in its raw format. Examples of data include images. It is capable of creating new designs. Some of the uses are as follows:

10.2.3.1 Applications in Microscopy

Electron and scanning probe microscopy produce a large number of high-resolution images needed for nanoscale research (Xu and Chen, 2020). These images can be analyzed with high precision by applying deep learning models, but from this chapter, CNNs have been determined to be most suitable. For instance, a CNN trained on atomic-resolution microscopy images for the recognition of definite crystal defects in nano-ceramics reached more than 98% accuracy (Cho et al., 2021). The above approach eliminates the need for making manual inspections, which, for the researchers, is a huge timesaver.

10.2.3.2 Generative Models for Nanostructure Synthesis

The generative models for nanostructure design include the Generative Adversarial Networks (GANs) and Variational Autoencoders (VAEs). They are becoming a net for designing nanostructures, and the components are evident in Figure 10.2 (Regenwetter et al., 2022). Such models can create their nano-material designs

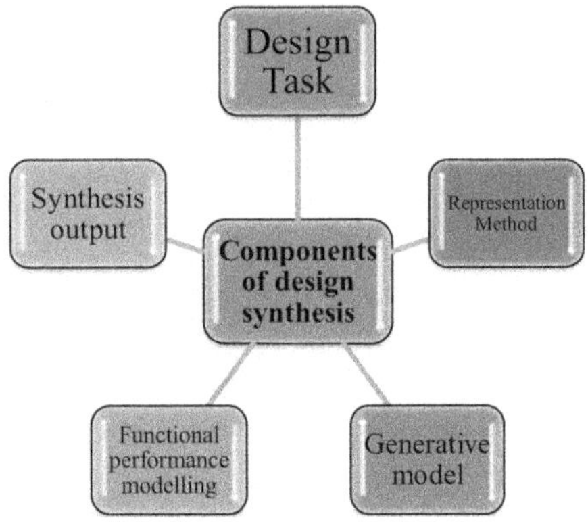

FIGURE 10.2 Typical components of design synthesis.

by looking at datasets, models of which are given below. For instance, a GAN fine-tuned for nanoparticle configuration was able to create new structures suitable for targeted dispatch of therapeutic cargo, boosting the stability of the load by 15% from conventional techniques (Figure 10.3). Such models enhance the advancement of nano-design and decrease the number of errors that could arise from experimentation (Figure 10.2).

An experiment conducted by a premier nanotechnology research center showed that supervised ML models cut the time to make the property prediction by 40% and achieved 90–95% accuracy compared to experimental methods (Chaikittisilp et al., 2022). Further, clustering analysis of 10,000 nano-compounds without prior labels for clustering purposes revealed several new material classes, two of which were patented for high-performance catalysts (Ahire et al., 2022). Moreover, deep learning-based defect detection systems also improved the accuracy of the manufacturing processes by 25% with more than $1 million worth of value in the production losses incurred by a nano-electronics firm annually (Park et al., 2022).

There is a clear indication that nanotechnology is being revolutionized. It is being done by ML through supervised learning. It is helpful for prediction and defect detection. It further helps in unsupervised learning for discovery and visualization. There are benefits in the field of deep learning for imaging and design. These algorithms improve efficiency and accuracy and open new opportunities for further development at the nano level. With the increasing availability of datasets and better algorithms, these tools will further bring transformation in the field.

10.3 INTEGRATION OF ML WITH NANO-SPECIFIC COMPUTATIONAL METHODS

ML is now revolutionizing the computational scenario in nanotechnology by solving issues with conventional approaches. ML speeds up and improves the accuracy of complete calculations. They would otherwise be impossible. They have to start with simulations of molecular dynamics (MD) and culminate with quantum physics.

10.3.1 Accelerating Molecular Simulations with ML

MD simulations play critical roles in investigating nano-scale processes, including the nanoparticles' interactions with organizing, living cells, or the processes of nano-materials' formation (Tian, 2008). However, conventional MD simulations are computationally expensive; they take days or even weeks to complete for large systems. However, ML gets over these restrictions with the help of surrogate models and better sampling techniques.

10.3.1.1 Example: Surrogate Models for Molecular Potentials

Neural networks or Gaussian process regressions serve as surrogate models of ML that can replace calculations of interatomic potentials, which, otherwise, would be computationally costly (Ceriotti, 2022). A neural network trained on quantum mechanical data was able to predict the potential energy surface of a system with 99% accuracy and required an order of magnitude less computational time than conventional approaches (Ceriotti, 2022).

10.3.1.2 Enhanced Sampling Techniques

Procedures such as reinforcement learning (RL) algorithms are also applied to steer simulations in the phase space of interest (Aradi, 2020). For instance, by using RL algorithms in MD simulations of nano-proteins, their time sampling was reduced by 50%, and the predictive analysis of the folding process that is so important in nano-bioengineering was also preserved (Aradi, 2020). The results of a comparative analysis of ML-accelerated simulations showed that the time for performing high-precision computer modeling is cut to the minimum—up to 60% less than in control, which allows integrating high-fidelity simulations into smaller-scale research setups.

10.3.2 Machine Learning and Quantum Mechanics Synergies

According to Kisielowski et al. (2023), nano-scale science is based on quantum mechanics (QM), which offers a base knowledge on the relation between electron and material response. However, the execution of QM calculations, including Density Functional Theory (DFT), is computationally costly, especially for large systems (Nakata et al., 2022). Concerning this, ML also offers attributes for increasing the velocity and scope of QM computations without compromising precision.

10.3.2.1 Example: Enhancing DFT Calculations

ML models, such as kernel ridge regression and deep neural networks, are used to predict electronic properties directly from atomic configurations. Kolluru et al. (2022) demonstrated that an ML-augmented DFT model predicted band gaps of nano-materials 1000 times faster than conventional methods with an error margin of less than 0.1 eV.

10.3.2.2 Hybrid ML-QM Approaches

Hence, hybrid models complement the capability of QM protocols to interpret results with the possibility of using ML. For instance, through QM-derived data, an ML model provided an almost perfect estimate for the adsorption energy of individual molecular species on nano-catalyst surfaces. These premises of the hybrid approach enabled the authors to identify novel catalyst designs for increasing the efficiency of hydrogen generation by an average of 20%. Comparing the hybrid ML-QM studies revealed that it was possible to reduce computation time by up to 70% for large-scale simulation, including more than 500 atoms, enabling the analysis of nano-systems at never-before-achieved levels.

10.3.3 Case Studies: Successful Applications in Nano-science

Examples are given of how successful ML implementation has been in combination with other nano-specific computational techniques.

Case Study 1: Nano-drug Delivery Optimization
Researchers integrated ML with molecular modeling to develop nano-carriers for drug delivery (Patra et al., 2018). According to Patra et al. (2018), they employed a deep learning model to determine the particle release profile of drugs as influenced by size, surface modification, and environmental effects. The tested formulations showed a 30% increase in drug effectiveness and 25% reduction in side effects during testing (Muldoon et al., 2022).

Case Study 2: Photovoltaic Nano-materials
A combined work of material scientists and AI developers showed that the application of the ML algorithm could accurately estimate the characteristics of perovskite-based nano-materials for solar cells (Hui et al., 2023). This allowed for much lower times in screening hundreds of different materials to be reduced from monthly to an average of one week when ML was combined with QM simulations. Thus, the formed nano-materials increased the level of photocurrent conversion by 5% and exceeded the indicators of similar materials.

Case Study 3: Corrosion-Resistant Nano-coatings
In the automotive industry, molecular simulations through machine learning were used in the design of nano-coating with better corrosion characteristics (Gu et al., 2020). By simulating 10,000 combinatorial possibilities, the ML model discovered the material configuration parameterizations, resulting in a doubling of the coating's useful life under high wear stress trials (Gu et al., 2020).

Case Study 4: Nano-scale 3D Printing
The stability of areas subjected to different stress levels was predicted using algorithms as the ML models for nano-scale 3D-printed component materials (Muldoon et al., 2022). The combination of ML predictions with finite element analysis resulted in enhanced accuracy of simulations and better means for assuring the safety and durability of materials in aerospace structures by as much as 15% (Muldoon et al., 2022).

The interaction of ML with nano-specific computational methods augments various nanotechnology applications, including reduced simulation times, scaling up QM applications, and driving nanotechnology breakthroughs driven by success stories. From surrogate models of MD simulations to hybrid methods of QM, ML offers an accurate, scalable, and low-cost solution. It has the realistic capability to transform industries such as healthcare, energy, and manufacturing, with case studies. Therefore, this interdependence will continue to be for progress in nano-science as computational methods advance.

10.4 CHALLENGES AND ETHICAL CONSIDERATIONS

The combination of ML with nanotechnology features possible breakthroughs, with prospects and threats reviewed below. All these problems need to be solved to achieve more efficient and permissive further development.

10.4.1 Data Scarcity and Quality Challenges

ML models require large and rich datasets for their training, as they work based on probability. Such datasets are rare in nanotechnology since the methods for obtaining the information are costly, and there are difficulties in employing experiments. Nano-specific data, in most cases, are scarce, messy, or contradictory, which makes model construction difficult.

10.4.1.1 Example: Sparse Data in Nano-material Research

For example, information concerning the mechanical properties of the nanocomposites can contain lacunae resulting from unfinished tests and/or limited measurement capabilities. The values of the attributes in such a dataset may lead to an ML model that are biased or produces unreliable predictions.

10.4.1.2 Solutions

Approaches such as transfer learning, whereby models are developed by overfitting from related sets and then fine-tuning them for nano-tasks, are being developed. Another interesting approach to building up the datasets is the usage of simulations or generative models, including GANs. Moreover, the development of Open Access Nano data repositories can help create data sharing and standardization. Gracia Moisés et al. (2023) compared several data augmentation methods, showing that they increase the accuracy by up to 15% in predicting nano-material properties, particularly from sample sizes of less than 1000.

The integration of ML and nanotechnology raises various ethical questions and consequences for society and the environment.

10.4.1.3 Environmental Concerns

Nano-materials are used in the production of manufactured goods. The distribution and installation of the nanoparticles themselves can produce waste products. They can introduce toxic elements into the environment. When complemented with the present and accelerated production and application of ML, these problems will become worse if more nano-products are produced. Technological advances in green synthesis processes and guaranteeing the life cycle assessment of the generated nano-products are significant in reducing their impact on the environment (Figure 10.3).

10.4.2 Ethical and Environmental Implications of Nano-AI Integration

10.4.2.1 Ethical Risks in Autonomous Systems

The use of nano-robots and devices through the help of AI can result in self-controlled systems, hence increasing chances of passiveness in case of failure or incidence of undesired results (Giri et al., 2021). For instance, medical nano-robots might pose a safety risk if algorithms cannot respond to the dynamic environment. It is important to set regulatory policies needed to control these risks of AI-nano devices.

10.4.2.2 Equity and Access

Some of the preparations for ML and nanotechnology implementation are so expensive that they may limit availability to only more affluent institutions or even nations, exacerbating the technology gap (Giri et al., 2021). Partnership work and grants for worldwide availability can assist in countering these inequalities. According to Stahl and Leach (2022), more than 60% of workers in the nano-field are of the view that ethics can become a barrier in the acceptance of nano-applications driven by artificial intelligence if not addressed.

The issues concerning data unavailability and data quality, as well as ethical concerns, all indicate that further development of nanotechnology that incorporates ML requires preventive action. Through advancing data access, securing sustainable mechanisms, and setting strong ethical standards, the promise of such innovations will be felt by the researchers and policymakers while potential harms are avoided.

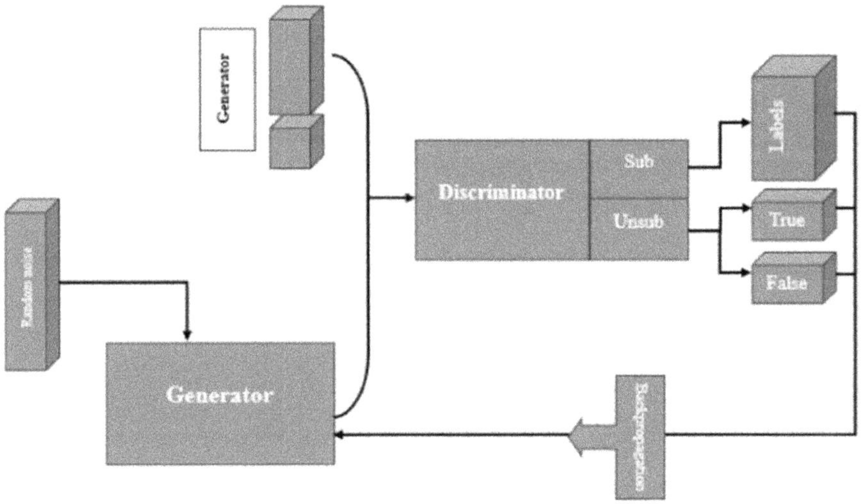

FIGURE 10.3 GAN model.

10.5 CONCLUSION

The combination of ML with nanotechnology is the most recent shift in the design and application of nano-scale advances. ML helps in solving the major problems that nanotechnology faces, including data constraints and computational limitations. Other trends, including self-supervised learning and physics-informed learning, are some of the trends that are emerging and are poised to take the field even further ahead. These technologies promise to solve problems of global proportions in medicine, energy, and the environment through disciplinary integration. Likewise, the blending of ML and nanotechnology will create exciting opportunities to propose novel opportunities and prospects for both science and industry.

10.6 FUTURE DIRECTIONS

A fascinating synergy is to point at the integration of the mechanics of ML and nanotrending for further development.

10.6.1 Emerging Trends in ML and Nanotechnology

10.6.1.1 Self-Supervised and Federated Learning

Self-supervised learning, which leverages unlabeled data, is gaining attention for addressing data scarcity in nanotechnology (Mamun et al., 2024). Federated learning, allowing collaborative model training across institutions without sharing sensitive data, can promote global innovation while preserving data privacy.

10.6.1.2 AI-Driven Discovery Pipelines

Self-driving ML processes are emerging to discover new nano-materials and optimize operations. For instance, smart experimentation technologies link the robotic systems and the ML to enhance the efficiency of discovering the best catalysts for nano-coating (Cho et al., 2021).

10.6.1.3 Integration of Physics-Informed ML Models

Physics-constrained ML models use prior knowledge from physics in the algorithms to guarantee that they work and produce correct data (Karniadakis et al., 2021). These models are particularly useful for the imitation of quantum characteristics and interactions by nano-nodes.

10.6.2 Opportunities for Interdisciplinary Collaboration

The complexity of nano-scale challenges requires collaboration between diverse fields, for example, AI, materials science, physics, biology, and engineering.

10.6.2.1 Nano-bio-Interfaces

Collaboration between biologists and AI engineers can create nano-bio systems that have never been seen before, including applications such as drug delivery devices and biosensors (Liang and Liang, 2021). Synergism could help push the development of nanobots. They can map out the ways of the biological world (Figure 10.4).

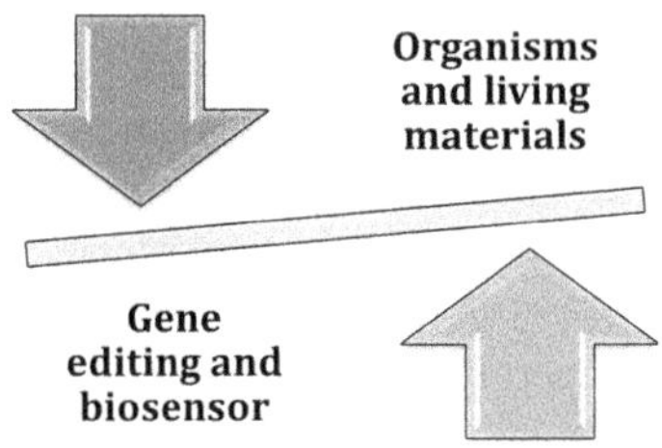

FIGURE 10.4 Impact.

10.6.2.2 Quantum-AI Synergies

The collaborations between quantum physicists and AI experts can directly lead to the developments in the integration of quantum computing with ML to yield highly accurate simulations and predictions at the nano level.

10.6.2.3 Global Research Networks

Resource disadvantages that exist between developing and developed nations can be effectively addressed by designing global research consortia and data sharing models, thereby providing equal opportunities for nano-innovation through ML across nations (Wang et al., 2023). New directions and interdisciplinarity open new opportunities in ML for nanotechnology. By adopting these directions, the researchers shall be able to tackle these large problems and enable the achievement of an environmentally friendly revolution that has a knock-on effect on society and industries.

REFERENCES

Ahire, S.A., Bachhav, A.A., Pawar, T.B., Jagdale, B.S., Patil, A.V. and Koli, P.B. 2022. The Augmentation of nanotechnology era: A concise review on fundamental concepts of nanotechnology and applications in material science and technology. *Results in Chemistry*, 4, 100633.

Aradi, S. 2020. Survey of deep reinforcement learning for motion planning of autonomous vehicles. *IEEE Transactions on Intelligent Transportation Systems*, 23 (2), 740–759.

Ceriotti, M. 2022. Beyond potentials: Integrated machine learning models for materials. *MRS Bulletin*, 47(10), 1045–1053.

Chaikittisilp, W., Yamauchi, Y. and Ariga, K. 2022. Material Evolution with Nanotechnology, Nanoarchitectonics, and Materials Informatics: What will be the Next Paradigm Shift in Nanoporous Materials? *Advanced Materials*, 34(7), 2107212.

Cho, P., Wood, A., Mahalingam, K. and Eyink, K. 2021. Defect detection in atomic resolution transmission electron microscopy images using machine learning. *Mathematics*, 9(11), 1209.

Contera, S. 2019. *Nano Comes to Life: How Nanotechnology is Transforming Medicine and the Future of Biology*. Princeton University Press.

Dananjaya, V., Marimuthu, S., Yang, R., Grace, A.N. and Abeykoon, C., 2024. Synthesis, properties, applications, 3D printing and machine learning of graphene quantum dots in polymer nanocomposites. *Progress in Materials Science*, 144, 101282.

Giri, G., Maddahi, Y. and Zareinia, K., 2021. A brief review on challenges in design and development of nanorobots for medical applications. *Applied Sciences*, 11(21), 10385.

Gracia Moisés, A., Vitoria Pascual, I., Imas González, J.J. and Ruiz Zamarreño, C., 2023. Data augmentation techniques for machine learning applied to optical spectroscopy datasets in agrifood applications: A comprehensive review. *Sensors*, 23(20), 8562.

Gu, Y., Xia, K., Wu, D., Mou, J. and Zheng, S. 2020. Technical Characteristics and Wear-Resistant Mechanism of Nano Coatings: A Review. *Coatings*, 10(3), 233.

Hui, Z., Wang, M., Yin, X., Wang, Y. and Yue, Y. 2023. Machine learning for perovskite solar cell design. *Computational Materials Science*, 226, 112215.

Karniadakis, G.E., Kevrekidis, I.G., Lu, L., Perdikaris, P., Wang, S. and Yang, L. 2021. Physics-informed machine learning. Nature Reviews *Physics*, 3(6), 422–440.

Kaushik, M. and Mathur, B., 2014. Comparative study of K-means and hierarchical clustering techniques. *International Journal of Software & Hardware Research in Engineering*, 2(6), 93–98.

Kisielowski, C., Specht, P., Stig Helveg, Chen, F.-R., Freitag, B., Jinschek, J. and Dyck, D.V. 2023. Probing the boundary between classical and quantum mechanics by analyzing the energy dependence of single-electron scattering events at the nanoscale. *Nanomaterials*, 13(6), 971.

Kolluru, A., Shuaibi, M., Palizhati, A., Shoghi, N., Das, A., Wood, B., Zitnick, C.L., Kitchin, J.R. and Ulissi, Z.W., 2022. Open challenges in developing generalizable large-scale machine-learning models for catalyst discovery. *ACS Catalysis*, 12(14), 8572–8581.

Liang, J. and Liang, K. 2021. Nano-bio-interface engineering of metal-organic frameworks. *Nano Today*, 40, 101256.

Madkour, L. H., 2019a. Environmental impact of nanotechnology and novel applications of nano materials and nano devices. In: *Nanoelectronic Materials: Advanced Structural Materials*, 116, 605–699.Springer, Cham.

Madkour, L.H. 2019b. Introduction to Nanotechnology (NT) and Nanomaterials (NMs). In: *Nanoelectronic Materials. Advanced Structured Materials*, 116, 1–47, Springer, Cham. https://doi.org/10.1007/978-3-030-21621-4_1

Malek, I.I., Sarkar, K. and Zubair, A., 2024. Predicting the mechanical properties of pristine and defective carbon nanotubes using a random forest model. *Nanoscale Advances*, 6(20), 5112–5132.

Mamun, A.A., Ahmedt-Aristizabal, D., Zhang, M., Hossen, Md I., Hayder, Z. and Awrangjeb, M. 2024. Plant disease detection using self-supervised learning: A systematic review. *IEEE Access*, 12, 171926–171943.

Muldoon, K., Song, Y., Ahmad, Z., Chen, X. and Chang, M.-W. 2022. High precision 3d printing for micro to nano scale biomedical and electronic devices. *Micromachines*, 13(4), 642.

Mutha, N., Olukayode, F. and Desai, S. 2024. Bridging Nanomanufacturing and Artificial Intelligence: A Comprehensive Review. *Materials*, 17(7), 1621.

Nakata, A., Bowler, D.R. and Miyazaki, T. 2022. Large-Scale DFT methods for calculations of materials with complex structures. *Journal of the Physical Society of Japan*, 91(9), 091011.

Park, S.-H., Lee, K.-H., Park, J.-S. and Shin, Y.-S. 2022. Deep learning-based defect detection for sustainable smart manufacturing. *Sustainability*, 14(5), 2697.

Patra, J.K., Das, G., Fraceto, L.F., Campos, E.V.R., Rodriguez-Torres, M.D.P., Acosta-Torres, L.S., Diaz-Torres, L.A., Grillo, R., Swamy, M.K., Sharma, S., Habtemariam, S. and Shin, H.-S. 2018. Nano based drug delivery systems: Recent developments and future prospects. *Journal of Nanobiotechnology*, 16(1), 71.

Pizzagalli, L., Brochard, S., Godet, J. and Durinck, J. (2024). Structural stability of tungsten nanoparticles. *Physical Review Materials*, 8(8), 086001.

Regenwetter, L., Nobari, A. H. and Ahmed, F. 2022. Deep generative models in engineering design: A review. *Journal of Mechanical Design*, 144(7), 071704.

Sarmina, B.G., Sun, G.H. and Dong, S.H., 2023. Principal component analysis and t-distributed stochastic neighbor embedding analysis in the study of quantum approximate optimization algorithm entangled and non-entangled mixing operators. *Entropy*, 25(11), 1499.

Stahl, B.C. and Leach, T. 2022. Assessing the ethical and social concerns of artificial intelligence in neuroinformatics research: An empirical test of the European Union Assessment List for Trustworthy AI (ALTAI). *AI and Ethics*, 3, 745–767.
Tian, P. 2008. Molecular dynamics simulations of nanoparticles. *Annual Reports Section 'C' (Physical Chemistry)*, 104(0), 142–164.
Wang, J., Gao, F., Zhou, Y., Guo, Q., Tan, C.-W., Song, J. and Wang, Y. 2023. Data sharing in energy systems. *Advances in Applied Energy*, 10, 100132.
Xu, K. and Chen, J. 2020. High-resolution scanning probe lithography technology: A review. *Applied Nanoscience*, 10(4), 1013–1022.
Zheng, B. and Gu, G.X. 2020. Machine learning-based detection of graphene defects with atomic precision. *Nano-Micro Letters*, 12(1), 181.

11 A Comprehensive Review of Diverse Fabrication and Optimization Techniques for Metal Matrix Composites (MMCs)

Vaibhav Chandra, Praveen Kumar, Umesh Khandey, Pankaj Kumar Singh, Gaurav Arora, and Kanwarjeet Singh

11.1 INTRODUCTION

The primary objective of the manufacturing sector these days is to manufacture higher-quality goods at a faster rate of production. The need for lightweight components with improved physical qualities has grown as a result of developments in many industries. Metal matrix composites (MMCs) possess good mechanical and physical properties: good strength, stiffness, and thermal conductivity. Due to these favorable properties, the MMCs are used in various fields, especially in the aerospace, aircraft, electronics, automobile, military, and recreational (sports) industries (Muguthu, Dong, and Ikua 2015) (Kaczmar, Pietrzak, and Włosiński 2000). The industries will be more likely to reduce manufacturing costs without compromising the productivity and quality of the components if even one optimization is carried out. However, due to their anisotropic, non-homogeneous features, composite materials are difficult to machine. The cutting tool may sustain severe damage and wear out quickly when a composite workpiece is machined. After all, a number of factors or elements, such as the material's characteristics, the relative amounts of reinforcement and matrix materials, the machining parameters, and the reaction to the machining process, affect how composite materials are machined (Arun Ramnath et al. 2018). Particulate-reinforced MMCs and fiber-reinforced polymer matrix composites have both undergone extensive experimental machining in the past.

DOI: 10.1201/9781003620570-11

These composite materials are shown to result in excessive tool wear, which causes damage phenomena like fiber pullout, particle fracture, delamination, and debonding at the interface between the fiber or particle and matrix. The type and orientation of the reinforcement, the type and geometry of the tool, and the machining parameters are the main factors that affect these composites' machinability. As a result, these characteristics classify such materials as difficult to machine (Dandekar and Shin, n.d.; Pramanik, Zhang, and Arsecularatne 2006). MMCs consist of two phases: the reinforcing phase and the matrix phase, which is the base material. The matrix, the most important component, can be made of metals, polymers, or ceramics. Constituents can benefit from the reinforcement's ability to improve their mechanical, electrical, thermal, wear, and corrosion-resistant qualities. It can be composed of continuous or irregular fibers or particles (Kim, Qaiser, and Hwang 2023). A metallic matrix and a dispersed ceramic or metallic phase are the usual components of MMCs. Examples of composite materials with a low density and a high strength-to-weight ratio that are reinforced with fibers, whiskers, or particles include titanium (Ti), magnesium (Mg), and aluminum (Al). Reinforcement materials can be made of metal, ceramic, or inorganic components (Shamim, Dvivedi, and Kumar 2022). The various kinds of MMCs with different matrices and reinforcement particles are depicted in Figure 11.1 (Jenczyk et al. 2021) (Sarmah and Gupta 2024a).

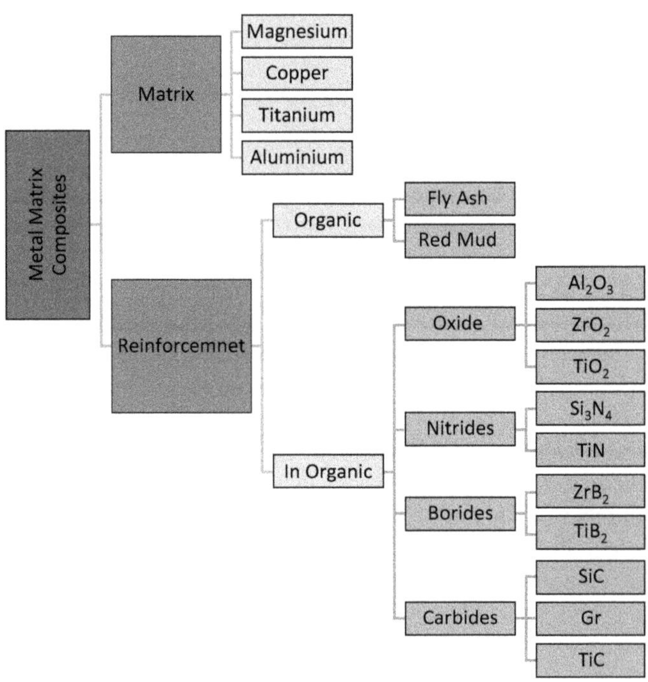

FIGURE 11.1 Different types of metal matrix composites.

11.2 PRODUCTION PROCESSES COMMONLY USED FOR MMCS

The matrix phase, the primary component of composites, is usually made of a soft metal that has lower thermal expansion and higher toughness, tensile strength, and shear modulus. The most commonly used matrices in MMCs are Al, Mg, and Ti. Because of their unique characteristics, aluminum and magnesium are also frequently used (Satishkumar et al. 2011; Sarmah and Patowari 2023). A number of manufacturing techniques have been used to manufacture and synthesize MMCs. Three major methods of fabrication for MMCs are defined as follows:

1. Liquid-State Fabrication
2. Solid-State Fabrication
3. In Situ Fabrication

Different MMC fabrication techniques are shown in Figure 11.2.

11.2.1 Liquid-State Fabrication

The reinforcement phase gets incorporated into the molten metal during the liquid phase processing of MMCs. For enhanced mechanical characteristics, fine interfacial bonding should be achieved between molten metal and reinforcements. Enhancement of wetting may be accomplished by coating the reinforcement particles (fibers). In the stir casting method, the reinforcement phase is blended, and the base metal is kept in a crucible and heated in a furnace above the melting point. To stir and evenly distribute the heat throughout the liquid metal, we use a stirrer. Then, fibers or reinforcement particles are added, and a stirring process is done for the uniform dissemination of reinforcement material. And after that, the solidification of the preferred MMCs is obtained with more homogeneity and fewer defects (Ozben, Kilickap, and Çakır 2008). The improved mechanical characteristics of MMCs produced using this process make them significant for a range of engineering applications (Rajmohan, Ranganathan, and Suryakumari 2014). Under the category of liquid-state fabrication, one process is infiltration. In this method, the preformed reinforcement phase

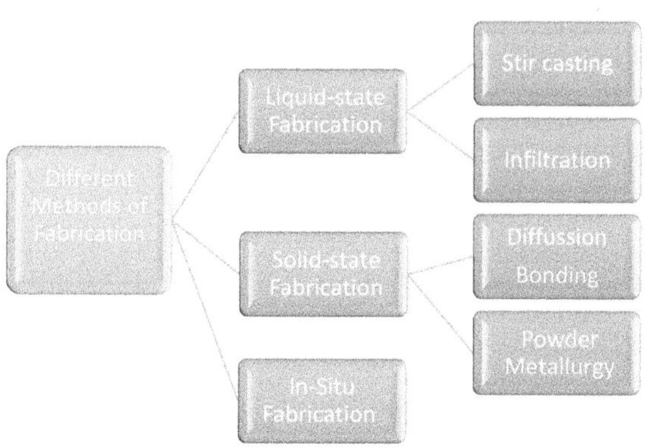

FIGURE 11.2 Different methods of fabrication used for MMC materials.

(fibers and particles) is submerged in a matrix of liquid metal. The space between the completed reinforcement phase inclusions is occupied by the liquid metal. Some limitations of this approach, including void formation during the infiltration of particle-reinforced metals, were discovered by Schöbel et al. (2014). High volume fractions of particulates and variations in the matrix and reinforcement's coefficients of thermal expansion result in the formation of voids. Normally, there are three types of infiltration methods, such as pressure die infiltration, squeeze casting infiltration, and gas pressure infiltration. The Massachusetts Institute of Technology was the first to develop the basic gas pressure infiltration technology as a means of investigation because it allowed for accurate regulation of processing parameters like cooling rate, metal temperature, and preform temperature (Cornie 1995). This process involves forcing molten metal through gas pressure into fiber, whisker, or particulate preforms that are placed in cheap, disposable containers. The system requires very little strength because gas pressure creates a quasi-hydrostatic pressure condition in the container (Blucher 1992) (Yang and Chung 1989; Evans, San Marchi, and Mortensen 2003). Figure 11.3 shows the schematic diagram of gas pressure infiltration. Additionally, there is the squeeze casing method, which is a form of pressure infiltration for the liquid state fabrication of MMCs. This method uses a movable ram to apply mechanical pressure to the molten metal, pushing it into a preform phase that is placed at the lower fixed part of the mold. The pressure is quite high compared to gas pressure infiltration, ranging from 10 to 100 MPa or even higher (Chadwick 1991) (Yue et al. 1996; Schultz, Ferguson, and Rohatgi 2011).

As it is frequently seen in experiments, applying high pressure can cause preform deformation or breakage during infiltration. This procedure typically has two steps in order to avoid this issue. To prevent excessively rapid infiltration that could harm the preform, the molten metal is pressed into it at a lower pressure in the first stage. For the purpose of producing a fine-grained, superior structure with the least amount of porosity, a higher pressure is then applied during the solidification phase (Fukunaga 1988; Schultz, Ferguson, and Rohatgi 2011; Yue et al. 1996). The quality of squeeze casting of composite materials is influenced by a number of process variables,

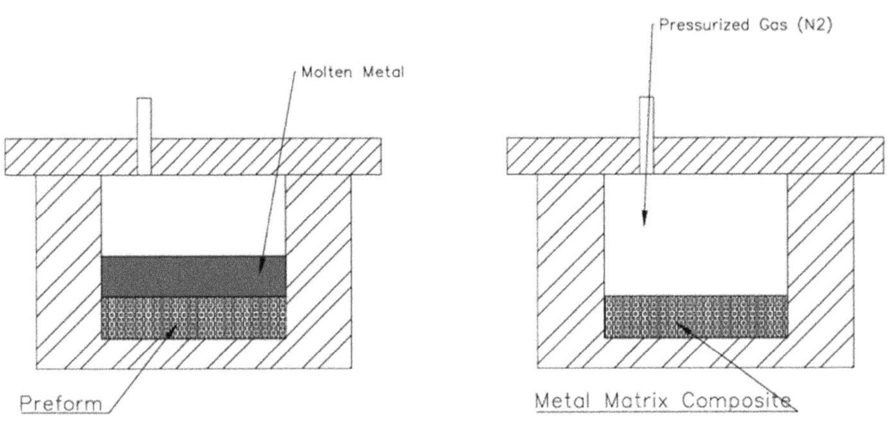

FIGURE 11.3 Gas pressure infiltration.

including pouring temperature, filling velocity, mold temperature, and time delay in pressurizing the mold. There are currently no particular guidelines in place to regulate these process variables (Panwar and Chauhan 2018; Nishida 2013). Figure 11.4 shows the schematic diagram of squeeze casting infiltration.

An additional mechanical infiltration technique for the liquid state fabrication of MMCs is pressure die infiltration. Molten metal enters the die under the pressure of a piston and passes via a gating system to infiltrate the preform. This process, which is adapted from the die-casting technique for unreinforced metals, involves placing a preformed dispersed phase (particles, fibers, or woven) into the die (Jarry et al. 1992). The basic structure of pressure die infiltration is shown in Figure 11.5.

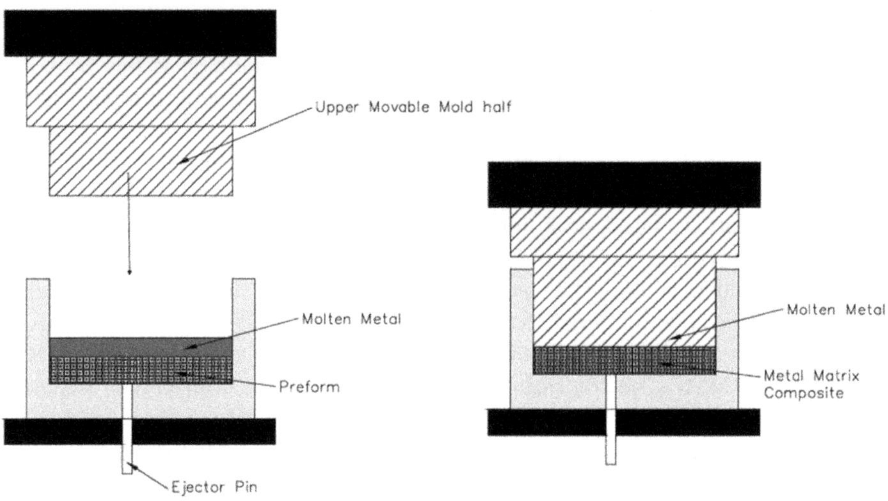

FIGURE 11.4 Squeeze casting infiltration method.

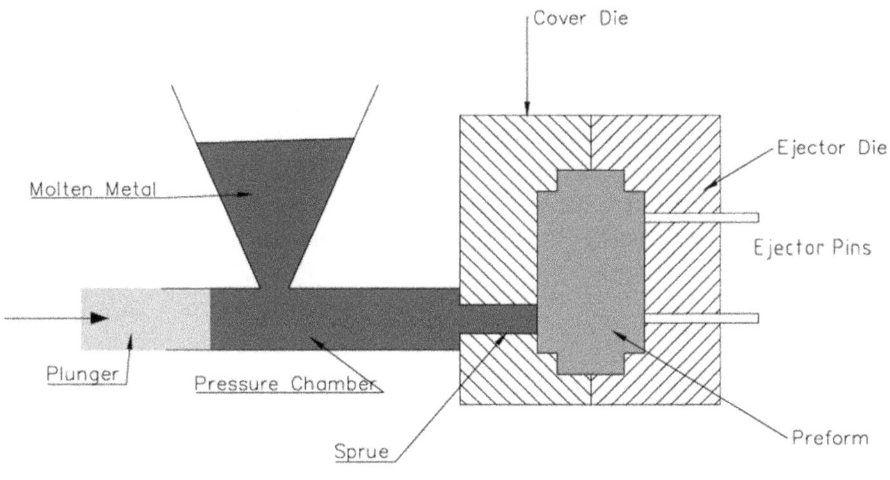

FIGURE 11.5 Pressure die infiltration method.

11.2.2 Solid-State Fabrication

The process for producing MMCs in solid states requires bonding between the matrix metal and dispersed phase through mutual diffusion, which takes place between them in solid states at high temperatures and pressures. When compared to liquid-state fabrication of MMCs, the low temperature of the solid-state fabrication process suppresses unwanted reactions at the boundary between the matrix and dispersed (reinforcing) phases. It is mainly classified under the following two categories:

- Powder metallurgy
- Diffusion bonding

Powder metallurgy, one of the multiple fabrication methods available, provides MMCs with flexibility and ease of fabrication. Powder metallurgy is free from too many chemical reactions, wettability challenges, and particle segregation caused by density, surface tension, and temperature gradient compared to procedures like stir casting (Brabazon, Browne, and Carr 2002). Reinforcement segregation, which is too difficult to prevent, is one of the major disadvantages of the stir casting process. The powder metallurgy process does not have this problem. In powder metallurgy, the three main procedures are compacting, post-processing, and powder mixing. To maximize the advantages of the powder metallurgy process, the mixing should consequently get the attention it requires. (X. Zhang and Tan 1997). Figure 11.6 shows a schematic representation of the powder metallurgy method used to create MMCs. When creating MMCs, silicon carbide (SiC) is the most often utilized reinforcement material. A matrix made of foils and a dispersed phase made of long fibers are stacked in a specific order and then compressed at a high temperature in the solid-state fabrication process known as diffusion bonding. Diffusion bonding is used for the fabrication of simple-shape parts (plates and tubes). The diffusion bonding process provides appropriate solutions for dealing with these problems so that the microstructure does not deteriorate (Pilling et al. 1984; Enjo, Ikeuchi, and Furukawa 1985). The schematic diagrams for diffusion bonding processes are shown in Figure 11.7.

11.2.3 In Situ Fabrication of MMC

To increase reinforcement adhesion with the interface, in situ reinforcement formation has been employed. These methods use exothermic reactions during the composite manufacturing process to create the reinforcements. It has even been discovered that certain reinforcement leads to excellent wettability and decreased lattice mismatch

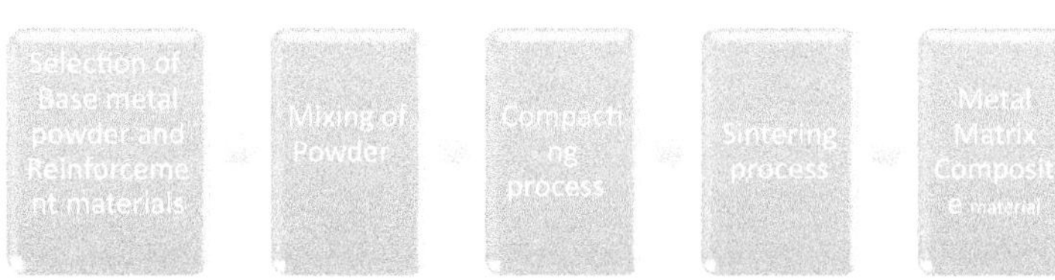

FIGURE 11.6 Schematic diagram for steps adopted in the powder metallurgy process.

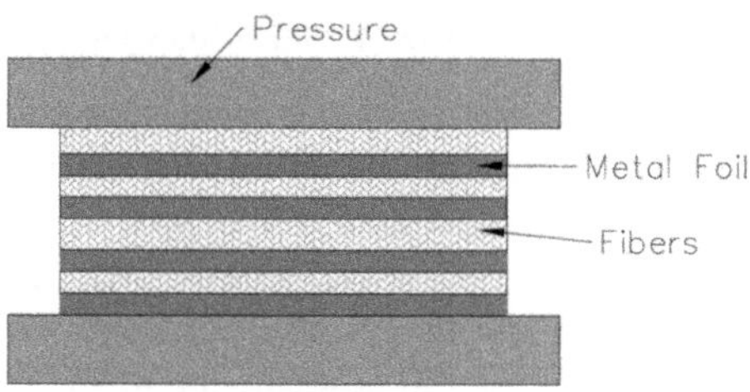

FIGURE 11.7 Diffusion bonding.

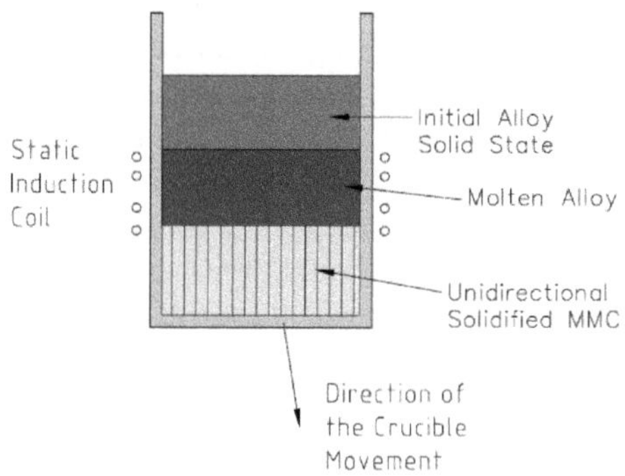

FIGURE 11.8 In-situ fabrication of MMCs. Composites.

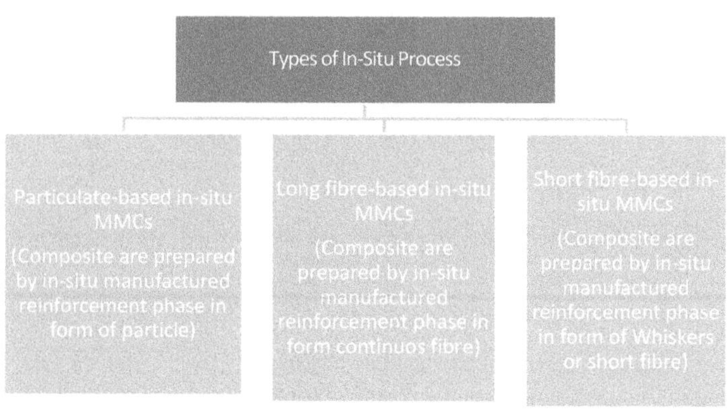

FIGURE 11.9 Types of in-situ processes of fabrication of MMCs.

(Wang et al. 2018; Qu et al. 2018). In this process, precipitation from the melt during cooling and solidification results in the dispersed phase in the form of a matrix, as illustrated in Figure 11.8. Figure 11.9 illustrates the in-situ procedure that can be used to prepare different kinds of MMCs.

Because it provides interesting alternatives to the limitations of other techniques, in situ powder metallurgy in particular has gained popularity in many sectors.

11.2.4 Some Advanced Methods of Fabrication of MMC Materials

Due to the increasing demand for products made from MMCs, a number of state-of-the-art processes are currently being used to fabricate them. The objective of cutting-edge and innovative techniques like wire arc additive manufacturing (WAAM), friction stir processing, and selective laser melting (SLM) is to enhance MMCs' mechanical capabilities and processability (Altıparmak et al. 2021). Some other modern techniques that are used to fabricate the MMC's material are continuous binder powder coating, metal injection molding (MIM), and mechanical alloying (A. Kumar et al. 2021). In this method, the matrix powders continuously coat the reinforcing fibers. After cleaning with acetone, the fibers are re-dipped in the binding solution. Atmospheric temperature is maintained for both the acetone and the binder solution. The fibers are now passed through a matrix powder vessel, and the matrix grains stick to the binder coating. The coated fibers are then rolled out and allowed to dry. This method is primarily used to prepare titanium-based MMC (Sanguinetti Ferreira et al. 2006). Figures 11.10 and 11.11 demonstrate the various advanced techniques used for the fabrication of MMCs.

A near-net-shape manufacturing technique for the creation of ceramic and metal powder is MIM. The MIM process consists of four stages: mixing, injection molding, debinding, and sintering. The primary objective of this process is to mass-produce

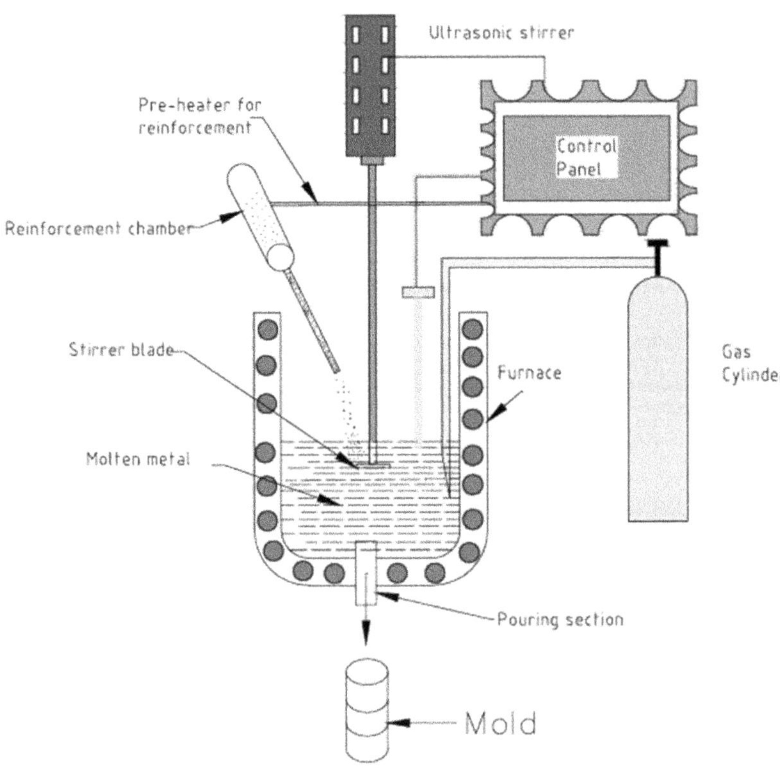

FIGURE 11.10 Ultrasonic-assisted stir casting.

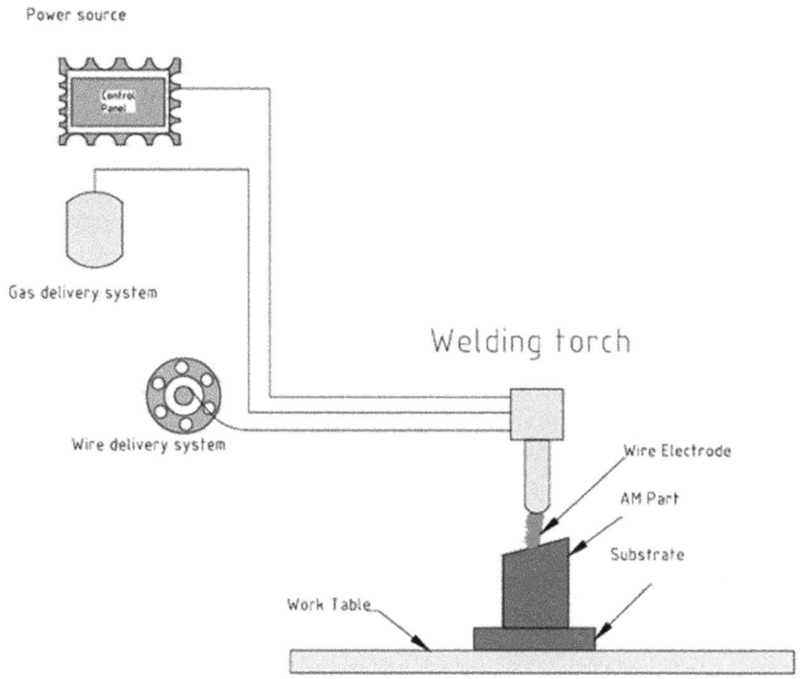

FIGURE 11.11 Wire and arc additive manufacturing (WAAM). Various advanced fabrication techniques for MMCs (Sarmah and Gupta 2024b).

small, intricate components. When used in feedstock, very fine powders improve compaction during the final sintering process. This MMC fabrication process can greatly lower the cost of composite materials used for commercial purposes (Ye, Liu, and Hong 2008). Additive manufacturing and laser additive manufacturing have numerous benefits for metallic materials. The advantages include quick prototypes, good material used, customization, and complex structure, whereas LMD offers higher production efficiency.

The proper selection of process parameters in laser deposition-additive manufacturing (LD-AM) can raise the melting point and uniformize the properties of the produced parts while reducing the detrimental effects of excessive heat stress and component deformation (Song and Jinliang 2022; Sun et al. 2023). Recent studies have examined the advantages of ultrasonic vibrations on the uniform distribution of reinforcements and improving tribological and mechanical properties in the production of magnesium-based MMCs (MMMCs) using ultrasonic-assisted stir casting (D. Kumar and Thakur 2019). This method was created to produce MMCs for use in the aerospace sector. Mechanical alloying is essentially broken down into three steps:

- In a ball mill, matrix and reinforcement materials are added and transformed into fine powders.
- A hot isostatic process is used to compress the powders.
- Finally, sintering techniques are used to reduce the stresses generated during compaction.

11.3 SOME ADVANCED TECHNIQUES ADOPTED FOR FABRICATION OF MMCS

A summary of the latest techniques applied for the fabrication of MMCs, with the findings, is depicted in Table 11.1.

TABLE 11.1
Some Fabrication Methods Adopted to Produce MMC Materials

S. No.	Materials	Fabrication Method	References
1.	TiB_2) is reinforced Al to produce Al-TiB_2	Hybrid ultrasonic casting	Sheelwant, Jadhav, and Narala (2021)
2.	Aluminum-based composite AA7075/B4C and FSP	Friction stir-processed	Sharma et al. (2021)
3.	Boron carbide over AA6063	Friction stir processing	Pratap et al. (2022)
4.	B4C addition into an AA6061-B4C composite	Ultrasonic-assisted stir casting	Gudipudi et al. (2020)
5.	3D-printed ($Ni_{60}Nb_{20}Ta_{20}$)	Gas pressure infiltration	Dittmann, Trauth, and André Weidenmann (2024)
6.	Ni Al material reinforcement of Al, Mo	Reactive infiltration	Venkatesh and Dunand (2000), Hsu, Chou, and Tuan (2016)
7	Fabrication of Al/Ni/Cu	Diffusion bonding	Hosseini et al. (2017)
8.	Al2219 alloy is hybridized with nanoparticles of TiC, Al_2O_3, and Si_3N_4	Squeeze casting process	Adithiyaa, Chandramohan, and Sathish (2020)
9.	Metallic glass lattice structure ($Ni_{60}Nb_{20}Ta_{20}$)	Gas pressure infiltration process	Dittmann, Trauth, and André Weidenmann (2024)
10.	ZrB_2-fortified Inconel 718 (In718+ZrB_2) superalloy metal matrix composite (MMC)	Laser powder bed fusion	Tekoğlu et al. (2024)
11.	Fe–TiB_2-based high modulus steels (HMS)	In-situ method and vacuum induction furnace (VIM) under an argon (Ar) atmosphere	H. Zhang et al. (2016)
12.	Silicon carbide (SiC) and Palm Sprout Shell Ash (PSSA)	Ultrasonic-assisted bottom-poured stir casting technique.	Tenali, Ganesan, and Ravindra Babu (2024)
13.	Aluminum matrix composite contains TiO_2 reinforcement particles	Liquid powder metallurgy	Ravichandran and Dineshkumar (2014)
14.	AA7075 MMC	Friction stir processing technique	S. Kumar, Srivastava, and Singh (2020)
15.	Al- and Mg-based MMCs	Pressure infiltration	Contreras et al. (2000), Contreras, López, and Bedolla (2004), Bedolla, Lemus-Ruiz, and Contreras (2012)

11.4 OPTIMIZATION OF MMCS COMPOSITES

This review work's objective is to find composites that are machined employing various machining parameters, methods, and materials by applying a number of advanced optimization strategies. The turning, drilling, and milling procedures are the machining operations taken into consideration in this current work. This review work provides a basis for machining parameter optimization. One of the most common materials found on airplanes is composite. In order to provide insight into the development and trends of research in this area, a bibliometric analysis of composite material optimization was conducted nowadays (Vieira et al. 2024). Figure 11.12 illustrates the extended history of optimizing composite materials that goes back to 1918 till now. An increasing tendency in research effort over time is revealed by general bibliometric research on composite material optimization.

Composite materials, such as carbon fiber-reinforced polymers (CFRPs), have become an appropriate replacement for the aluminum alloys frequently utilized in the aviation industry. Composite materials, renowned for their remarkable design flexibility, have supplanted conventional metallic materials in numerous applications. This phenomenon results from the intrinsic properties of these materials—such as their high flexural modulus, high specific strength, low density, resistance to corrosion, and design flexibility—that are employed in structural design (Chen and Xu 2023; Tran-Ngoc et al. 2021). However, employing these materials presents additional difficulties, such as the need for more accurate optimization to enhance their structural performance. This is because composite materials exhibit anisotropic properties, necessitating the use of more sophisticated material design approaches (Bisagni and Lanzi 2002).

11.4.1 Some Optimization Techniques Implemented in MMCs Machining Processes

The review is conducted for several fiber matrix composite machining processes, and particle-reinforced fiber matrix composites can be manufactured using the same methodology and conditions. The turning technique came into existence many years

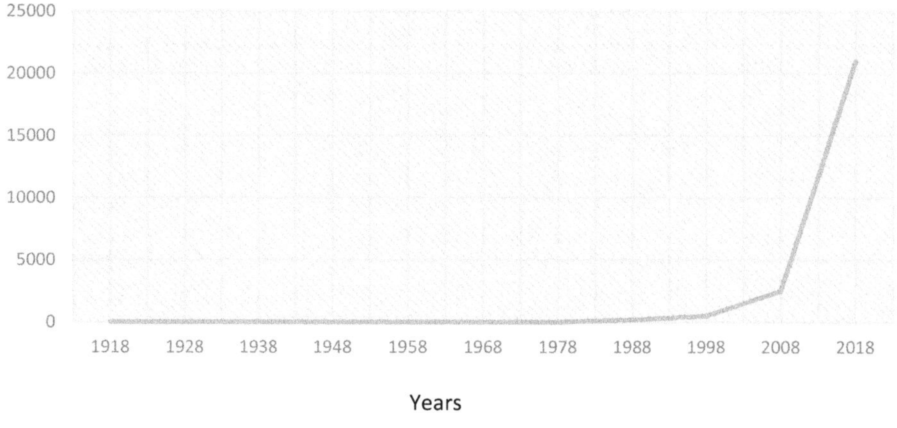

FIGURE 11.12 Composite material optimization research trend (Vieira et al. 2024).

ago and is currently the most popular and commonly used conventional machining method. For a variety of applications, the turning process of fiber matrix composite should yield a higher MRR, a good surface polish, and a low tool wear rate. The best process parameters must be found under such conflicting circumstances and multi-attribute decision-making without compromising the desired outcome. Only one goal may be achieved at a time by changing these parameters by trial and error, which could result in lower MRR and surface finish. Therefore, identifying the best process parameters is necessary to ensure that the process meets all of the desired objectives. Only suitable optimization approaches can be used to make these multi-attribute decisions. Various optimization techniques include Promethee, TOPSIS, AHP, Genetic Algorithm (GA), and others. Several investigations have been conducted with hybridization of algorithms such as fuzzy grey relation analysis, TOPSIS with grey fuzzy, Taguchi design with GA, and others. Similarly, a number of studies on the turning, drilling, and milling processes in fiber matrix composite materials have been conducted, which are presented in Table 11.2.

TABLE 11.2
Some Optimization Techniques Implemented on Composite Materials

S. No.	Process	Work Piece/Material	Optimization Technique Applied	References
1.	Turning	GFRP epoxy composites	Grey and Fuzzy coupled Taguchi methodology	Vasudevan, Deshpande, and Rajguru (2014)
2.	Turning	GFRP composites	TOPSIS method	A. K. Parida and Routara (2014)
3.	Turning	GFRP composite	Taguchi method	A. kumar Parida et al. (2014)
4.	Turning	LM25 Al/SiC composite	Taguchi and RSM	Palanikumar and Karthikeyan (2006)
5	Turning	Aluminum matrix composite	Genetic Algorithm	Davim and Conceição António (2001)
6	Drilling	GFRP composites	TOPSIS and Deng similarity method	Sonkar et al. (2014)
7	Drilling	GFRP composites	Group Method Data Handling algorithm (GMDH)	Arul et al. (2006)
8.	Drilling	GFRP composites	Fuzzy Logic	Palanikumar et al. (2008)
9	Drilling	Hybrid metal matrix composites	Taguchi method and regression analysis	Rajmohan and Palanikumar (2012)
10	Drilling	UD-GFRP composites	Taguchi Method	Singh, Bhatnagar, and Viswanath (2008)
11	Drilling	Boron carbide-reinforced metal matrix composites	Taguchi Gray relation analysis	Taşkesen and Kütükde (2014)
12	Drilling	Hybrid metal matrix composites	Taguchi method	Basavarajappa, Chandramohan, and Davim (2008)

(Continued)

TABLE 11.2 (Continued)

S. No.	Process	Work Piece/Material	Optimization Technique Applied	References
13	End Milling	MWCNT-reinforced epoxy composite	Taguchi method	Rao, Ansari, and Begum (2013)
14	Face Milling	AA7039/Al2O3 metal matrix composite	ANN and Taguchi method	Karabulut (2015)
15	End Milling	aluminum alloy (HNT/Al/Ep)	Taguchi method	Pang et al. (2014)

11.5 CONCLUSION

The purpose of this chapter was to present new information about developments in MMC fabrication. By analyzing the most recent developments in MMC fabrication techniques and pinpointing areas in need of further study, several techniques, including stir casting and powder metallurgy, have greatly improved the production of MMCs. After studying developments in MMC fabrication, the following conclusions can be made:

- Developments in solid-state methods for producing MMCs have immensely benefited the transportation, power generation, aerospace, and defense sectors. However, improvements in solid-state methods like friction stir processing ensure better reinforcement spread, higher densification, and enhanced interfacial bonding compared to methods like powder metallurgy, resulting in MMCs with superior physical properties.
- This chapter reviews the various liquid-state and solid-state processes used to fabricate MMCs, including stir casting, infiltration, sintering, and diffusion bonding.
- The primary goal is to increase the productivity, economy, and efficiency of these techniques. In order to fabricate MMCs and satisfy the needs of diverse industries, researchers are developing new, cutting-edge technologies such as mechanical alloying, metal injection molding, and continuous binder powder coating (CBPC).
- With a number of benefits over traditional sand casting, powder metallurgy is becoming more and more popular as a viable alternative.
- For the production of MMC, liquid-state processing techniques have produced significant improvements in material properties, cost savings, and ease of use. Despite the limitations of conventional liquid-state techniques, advancements in MMC fabrication have made it feasible to create MMCs with enhanced tensile, wear, and metallurgical characteristics.
- There are various techniques that are implemented to perform the optimization for different machining processes, especially for composite materials, such as Response Surface Methods (RSM), Artificial Neural Networks, the Taguchi method, and GA.s

REFERENCES

Adithiyaa, T., D. Chandramohan, and T. Sathish. 2020. "Optimal Prediction of Process Parameters by GWO-KNN in Stirring-Squeeze Casting of AA2219 Reinforced Metal Matrix Composites." *Materials Today: Proceedings, International Conference on Recent Trends in Nanomaterials for Energy, Environmental and Engineering Applications*, 21 (January):1000–7. https://doi.org/10.1016/j.matpr.2019.10.051

Altıparmak, S. C., V. A. Yardley, Z. Shi, and J. Lin. 2021. "Challenges in Additive Manufacturing of High-Strength Aluminium Alloys and Current Developments in Hybrid Additive Manufacturing." *International Journal of Lightweight Materials and Manufacture* 4 (2): 246–61. https://doi.org/10.1016/j.ijlmm.2020.12.004

Arul, S., D. Samuel Raj, L. Vijayaraghavan, S. K. Malhotra, and R. Krishnamurthy. 2006. "Modeling and Optimization of Process Parameters for Defect Toleranced Drilling of GFRP Composites." *Materials and Manufacturing Processes* 21 (4): 357–65. https://doi.org/10.1080/10426910500411587

Arun Ramnath, R., P. R. Thyla, N. Mahendra Kumar, and S. Aravind. 2018. "Optimization of Machining Parameters of Composites Using Multi-Attribute Decision-Making Techniques: A Review." *Journal of Reinforced Plastics and Composites* 37 (2): 77–89. https://doi.org/10.1177/0731684417732840

Basavarajappa, S., G. Chandramohan, and J. Paulo Davim. 2008. "Some Studies on Drilling of Hybrid Metal Matrix Composites Based on Taguchi Techniques." *Journal of Materials Processing Technology* 196 (1): 332–38. https://doi.org/10.1016/j.jmatprotec.2007.05.043

Bedolla, E., J. Lemus-Ruiz, and A. Contreras. 2012. "Synthesis and Characterization of Mg-AZ91/AlN Composites." *Materials & Design* 38 (June):91–98. https://doi.org/10.1016/j.matdes.2012.02.001

Bisagni, C., and L. Lanzi. 2002. "Post-Buckling Optimisation of Composite Stiffened Panels Using Neural Networks." *Composite Structures* 58 (2): 237–47. https://doi.org/10.1016/S0263-8223(02)00053-3

Blucher, J. T. 1992. "Discussion of a Liquid Metal Pressure Infiltration Process to Produce Metal Matrix Composites." *Journal of Materials Processing Technology* 30 (3): 381–90. https://doi.org/10.1016/0924-0136(92)90227-J

Brabazon, D, D. J. Browne, and A. J. Carr. 2002. "Mechanical Stir Casting of Aluminium Alloys from the Mushy State: Process, Microstructure and Mechanical Properties." *Materials Science and Engineering: A* 326 (2): 370–81. https://doi.org/10.1016/S0921-5093(01)01832-9

Chadwick, G. A. 1991. "Squeeze Casting of Metal Matrix Composites Using Short Fibre Performs." *Materials Science and Engineering: A, European Materials Research Society* 1990 Spring Meeting on Metal Matrix Composites, 135 (March):23–28. https://doi.org/10.1016/0921-5093(91)90530-Z

Chen, S., and N. Xu. 2023. "The Deep-Learning-Based Evolutionary Framework Trained by High-Throughput Molecular Dynamics Simulations for Composite Microstructure Design." *Composite Structures* 318 (August):117118. https://doi.org/10.1016/j.compstruct.2023.117118

Contreras, A., V. H. López, and E. Bedolla. 2004. "Mg/TiC Composites Manufactured by Pressureless Melt Infiltration." *Scripta Materialia* 51 (3): 249–53. https://doi.org/10.1016/j.scriptamat.2004.04.007

Contreras, A., M. Salazar, C. A. León, R. A. L. Drew, and E. Bedolla. 2000. "Kinetic Study of the Infiltration of Aluminum Alloys into TiC Preforms." *Materials and Manufacturing Processes* 15 (2): 163–82. https://doi.org/10.1080/10426910008912980

Cornie, J. 1995. "Advanced Pressure Infiltration Casting Technology Produces Near-Absolute Net-Shape Metal Matrix Composite Components Cost Competitively." *Materials Technology* 10 (3–4): 43–48. https://doi.org/10.1080/10667857.1995.11752588

Dandekar, C. R., and Y. C. Shin. n.d. "Multiphase Finite Element Modeling of Machining Unidirectional Composites: Prediction of Debonding and Fiber Damage." *Journal of Manufacturing Science and Engineering* 130(5): 051016.

Davim, J. Paulo, and C. A. Conceição António. 2001. "Optimisation of Cutting Conditions in Machining of Aluminium Matrix Composites Using a Numerical and Experimental Model." *Journal of Materials Processing Technology* 112 (1): 78–82. https://doi.org/10.1016/S0924-0136(01)00551-9

Dittmann, K., A. Trauth, and K. A. Weidenmann. 2024. "Manufacturing and Characterization of an Interpenetrating Metal Matrix Composite Reinforced with a 3D-Printed Metallic Glass Lattice Structure (Ni60Nb20Ta20)." *Composite Structures* 327 (January):117697. https://doi.org/10.1016/j.compstruct.2023.117697

Enjo, T., K. Ikeuchi, and K. Furukawa. 1985. "Behavior of Oxide Film in Diffusion Bonding of Aluminum(Materials, Metallurgy & Weldability)." *Welding Research Institute of Osaka University*. https://doi.org/10.18910/11121

Evans, A., C. S. Marchi, and A. Mortensen. 2003. "Processing Metal Matrix Composites." In Metal Matrix *Composites in Industry: An Introduction and a Survey, edited by Alexander Evans, Christopher San Marchi, and Andreas Mortensen*, 39–64. Boston, MA: Springer US. https://doi.org/10.1007/978-1-4615-0405-4_3

Fukunaga, H. 1988. "Processing Aspects of Squeeze Casting for Short Fiber Reinforced Metal Matrix Composites." *Advanced Materials and Manufacturing Processes* 3 (4): 669–87. https://doi.org/10.1080/10426918808953223

Gudipudi, S., S. Nagamuthu, K. S. Subbian, and S. P. R. Chilakalapalli. 2020. "Enhanced Mechanical Properties of AA6061-B4C Composites Developed by a Novel Ultra-Sonic Assisted Stir Casting." *Engineering Science and Technology, an International Journal* 23 (5): 1233–43. https://doi.org/10.1016/j.jestch.2020.01.010

Hosseini, M., N. Pardis, H. Danesh Manesh, M. Abbasi, and D. -Ik Kim. 2017. "Structural Characteristics of Cu/Ti Bimetal Composite Produced by Accumulative Roll-Bonding (ARB)." *Materials & Design* 113 (January):128–36. https://doi.org/10.1016/j.matdes.2016.09.094

Hsu, H.-C., J.-Y. Chou, and W.-H. Tuan. 2016. "Preparation of AlN/Cu Composites through a Reactive Infiltration Process." *Journal of Asian Ceramic Societies* 4 (2): 201–4. https://doi.org/10.1016/j.jascer.2016.03.003

Jarry, P., V. J. Michaud, A. Mortensen, A. Dubus, and R. Tirard-Collet. 1992. "Infiltration of Fiber Preforms by an Alloy: Part III. Die Casting Experiments." *Metallurgical Transactions A* 23 (8): 2281–89. https://doi.org/10.1007/BF02646021

Jenczyk, P., H. Grzywacz, M. Milczarek, and D. M. Jarząbek. 2021. "Mechanical and Tribological Properties of Co-Electrodeposited Particulate-Reinforced Metal Matrix Composites: A Critical Review with Interfacial Aspects." *Materials* 14 (12): 3181. https://doi.org/10.3390/ma14123181

Kaczmar, J. W., K. Pietrzak, and W. Włosiński. 2000. "The Production and Application of Metal Matrix Composite Materials." *Journal of Materials Processing Technology* 106 (1): 58–67. https://doi.org/10.1016/S0924-0136(00)00639-7

Karabulut, Ş. 2015. "Optimization of Surface Roughness and Cutting Force during AA7039/Al2O3 Metal Matrix Composites Milling Using Neural Networks and Taguchi Method." *Measurement* 66 (April):139–49. https://doi.org/10.1016/j.measurement.2015.01.027

Kim, H., N. Qaiser, and B. Hwang. 2023. "Electro-Mechanical Response of Stretchable PDMS Composites with a Hybrid Filler System." *Facta Universitatis, Series: Mechanical Engineering* 21 (1): 51–61. https://doi.org/10.22190/FUME221205002K

Kumar, A., O. Vichare, K. Debnath, and M. Paswan. 2021. "Fabrication Methods of Metal Matrix Composites (MMCs)." *Materials Today: Proceedings, ICFTMM-2020*, 46 (January):6840–46. https://doi.org/10.1016/j.matpr.2021.04.432

Kumar, D., and L. Thakur. 2019. "Recent Studies on the Fabrication of Magnesium Based Metal Matrix Nano-Composites by Using Ultrasonic Stir Casting Technique: A Review." *Materials Science Forum* 969:889–94. https://doi.org/10.4028/www.scientific.net/MSF.969.889

Kumar, S., A. K. Srivastava, and R. K. Singh. 2020. "Fabrication of AA7075 Hybrid Green Metal Matrix Composites by Friction Stir Processing Technique." *Annales de Chimie - Science Des Matériaux* 44 (4): 295–300. https://doi.org/10.18280/acsm.440409

Muguthu, J. N., G. Dong, and B. W. Ikua. 2015. "Optimization of Machining Parameters Influencing Machinability of Al2124SiCp (45%wt) Metal Matrix Composite." *Journal of Composite Materials* 49 (2): 217–29. https://doi.org/10.1177/0021998313516141

Nishida, Y.. 2013. "Fabrication by Squeeze Casting." In *Introduction to Metal Matrix Composites: Fabrication and Recycling*, edited by Yoshinori Nishida, 53–66. Tokyo: Springer Japan. https://doi.org/10.1007/978-4-431-54237-7_3

Ozben, T., E. Kilickap, and O. Çakır. 2008. "Investigation of Mechanical and Machinability Properties of SiC Particle Reinforced Al-MMC." *Journal of Materials Processing Technology* 198 (1): 220–25. https://doi.org/10.1016/j.jmatprotec.2007.06.082

Palanikumar, K., and R. Karthikeyan. 2006. "Optimal Machining Conditions for Turning of Particulate Metal Matrix Composites Using Taguchi and Response Surface Methodologies." *Machining Science and Technology* 10 (4): 417–33. https://doi.org/10.1080/10910340600996068

Palanikumar, K., S. Prakash, and K. Shanmugam. 2008. "Evaluation of Delamination in Drilling GFRP Composites." *Materials and Manufacturing Processes* 23 (8): 858–64. https://doi.org/10.1080/10426910802385026

Pang, J. S., M. N. M. Ansari, Omar S. Zaroog, Moaz H. Ali, and S. M. Sapuan. 2014. "Taguchi Design Optimization of Machining Parameters on the CNC End Milling Process of Halloysite Nanotube with Aluminium Reinforced Epoxy Matrix (HNT/Al/Ep) Hybrid Composite." *HBRC Journal* 10 (2): 138–44. https://doi.org/10.1016/j.hbrcj.2013.09.007

Panwar, N., and A. Chauhan. 2018. "Fabrication Methods of Particulate Reinforced Aluminium Metal Matrix Composite-A Review." *Materials Today: Proceedings, 7th International Conference of Materials Processing and Characterization*, March 17–19, 2017, 5 (2, Part 1): 5933–39. https://doi.org/10.1016/j.matpr.2017.12.194

Parida, A. K., R. Das, A. K. Sahoo, and B. C. Routara. 2014. "Optimization of Cutting Parameters for Surface Roughness in Machining of Gfrp Composites with Graphite/Fly Ash Filler." *Procedia Materials Science, 3rd International Conference on Materials Processing and Characterisation (ICMPC 2014)*, 6 (January):1533–38. https://doi.org/10.1016/j.mspro.2014.07.134

Parida, A. K., and B. C. Routara. 2014. "Multiresponse Optimization of Process Parameters in Turning of GFRP Using TOPSIS Method." *International Scholarly Research Notices* 2014 (1): 905828. https://doi.org/10.1155/2014/905828

Pilling, J., Livesey, D. W., Hawkyard, J. B., and N. Ridley. 1984. "Solid State Bonding in Superplastic Ti-6Al-4V." *Metal Science* 18 (3): 117–22. https://doi.org/10.1179/msc.1984.18.3.117

Pramanik, A., L. C. Zhang, and J. A. Arsecularatne. 2006. "Prediction of Cutting Forces in Machining of Metal Matrix Composites." *International Journal of Machine Tools and Manufacture* 46 (14): 1795–803. https://doi.org/10.1016/j.ijmachtools.2005.11.012

Pratap, C., P. Chandra, R. Butola, and A. Shukla. 2022. "Fabrication and Characterization of AA6063/B4C Metal Matrix Surface Nanocomposite Using Friction Stir Processing." *ECS Journal of Solid State Science and Technology* 11 (3): 033010. https://doi.org/10.1149/2162-8777/ac5c81

Qu, X., F. Wang, C. Shi, N. Zhao, E. Liu, C. He, and F. He. 2018. "In Situ Synthesis of a Gamma-Al2O3 Whisker Reinforced Aluminium Matrix Composite by Cold Pressing and Sintering." *Materials Science and Engineering: A* 709 (January):223–31. https://doi.org/10.1016/j.msea.2017.10.063

Rajmohan, T., and K. Palanikumar. 2012. "Optimization of Machining Parameters for Surface Roughness and Burr Height in Drilling Hybrid Composites." *Materials and Manufacturing Processes* 27 (3): 320–28. https://doi.org/10.1080/10426914.2011.585491

Rajmohan, T., S. Ranganathan, and T. S. A. Suryakumari. 2014. "Experimental Study on Fabrication of Hybrid (Al+Tio2+Gr) Metal Matrix Composites." *International journal of Advanced Engineering Applications.* 7 (2): 11–14.

Rao, G. H. K., M. N. M. Ansari, and Shahida Begum. 2013. "Effect of Cutting Parameters on the Surface Roughness of MWCNT Reinforced Epoxy Composite Using CNC End-Milling Process" *International Journal of Science and Research (IJSR)* 2 (11): 51–55

Ravichandran, M. and S. Dineshkumar 2014. "Synthesis of Al-TiO2 Composites through Liquid Powder Metallurgy Route." *International Journal of Mechanical Engineering* 1 (1): 12–15. https://doi.org/10.14445/23488360/IJME-V1I1P103

Sanguinetti Ferreira, R. A., C. Arvieu, B. Guillaume, and J. M. Quenisset. 2006. "Titanium Matrix Composites Processed by Continuous Binder-Powder Coating: An Alternative Fabrication Route." *Composites Part A: Applied Science and Manufacturing* 37 (10): 1831–36. https://doi.org/10.1016/j.compositesa.2005.10.004

Sarmah, P., and K. Gupta. 2024a. "A Review on the Machinability Enhancement of Metal Matrix Composites by Modern Machining Processes." *Micromachines* 15 (8): 947. https://doi.org/10.3390/mi15080947

Sarmah, P., and K. Gupta. 2024b. "Recent Advancements in Fabrication of Metal Matrix Composites: A Systematic Review." *Materials* 17 (18): 4635. https://doi.org/10.3390/ma17184635

Sarmah, P., and P. K. Patowari. 2023. "Mechanical and Tribological Analysis of the Fabricated Al 6063-Based MMCs with SiC Reinforcement Particles." *Silicon* 15 (6): 2781–96. https://doi.org/10.1007/s12633-022-02175-8

Satishkumar, D., M. Kanthababu, V. Vajjiravelu, R. Anburaj, N. Thirumalai Sundarrajan, and H. Arul. 2011. "Investigation of Wire Electrical Discharge Machining Characteristics of Al6063/SiCp Composites." *The International Journal of Advanced Manufacturing Technology* 56 (9): 975–86. https://doi.org/10.1007/s00170-011-3242-5

Schöbel, M., G. Requena, G. Fiedler, D. Tolnai, S. Vaucher, and H. P. Degischer. 2014. "Void Formation in Metal Matrix Composites by Solidification and Shrinkage of an AlSi7 Matrix between Densely Packed Particles." *Composites Part A: Applied Science and Manufacturing* 66 (November):103–8. https://doi.org/10.1016/j.compositesa.2014.07.011

Schultz, B. F., J. B. Ferguson, and P. K. Rohatgi. 2011. "Microstructure and Hardness of Al2O3 Nanoparticle Reinforced Al–Mg Composites Fabricated by Reactive Wetting and Stir Mixing." *Materials Science and Engineering: A* 530 (December):87–97. https://doi.org/10.1016/j.msea.2011.09.042

Shamim, F. A., A. Dvivedi, and P. Kumar. 2022. "Fabrication and Characterization of Al6063/SiC Composites Using Electromagnetic Stir Casting Process." *Proceedings of the Institution of Mechanical Engineers, Part E: Journal of Process Mechanical Engineering* 236 (1): 187–93. https://doi.org/10.1177/09544089211045796

Sharma, H., S. K. Tiwari, V. S. Chauhan, R. Kumar, and P. Gulati. 2021. "Wear Analysis of Friction Stir Processed Aluminum Composite Reinforced by Boron Carbide." *Materials Today: Proceedings*, ICMN-2020, 46 (January):2141–45. https://doi.org/10.1016/j.matpr.2021.02.468

Sheelwant, A., P. M. Jadhav, and S. K. R. Narala. 2021. "ANN-GA Based Parametric Optimization of Al-TiB2 Metal Matrix Composite Material Processing Technique." *Materials Today Communications* 27 (June):102444. https://doi.org/10.1016/j.mtcomm.2021.102444

Singh, I., N. Bhatnagar, and P. Viswanath. 2008. "Drilling of Uni-Directional Glass Fiber Reinforced Plastics: Experimental and Finite Element Study." *Materials & Design* 29 (2): 546–53. https://doi.org/10.1016/j.matdes.2007.01.029

Song, Bo, and Z. Jinliang. 2022. "Research Progress of Materials Design for Metal Laser Additive Manufacturing." *Acta Metallurgica Sinica* 59 (1): 1–15. https://doi.org/10.11900/0412.1961.2022.00026

Sonkar, V., K. Abhishek, S. Datta, and S. S. Mahapatra. 2014. "Multi-Objective Optimization in Drilling of GFRP Composites: A Degree of Similarity Approach." *Procedia Materials Science, 3rd International Conference on Materials Processing and Characterisation (ICMPC 2014)*, 6 (January):538–43. https://doi.org/10.1016/j.mspro.2014.07.068

Sun, J., D. Ye, J. Zou, X. Chen, Y. Wang, J. Yuan, H. Liang, H. Qu, J. Binner, and J. Bai. 2023. "A Review on Additive Manufacturing of Ceramic Matrix Composites." *Journal of Materials Science & Technology* 138 (March):1–16. https://doi.org/10.1016/j.jmst.2022.06.039

Taşkesen, A., and K. Kütükde. 2014. "Experimental Investigation and Multi-Objective Analysis on Drilling of Boron Carbide Reinforced Metal Matrix Composites Using Grey Relational Analysis." *Measurement* 47 (January):321–30. https://doi.org/10.1016/j.measurement.2013.08.040

Tekoğlu, E., A. D. O'Brien, J.-S. Bae, K.-H. Lim, J. Liu, S. Kavak, Y. Zhang, et al. 2024. "Metal Matrix Composite with Superior Ductility at 800°C: 3D Printed In718+ZrB2 by Laser Powder Bed Fusion." *Composites Part B: Engineering* 268 (January):111052. https://doi.org/10.1016/j.compositesb.2023.111052

Tenali, N., G. Ganesan, and P. Ravindra Babu. 2024. "A Study on the Effects of Hybrid (SIC-PSSA) Nano Sized Reinforcement on Mechanical and Tribological Behaviour of Al Alloy-Based Metal Matrix Composite Produced by Ultrasonic Assisted Stir Casting." *Materials Research Express* 11 (7): 075010. https://doi.org/10.1088/2053-1591/ad62bb

Tran-Ngoc, H., S. Khatir, H. Ho-Khac, G. De Roeck, T. Bui-Tien, and M. Abdel Wahab. 2021. "Efficient Artificial Neural Networks Based on a Hybrid Metaheuristic Optimization Algorithm for Damage Detection in Laminated Composite Structures." *Composite Structures* 262 (April):113339. https://doi.org/10.1016/j.compstruct.2020.113339

Vasudevan, H., N. C. Deshpande, and R. R. Rajguru. 2014. "Grey Fuzzy Multiobjective Optimization of Process Parameters for CNC Turning of GFRP/Epoxy Composites." *Procedia Engineering, "12th Global Congress on Manufacturing and Management" GCMM - 2014*, 97 (January):85–94. https://doi.org/10.1016/j.proeng.2014.12.228

Venkatesh, T. A., and D. C. Dunand. 2000. "Reactive Infiltration Processing and Secondary Compressive Creep of NiAl and NiAl-W Composites." *Metallurgical and Materials Transactions A* 31 (3): 781–92. https://doi.org/10.1007/s11661-000-0020-7

Vieira, A. F. M. R. Costa T. Filho, J. P. Eguea, and M. L. Ribeiro. 2024. "Optimization of Structures and Composite Materials: A Brief Review." *Eng* 5 (4): 3192–211. https://doi.org/10.3390/eng5040168

Wang, F., J. Li, C. Shi, N. Zhao, E. Liu, C. He, and F. He. 2018. "Preparation and Mechanical Properties of In-Situ Synthesized Nano-MgAl2O4 Particles and MgxAl(1-x)B2 Whiskers Co-Reinforced Al Matrix Composites." *Materials Science and Engineering: A* 735 (September):236–42. https://doi.org/10.1016/j.msea.2018.08.054

Yang, J., and D. D. L. Chung. 1989. "Casting Particulate and Fibrous Metal-Matrix Composites by Vacuum Infiltration of a Liquid Metal under an Inert Gas Pressure." *Journal of Materials Science* 24 (10): 3605–12. https://doi.org/10.1007/BF02385746

Ye, H., X. Y. Liu, and H. Hong. 2008. "Fabrication of Metal Matrix Composites by Metal Injection Molding: Review." *Journal of Materials Processing Technology* 200 (1): 12–24. https://doi.org/10.1016/j.jmatprotec.2007.10.066

Yue, T. M., Dai, Y., and W. S. Lau. 1996. "An Examination of Wire Electrical Discharge Machining (WEDM) of Al_2O_3 Particulate Reinforced Aluminum Based Composites." *Materials and Manufacturing Processes* 11 (3): 341–50. https://doi.org/10.1080/10426919608947490

Zhang, H., H. Springer, R. Aparicio-Fernández, and D. Raabe. 2016. "Improving the Mechanical Properties of Fe – TiB2 High Modulus Steels through Controlled Solidification Processes." *Acta Materialia* 118 (October):187–95. https://doi.org/10.1016/j.actamat.2016.07.056

Zhang, X., and M. J. Tan. 1997. "Selection of Particulate Reinforcement in P/M Metal Matrix Composites." *Journal of Materials Processing Technology* 63 (1): 913–17. https://doi.org/10.1016/S0924-0136(96)00013-1

Index

For Product Safety Concerns and Information please contact our EU
representative GPSR@taylorandfrancis.com
Taylor & Francis Verlag GmbH, Kaufingerstraße 24, 80331 München, Germany

www.ingramcontent.com/pod-product-compliance
Ingram Content Group UK Ltd.
Pitfield, Milton Keynes, MK11 3LW, UK
UKHW021514250926
475702UK00043B/496

* 9 7 8 1 0 4 1 0 2 2 6 8 8 *